Chemistry
for *CSEC*®

NEW EDITION

Keane Campbell, Nigel Jalsa, Leon Chin, Cheryl Remy,
Louise Mason, Norman Lambert, Marine Mohammed

HODDER
EDUCATION
AN HACHETTE UK COMPANY

Orders: please contact Hachette UK Distribution, Hely Hutchinson Centre, Milton Road, Didcot, Oxfordshire, OX11 7HH. Telephone: +44 (0)1235 827827. Email education@hachette.co.uk. Lines are open from 9 a.m. to 5 p.m., Monday to Friday. You can also order through our website: www.hoddereducation.com

First published in 2014
by Pearson Education Limited
Published from 2015 by Hodder Education,
An Hachette UK Company
Carmelite House
50 Victoria Embankment
London EC4Y 0DZ
www.hoddereducation.com

22
IMP 14

ISBN: 978 14479 5220 6

Produced by DTP Impressions
Original illustrations © Hodder & Stoughton Limited, 2007 and Robert Hichens
Cover design by Firelight Studio

Cover photo: © Alamy

Acknowledgements
The author and publisher would like to thank the following individuals and organisations for permission to reproduce photographs:
p1 left,Harcourt Education Ltd / Trevor Clifford; right, Bigstock; **p2** Corbis; **p3** Pearson photolibrary; **p12** all, Pearson photolibrary; **p17** Alamy Images / Zgraggen; **p19** Pearson photolibrary; **p28** left, both, Bigstock; right; Science Photo Library / ZEPHYR; **p29** top, Bigstock; bottom, Science Photo Library / ISM; **p30** left, Bigstock; right, iStockphoto; **p31** Corbis / Reuters; **p35** left, Science Photo Library / Dirk Wiersma, centre, Science Photo Library / John Carter, right, Science Photo Library / Andrew Lambert Photography; **p37** Corbis / Steve Raymer; **p39** right, Harcourt Education Ltd, left, Science Photo Library / Charles D Winters; **p45** Getty Images / PhotoDisc; **p58** left, Science Photo Library / Charles D Winters; right, Bigstock; **p61** Pearson photolibrary; **p63** Getty images; **p65** left, Getty Images / PhotoDisc; right, Science Photo Library / Russ Lappa; **p64** all, Science Photo LibraryCharles D Winters; all, Pearson photolibrary; **p75** top left, Harcourt Education Ltd / Devon Shaw; bottom right, Harcourt Education Ltd / Devon Shaw; top right, Bigstock; bottom left, Pearson photolibrary; **p78** Science Photo Library / Steve Allen; **p79** Pearson photolibrary; **p81** Harcourt Education Ltd / Trevor Clifford; **p87** all, Pearson photolibrary; **p88** Alamy; **p90** Alamy; **p92** both, Harcourt Education Ltd / Trevor Clifford; **p95** Pearson photolibrary; **p97** Alamy; **p105** all, Pearson photolibrary; **p109** all, Pearson photolibrary; **p111** all, Pearson photolibrary; **p114** all, Science Photo Library / Andrew Lambert Photography; **p119** from left, Harcourt Education Ltd / Gareth Boden, Science Photo Library / Russ Lappa, Ginny Stroud-Lewis, Science Photo Library / John Carter, Science Photo Library / Andrew Lambert Photography, Science Photo Library / Charles D Winters, Science Photo Library / Andrew Lambert Photography, Science Photo Library / Andrew Lambert Photography; **p129** both, Harcourt Education Ltd / Trevor Clifford; **p133** all, Harcourt Education Ltd / Trevor Clifford; **p167** all, Pearson photolibrary; **p1679** Pearson photolibrary; **p175** Science Photo Library / Andrew Lambert; **p188** left, Corbis / Jonathon Blair, centre and right, Harcourt Education Ltd / Devon Shaw; **p189** left and right, Science Photo Library / Martyn F Chillmaid, bottom, Science Photo Library / Charles D Winters; **p190** all, Alamy; **p192** all, Pearson photolibrary; **p194** left and right, Pearson photolibrary; centre, Bigstock; **p199** left, Bigstock; centre and right, Pearson photolibrary; **p200** all, Pearson photolibrary; **p210** Bigstock; **p214** all, Harcourt Education Ltd / Trevor Clifford; **p217** both, Harcourt Education Ltd / Trevor Clifford; **p232** all, Bigstock; **p233** both, Harcourt Education Ltd / Trevor Clifford; **p234** left and middle, Bigstock; right, Pearson photolibrary; **p248** Bigstock; **p249** Pearson photolibrary; **p251** both, Bigstock; **p256** left, KPT Power Photos, top right, Corbis; bottom right, Corbis / M. Dillon; **p257** all, Harcourt Education Ltd / Trevor Clifford; **p259** Getty images; **p268** Pearson photolibrary; **p269** left, Pearson photolibrary; centre and right, iStockphoto; **p274** left and bottom right, Harcourt Education Ltd / Trevor Clifford; top right, Bigstock; **p279** Harcourt Education Ltd / Trevor Clifford; **p283** Science Photo Library; **p318** Corbis / Morton Beebe; **p334** Corbis / Julia Waterlow Eye Ubiquitous; **p345** both Pearson photolibrary; **p348** Pearson photolibrary; **p350** Alamy; **p351** right, Alamy; **p352** top left and right, Pearson photolibrary; bottom, Alamy; **p359** Science Photo Library; **p359** Bigstock; **p368** Bigstock; **p378** Bigstock; **p380** Alamy; **p382** starting from top, Bigstock, Bigstock, Bigstock, Pearson photolibrary, Bigstock, Bigstock; **p385** Harcourt Education Ltd / Devon Shaw; **p389** both, Harcourt Education Ltd / Trevor Clifford; **p399** Science Photo Library; **p403** Pearson photolibrary; 404 Science Photo Library; **p410** Science Photo Library; **p411** Science Photo Library; **p419** Alamy; **p422** Alamy; **p423** Pearson photolibrary; **p425** Alamy; **p442** Reuters; **p442** top, Pearson photolibrary, bottom, Bigstock; **p443** Bigstock: **p450** Science Photo Library / John Carter; **p453** left, Science Photo Library / Gary Retherford, right, Science Photo Library / Andrew Lambert Photography; **p481** Steve O'Meara; **p489** Damion Mitchell; **p496** Pearson photolibrary; **p504** Pearson photolibrary; **p501** Pearson photolibrary; **p506** Science Photo Library; **p510** Getty Images / PhotoDisc

Printed and bound by CPI Group (UK) Ltd, Croydon, CR0 4YY

Contents

Introduction   vii

Section A Principles of Chemistry

Chapter 1 States of matter   1
1.1   States of matter   2
1.2   Evidence for Particle Theory   3
1.3   Using Particle Theory   4
1.4   Properties of different states   7
1.5   Changing states   11
Summary   14
End-of-chapter questions   14

Chapter 2 Structure of the atom   17
2.1   The idea of atoms   17
2.2   Arrangement of electrons in atoms   19
2.3   Protons, neutrons and electrons   21
2.4   Different types of atoms (elements)   22
2.5   Relative atomic mass of elements   24
2.6   Isotopes   26
2.7   A brief look at radioactivity   27
Summary   32
End-of-chapter questions   32

Chapter 3 Introducing the Periodic Table   35
3.1   The modern Periodic Table   35
3.2   Position of elements in the Periodic Table   37
3.3   Patterns in the Periodic Table   40
Summary   42
End-of-chapter questions   42

Chapter 4 Chemical bonding   45
4.1   What changes when elements combine?   45
4.2   The formation of bonds   46
4.3   The ionic bond and the transfer of electrons   46
4.4   The covalent bond   49
4.5   Non-polar and polar covalent molecules   53
4.6   Metallic bonding   55
4.7   Structure and properties   57
4.8   Structure and properties of ionic compounds   58
4.9   Structure and properties of simple covalent
      (molecular) substances   59
4.10 Intermolecular forces   61
4.11 Structure of simple covalent solids   62
4.12 Giant covalent solids   63
4.13 Differences in physical properties of ionic and
      molecular solids   65
4.14 Properties and uses of sodium chloride, diamond and
      graphite   68
4.15 Allotropy   70
Summary   72
End-of-chapter questions   73

Chapter 5 Mixtures and separations   75
5.1   How can we tell when a substance is pure?   76
5.2   Mixtures   78
5.3   Separating mixtures   85
5.4   The extraction of sucrose from cane sugar   86
Summary   88
End-of-chapter questions   89

Multiple-choice questions for Chapters 1–5   101

Chapter 6 Groups II and VII, and Period 3   103
6.1   Group I and Group II elements   103
6.2   The Group VII elements   112
6.3   Period 3 elements   118
Summary   124
End-of-chapter questions   125

Chapter 7 Writing formulae and equations   129
7.1   Molecular formulae   129
7.2   Chemical equations   133
Summary   141
End-of-chapter questions   141

Chapter 8 Introducing the mole   144
8.1   What is one mole of substance?   144
8.4   Empirical and molecular formulae   150
Summary   153
End-of-chapter questions   154

Chapter 9 The mole concept applied to gases
             and solutions   155
9.1   The mole concept applied to gases   155
9.2   Applications of Avogadro's Law   157
9.3   The Law of Conservation of Matter   159
9.4   The mole and solutions   159
9.5   The standard solution and its preparations   164
9.6   Volumetric analysis   167
Summary   172
End-of-chapter questions   173

Chapter 10 Calculations involving equations
             and moles   175
10.1 Quantitative relationships between reactants and
      products   176
10.2 The concept of the limiting reagent   180
10.3 Using chemical equations to find the concentration of
      solutions   182
Summary   184
End-of-chapter questions   184

Multiple-choice questions for Chapters 7–10   187

Contents

Chapter 11 Acids, bases and salts   188

11.1   Acids and bases – what is the difference?   188
11.2   Acids   191
11.3   Bases   200
11.4   Salts   202
Summary   211
End-of-chapter questions   212

Chapter 12 Types of chemical reactions   214

12.1   Direct combination reactions   214
12.2   Decomposition reactions   215
12.3   Substitution (displacement) reactions   216
12.4   Ionic precipitation reactions   217
12.5   Neutralisation reactions   218
12.6   Reversible reactions   219
12.7   Redox reactions or oxidation–reduction reactions   220
Summary   229
End-of-chapter questions   230

Chapter 13 Electrolysis and its applications   232

13.1   Electrical conductors   233
13.2   Electrolysis   236
13.3   The Faraday constant and related calculations   245
13.4   Electrolysis in industry   247
Summary   254
End-of-chapter questions   254

Chapter 14 The rates of chemical reactions   256

14.1   Measuring the rates of chemical reactions   257
14.2   Calculate the rate of reaction   258
14.3   Conditions required for chemical reactions
         to occur   260
14.4   Catalysts in industry   270
Summary   272
End-of-chapter questions   272

Chapter 15 Chemical reactions and energy changes   274

15.1   Energy changes during chemical reactions   275
15.2   A closer look   278
15.3   Measuring heat (enthalpy) changes in the
         laboratory   279
Summary   286
End-of-chapter questions   286

Multiple-choice questions for Chapters 11–15   288

Section B Organic chemistry

Chapter 16 Introducing organic chemistry   291

16.1   The bonding properties of carbon   292
16.2   Formulae of organic compounds   294
16.3   Homologous series and their functional groups   297

16.4   The IUPAC system of naming organic compounds   300
16.5   Deducing the homologous series from structural
         formulae of organic compounds   306
16.6   Determining the homologous series when given
         themolecular formula   307
16.7   Distinguishing an alkane and an alkene using a
         general formula approach   309
Summary   310
End-of-chapter questions   311

Chapter 17 Hydrocarbons: Alkanes and
         alkenes, and their derivatives   315

17.1   Natural sources of hydrocarbons   316
17.2   Fractional distillation of crude oil   316
17.3   Structural isomerism and nomenclature   321
17.4   The alkanes   326
17.5   Physical properties of alkanes   327
17.6   Chemical properties of alkanes   328
17.7   Uses of alkanes and their derivatives   330
17.8   The alkenes   331
17.9   Physical properties of alkenes   332
17.10 Chemical properties of alkenes   332
17.11 Distinguishing between alkanes and alkenes   335
17.12 Uses of alkenes   336
17.13 The alkanols   337
17.14 Physical properties of alkanols   338
17.15 The alkanoic acids   338
17.16 Physical properties of alkanoic acids   339
Summary   339
End-of-chapter questions   340

Chapter 18 The alkanols   344

18.1   The reactions of alkanols   344
18.2   Properties and uses of alkanols   349
18.3   The effects of ethanol on the body   354
Summary   355
End-of-chapter questions   355

Chapter 19 Alkanoic acids and esters   359

19.1   Introduction   359
19.2   Alkanoic acids   360
19.3   Hydrolysis of esters   366
19.4   Uses of alkanoic acids and esters   368
19.5   Detergents   370
19.6   Structure of a detergent molecule   371
19.7   The action of detergents   371
19.8   Chemicals used in laundering   373
Summary   374
End-of-chapter questions   375

Chapter 20 Synthetic polymers   378

20.1   What are polymers?   379
20.2   Forming addition polymers   383
20.3   Condensation polymerisation   386

20.4   Polysaccharides   391
Summary   395
End-of-chapter questions   396

Section C Inorganic Chemistry
Chapter 21 Metals and their compounds
                                                      398
21.1   Properties of metals   398
21.2   The reactions of metals with oxygen, water and
       non-oxidising acids   400
21.3   Displacement reactions of metals with solutions
       of salts   403
21.4   The ease of decomposition of metal nitrates,
       carbonates and hydroxides   407
21.5   The reaction of metal oxides, hydroxides and
       carbonates with acids and alkalis   411
21.6   The reactivity series of metals   413
21.7   How to deduce the order of reactivity of metals
       based on data   414
Summary   415
End-of-chapter questions   416

Chapter 22 Metals: Extraction and uses   418
22.1   Principles underlying the extraction of metals 418
22.2   Using the metal reactivity series to select an
       appropriate method for extracting a metal   421
22.3   Extraction and production of aluminium   422
22.4   Extraction of iron   424
22.5   Uses of some metals and their alloys   426
Summary   429
End-of-chapter questions   429

Chapter 23 Metals in the environment   431
23.1   Conditions necessary for corrosion of metals   431
23.2   Metals important to living systems   433
23.3   Harmful effects of metals and their compounds
       on living systems and the environment   436
Summary   443
End-of-chapter questions   443

Multiple-choice questions for Chapters 16–23   445

Chapter 24 Non-metals and their compounds 448
24.1   Non-metals   449
24.2   The uniqueness of hydrogen and silicon   450
24.3   Occurrence of the elements   451
24.4   How can non-metals be identified?   452
24.5   Physical properties of the gaseous non-metals 455
24.6   Chemical reactions of non-metals   456
24.7   Relative reactivity of the non-metals   459
24.8   Uses of the non-metals   460
Summary   464
End-of-chapter questions   465

Chapter 25 Preparation of gases and acids   467
25.1   Laboratory preparation of gases   467
25.2   Industrial production of gases and acids   475
25.3   The common tests for gases, anions and cations 481
Summary   485
End-of-chapter questions   485

Chapter 26 Non-metals in the
               environment   488
26.1   The environment   488
26.2   The composition of non-metals and their
       compounds in the atmosphere   490
26.3   Non-metals and the environment   491
26.4   Non-metals in living systems   493
26.5   Three important natural cycles   495
26.6   Green chemistry   500
Summary   504
End-of-chapter questions   505

Chapter 27 Water   507
27.1   The quality of water   507
27.2   Water: A unique commodity   508
27.3   Hardness of water   510
27.4   The treatment of water   512
27.5   Removal of hardness   512
Summary   517
End-of-chapter questions   518

Index   520

Introduction

Chemistry and you

Chemistry is the branch of Science that deals with matter – the forms it takes, its composition, its properties, its changes and its relationship with energy.

Whatever your reasons for choosing to study Chemistry, your immediate goal is to pass your CXC CSEC examinations. This textbook is designed to help you do just that. The text covers the syllabus comprehensively and is organised in the same way as the syllabus. In addition, the book has the following features shown in Table 1.

Features	Purpose
Chapter objectives at the start of each chapter	Helps ensure that you understand what topics are to be explores in each chapter
Understand it better boxes	Give you extra help with some difficult concepts
Find out more boxes	Provide extra information on some topics that may interest you.
Practice sections	Allow you to continuously check your understanding of topics
End-of-chapter questions and Multiple-choice questions	Provide practice in CXC CSEC-type questions
Many examples of the uses and importance of Chemistry in the Caribbean region	Helps understanding of the importance and relevance of Chemistry
More coloured photographs	Makes the chemistry topics clearer, easier to appreciate and improves relevance
A supporting website is available	Provides additional activities, photographs and videos, plus examination practice and answers to all questions in the book

Syllabus coverage

The order of the chapters follows the CSEC Chemistry Syllabus for examinations from May 2015. The book is divided into three sections covering the Principles of Chemistry, Organic Chemistry and Inorganic Chemistry. The learning objectives covered in each chapter are stated at the beginning of the chapter.

In addition to learning about the facts of Chemistry, it is also important that you develop the scientific and experimental skills of someone who is scientifically literate and also understand the uses to which your knowledge of Chemistry can be put. The CSEC examination assesses your abilities in all three of these areas:
1. Knowledge and comprehension
2. Uses of knowledge
3. Experimental skills

The CSEC examination will assess your abilities using three types of test:
1. Paper 1 is an objective test to 60 multiple choice questions
2. Paper 2 is a written examinations with data analysis questions, structured answer questions and extended questions

3. There is also School Based Assessment, which is an opportunity to assess your achievement in Experimental and Practical Skills, and Analysis and Interpretation of results. This is an important part of the examination worth up to a maximum of 20% of the marks.

This book provides multiple choice questions at the end of each Chapter to help you to practice for Paper 1. It also provides structured and extended answer questions for Paper 2. The practical and experimental activities will enable you to practise your skills for the SBA component of the examination.

Introduction to the School-Based Assessment (SBA)

As you study the topics, your knowledge and understanding of chemistry will improve. The experiments and practice questions are designed to help you develop theoretical as well as practical skills and to sharpen your awareness of and concern for things chemical in your environment.

There are 65 experiments in this book, and each one has been tagged with one or more SBA icons. These icons, which are illustrated below, give suggestions as to the skills that you may be assessed on, if this or some similar experiment is used either for your internal or continuous (SBA) assessment.

The SBA experimental skills are clearly stated in the CXC syllabus:

P&D • Planning and Designing

M&M • Manipulation and Measurement

ORR • Observation, Recording and Reporting

A&I • Analysis and Interpretation

Each skill will be assessed for SBA at least twice per year over a two-year period leading up to the examination.

You should become familiar with the sections in the CXC CSEC Chemistry examination syllabus on School-Based Assessment. Here CXC provides information on the kinds of investigative practical projects that you will have to carry out and how these will be assessed.

It is important for you to understand that SBA tests not only your **experimental skills** but also your **use of knowledge;** such as your ability to:

• interpret observations;

• analyse collected data;

• do any relevant calculations;

• appreciate that the types of instruments used in experiments, and how well you read them, can impose limits on experiments.

There is also a data analysis question in **Paper 2**, which tests the same range of skills as the SBA and, like your SBA mark, contributes to the third profile on the CXC CSEC certificate.

We have given consideration to the concerns expressed in the CXC reports on CSEC Chemistry over the years, especially those related to School-Based Assessment. These concerns have prompted us to encourage you to pay attention to the following sub-skills as you carry out and report on experiments.

• Accurate positioning and reading of instruments, for example the need to keep thermometers vertical and to read them at eye level at the same time having the bulb completely submerged in the material you are testing.

- The need to make clear distinctions between an **observation** (the changes you see or the readings on an instrument) and an **inference** (the chemical explanation for or interpretation of that observation).
- Observations should be clearly and concisely written preferably without repeating the test that was already described.
- Drawings should be relevant, drawn to scale, graduations should be shown on instruments, stoppers should be appropriately positioned.
- When drawing graphs, you should aim for clear and accurate plotting of points, label the axes and show the units, use best fit lines rather than joining point to point and choose a scale that allows use of most of the graph paper.
- The format of your lab report should be appropriate for the type of experiment and the particular skill being tested. In P & D experiments, for example, the past tense and active voice are used (for example 'we looked ...', 'I saw ...'). Also, subheadings such as 'expected results' and 'sample calculations' replace the actual results and calculations in other experiments.
- Careful tabulation of results, including headings and units, are important in experiments where measurements are made.

The experiments in this book and the suggested skills and topics are shown in Table 2.

▼ **Table 2**

Page	Experiment	Category	Experiment skill			Use of knowledge
			ORR	M&M	P&D	A&I
3	1.1 Diffusion in liquids	States of matter	✓			✓
3	1.2 Diffusion in gases, 1*	States of matter	✓			✓
4	1.3 Diffusion in gases, 2*	States of matter	✓			✓
5	1.4 Demonstrating Brownian motion	Evidence for particles in matter	✓			✓
6	1.5 Osmosis	Evidence for particles in matter	✓			✓
7	1.6 Osmosis in living tissue	Evidence for particles in matter	✓	✓		✓
10	1.7 Properties of different states	Evidence for particles in matter	✓			✓
40	3.1 Build models of elements	Evidence for particles in matter		✓	✓	
54	4.1 Testing liquids for polarity	Bonding and structure	✓			
64	4.2 Make models of ionic and covalent structures	Bonding and structure		✓	✓	
68	4.3 Differences in the physical properties of ionic and simple molecular solids	Bonding and structure	✓			✓
76	5.1 To determine the melting point of a solid	Bonding and structure		✓		✓
77	5.2 To determine the boiling point of a pure liquid	Bonding and structure		✓		✓
81	5.3 Determining the solubility of potassium nitrate	Separation / Acids, bases and salts	✓	✓		✓

▼ **Table 2** *continued*

Page	Experiment	Category	Experiment skill			Use of knowledge
			ORR	M&M	P&D	A&I
92	5.4 To compare the solubility of iodine in two solvents	Separation				
93	5.5 Removing caffeine by solvent extraction*	Separation	✓	✓		✓
116	6.1 Displacement reactions of halogens	Redox and electrolysis	✓	✓		✓
150	8.1 To work out the percentage composition by mass	Quantitative/volumetric analysis	✓			✓
165	9.1 To prepare a standard solution of sodium carbonate	Quantitative/volumetric analysis		✓		✓
168	9.2 Practising titrations	Quantitative/volumetric analysis	✓	✓		✓
170	9.3 An acid–base titration	Quantitative/volumetric analysis				✓
190	11.1 Testing toothpaste	Acids, bases and salts		✓		✓
198	11.2 Testing acids	Acids, bases and salts	✓	✓		✓
206	11.3 To prepare copper(II) sulphate from copper(II) oxide	Acids, bases and salts	✓	✓		
207	11.4 To prepare a sample of sodium chloride – a neutralisation reaction	Acids, bases and salts	✓	✓		
218	12.1 Thermometric titration	Energy and energetics	✓	✓		✓
233	13.1 Distinguishing conductors from non-conductors	Redox and electrolysis		✓	✓	✓
252	13.2 Working out the conditions for rusting	Redox and electrolysis		✓	✓	
263	14.1 The effect of concentration on reaction rate: the disappearing cross	Rates		✓		✓
265	14.2 The effect of temperature on the rate of reaction	Rates			✓	
267	14.3 Investigating the catalytic decomposition of hydrogen peroxide	Rates			✓	✓
269	14.4 The darkening of silver salts	Redox and electrolysis	✓			✓
275	15.1 Classifying reactions as exothermic or endothermic	Energy and energetics	✓	✓		✓
281	15.2 Measuring $\Delta H_{neutralisation}$	Energy and energetics	✓	✓		✓
281	15.3 Measuring $\Delta H_{solution}$	Energy and energetics	✓	✓		✓

▼ **Table 2** *continued*

Page	Experiment	Category	Experiment skill			Use of knowledge
			ORR	M&M	P&D	A&I
282	15.4 To determine heat change for a displacement reaction	Energy and Energetics/ Redox and electrolysis		✓		✓
336	17.1 To distinguish between an alkane and an alkene	Organic chemistry	✓	✓		✓
346	18.1 Some typical reactions of alkanols	Organic chemistry	✓	✓		✓
347	18.2 Preparation of ethane from ethanol*	Organic chemistry	✓			✓
350	18.3 Investigating the solubility of ethanol in water	Organic chemistry/ Bonding and structure		✓		✓
353	18.4 Making wine at home†	Organic chemistry	✓			✓
365	19.1 Preparing ethyl ethanoate*	Organic chemistry	✓			✓
367	19.2 Saponification	Organic chemistry				
368	19.3 Finding the concentration of ethanoic acid in vinegar	Quantitative/volumetric analysis	✓		✓	✓
389	20.1 Preparing a sample of nylon – the nylon rope trick*	Organic chemistry		✓		
402	21.1 To determine the reactivity of some metals	Organic chemistry	✓			✓
403	21.2 Displacement reactions, used to determine the relative reactivity of zinc, magnesium and copper	Reactivity series	✓			
408	21.3 To determine the relative stability of metallic nitrates, carbonates and hydroxides	Reactivity series	✓	✓		
411	21.4 Testing the oxides, hydroxides and carbonates of selected metals with acids and alkalis	Reactivity series	✓	✓		
420	22.1 Looking at the ease of reduction of metal oxides with carbon and hydrogen	Reactivity series	✓			✓
453	24.1 Making the two crystal forms of sulphur	Non-metals				

The experiments marked * are suggested as teacher demonstrations. The experiments marked † may be used as home or group projects.

1 Section A Principles of chemistry
States of matter

→ ## In this chapter, you will study the following:

- evidence that supports the idea that matter is made of particles;
- the three states of matter and their properties;
- changes from one state of matter to another.

This chapter covers
Objectives 1.1, 1.2 and 1.3 of Section A of the CSEC Chemistry Syllabus.

You already know that matter is anything that has mass and occupies space. Objects such as a chair or bottle, and materials such as salt or water, are examples of matter.

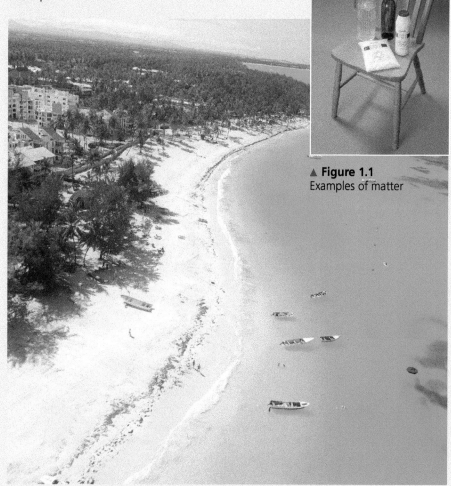

▲ Figure 1.1
Examples of matter

▲ Figure 1.2
A coastline

There are three states of matter: solid, liquid and gas. Matter can change from one state to another and back again. Think about taking an ice cube from a fridge.

1.1 States of matter

Physical properties
are characteristics that can be observed without changing the composition of matter. Physical properties include colour, odour, taste, density, solubility, melting point and boiling point. Can you think of any others?

The Particle Theory
says that matter is made up of particles.

The photograph in Figure 1.3 shows many examples of matter.

Matter can be solid (such as the land), liquid (such as the sea) or gaseous (such as the air). We can distinguish between these states because of their **physical properties** (or attributes) such as shape, volume and ease of flow. These properties are easily observed.

These properties can be linked to the theory that matter is made up of particles (this theory is known as **Particle Theory**). Whether matter exists as a solid, a liquid or a gas is determined by the arrangement of its particles and the strength of the attraction between the particles. We will first look at evidence that supports Particle Theory. Later in the chapter, we will look at differences between the three states of matter.

Figure 1.3
How many examples of the three states of matter can you see in this photograph? ▶

Find out more

Linking evidence and theory
You soon notice if someone spills a bottle of alcohol or sprays perfume at the front of a room. Very soon you will smell alcohol or perfume in all parts of the room. How can you explain this?

If we assume that the perfume is made of very tiny particles that can move, and that the air in the room is also made of particles, then a reasonable explanation is that the particles of perfume were able to spread throughout the room between the particles of the air. When we smell the perfume all over the room, that is the evidence that it has spread throughout the room. Although we cannot see the particles spreading, not even with a microscope, our explanation appears reasonable. A theory is an idea that can explain our observations.

We can also use theories to predict what will happen in an experiment. If the results of our experiment match our predictions, then we can be confident that our theory is accurate.

Q1 Find out all you can about Particle Theory and the Kinetic Theory of Gases.

1.2 Evidence for Particle Theory

Here are some experiments that we can use to gain more evidence for Particle Theory.

ORR

A&I

Experiment 1.1 Diffusion in liquids

Procedure

Diffusion is the movement of particles of a substance from where there is a lot of it to where there is less or none of it. Diffusion results in a substance becoming evenly spread out.

1 Place 50 cm³ of saturated sucrose solution in a tall measuring cylinder. With care, add 50 cm³ of copper(II) sulphate solution. Make sure that you disturb the lower layer as little as possible. Add 50 cm³ of water, again disturbing the previous layers as little as possible.

2 Set the measuring cylinder aside and observe it at regular intervals over several days.

3 Write up what you observe. What did you see, smell or hear?

4 Now try to explain what you observed using Particle Theory. These are your inferences.

Figure 1.4
A simple demonstration of diffusion ▶

A&I

ORR

Experiment 1.2 Diffusion in gases 1 (Teacher demonstration)

⚠ Observe carefully as your teacher conducts the following experiment in a fume cupboard.

Procedure

1 Liquid bromine is placed in a jar, which is then covered.

2 Observe what happens.

3 Write up what you observed.

4 Use Particle Theory to explain your observations.

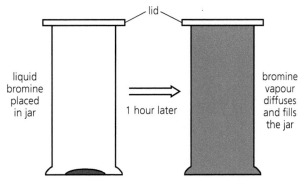

lid

liquid bromine placed in jar

1 hour later

bromine vapour diffuses and fills the jar

Figure 1.5
The diffusion of bromine vapour ▶

From Experiments 1.1 and 1.2, you can conclude that gases diffuse faster than liquids. In fact, the less dense the gas, the faster it diffuses.

You will need to use this conclusion in Experiment 1.3.

1.3 Using Particle Theory

We can use Particle Theory to make predictions. As you work through Experiments 1.3 to 1.6, you need to predict the results using Particle Theory. You can then check your predictions with the actual results of the experiments. Was Particle Theory useful in predicting and explaining your observations?

ORR

A&I

Experiment 1.3 Diffusion in gases 2 (Teacher demonstration)

⚠ Do not inhale either ammonia gas or hydrogen chloride gas.

Procedure

1 Your teacher will place small cotton wool pads soaked with concentrated aqueous solutions of ammonia and hydrogen chloride simultaneously at the ends of a glass tube at least 100 cm long (see Figure 1.6).

2 Allow the vapours to diffuse in the tube.

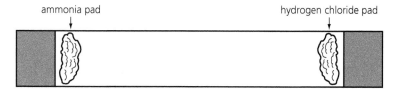

ammonia pad   hydrogen chloride pad

Figure 1.6
Diffusion of ammonia and hydrogen chloride ▶

Questions

Q1 Predict where the white ring will form. Use Particle Theory and the information given below.
  • Ammonia gas and hydrogen chloride gas are colourless. Where they meet, a white ring will form.
  • The particles of ammonia gas travel faster than those of hydrogen chloride gas.

  Below are results obtained by some students.

ammonia pad   white ring   hydrogen chloride pad

◀————————— 59 cm —————————▶◀——— 44 cm ———▶

Figure 1.7
Ammonia and hydrogen chloride diffuse and meet to form a white ring. ▶

Q2 Was your prediction accurate?

Q3 Can you explain why the white ring formed closer to the hydrogen chloride pad?

$$rate = \frac{distance\ travelled}{time}$$

Imagine that it took 10 minutes for the white ring in Figure 1.7 to form.

Q1 Calculate the rate of diffusion of each gas.

Q2 Which gas is denser? Use the results in Figure 1.7 and the conclusion from Experiment 1.3.

Q3 If all the air had been sucked out of the glass tube before the experiment, would the rates be affected? Explain your answer.

The results from Experiment 1.3 demonstrate that the less dense gas, ammonia, travelled (diffused) faster than the more dense gas, hydrogen chloride. The white fumes are formed as a result of the reaction between gaseous ammonia particles and gaseous hydrogen chloride particles.

$$ammonia + hydrogen\ chloride \rightarrow ammonium\ chloride$$
$$NH_3(g) + HCl(g) \rightarrow NH_4Cl(s)$$

The abbreviations (g) and (s) refer to the states of the substances: (g) indicates gas and (s) indicates solid.

Brownian motion is the visible movement of small pieces of solid that are large enough to be seen under a microscope. The small pieces of solid are bombarded by invisible particles of the liquid or gas in which they are suspended.

Brownian motion

The phenomenon of **Brownian motion** takes its name from the British botanist Robert Brown (1773–1858). While Brown was examining pollen grains in water using a microscope, he noticed the random but vigorous movement of the grains. The same haphazard movements are observed if a beam of sunlight shines through dust particles in the air in a room. Experiment 1.4 shows Brownian motion.

ORR

A&I

Experiment 1.4   Demonstrating Brownian motion

Procedure

1 Shake some pollen grains from a hibiscus flower onto a drop of water on a microscope slide.

2 View the slide under a microscope.

Observations

You see random, vigorous movements of the pollen.

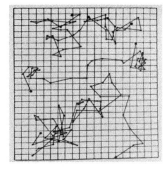

▶ **Figure 1.8**   Random motion of the particles in Brownian motion

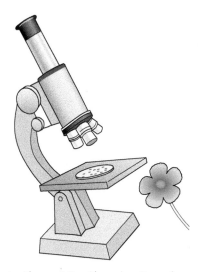

▶ **Figure 1.9**   Observing Brownian motion

Particle Theory explains why these visible pieces of solid (dust or pollen) dart about, changing direction in the air or water.

We can see the pollen grains under a microscope, because a piece of pollen is made up of millions of particles joined together. Air and water are also made up of particles, but these particles remain free and are too small to be seen under a microscope. A reasonable explanation for the motion of the visible particles is that they are bombarded by the invisible particles in the water (or air), changing direction as they do so.

Find out more

What does Brownian motion tell us about the invisible particles in liquids and gases?

ORR

A&I

Osmosis

Experiment 1.5 Osmosis

Osmosis is the diffusion of water (or any other solvent) through a semi-permeable membrane.

Osmosis can be demonstrated with the apparatus shown in Figure 1.10.

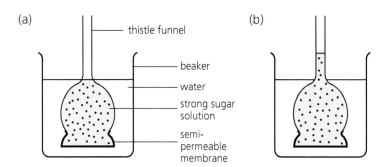

Figure 1.10
A demonstration of osmosis ▶

A **semi-permeable membrane** allows some particles (small ones) to pass through it, but not others (larger ones).

Procedure

A **semi-permeable membrane** is placed over a thistle funnel, as shown in Figure 1.10. The thistle funnel is filled with sugar solution and coloured water, and then immersed in a beaker of pure water (Figure 1.10a).

Observations

The level of liquid in the funnel rises initially, and then stays steady (Figure 1.10b).

Explanation

Osmosis is really a special case of diffusion, because we can think of it as diffusion in one direction. The small water molecules can pass through the semi-permeable membrane, and so they diffuse into the sugar solution. The larger sugar particles cannot move out of the funnel. Eventually, the pressure inside the funnel pushes some water particles out of the funnel while other water particles move in. At this point, the level of the sugar solution stops rising.

ORR
M&M
A&I

Experiment 1.6 Osmosis in living tissue

Procedure

1 Place 100 cm³ of water in each of two identical beakers.
2 Add six spatulas full of common salt to one of the beakers. Label the two beakers carefully.
3 Cut four strips (4 cm × 1 cm × 1 cm) of green pawpaw (or potato or yam).
4 Place two of the strips into each beaker. Note the time.
5 Leave the strips in the liquid for at least 1 hour.
6 Remove the strips from the liquids. Feel the strips and record their texture.
7 Measure the dimensions of the strips. Record your observations.
8 Explain your observations.

Questions

Q1 What is the semi-permeable membrane in this experiment?

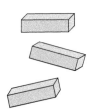

▲ **Figure 1.11**
Use pawpaw, potato or yam in this experiment.

Find out more

You should have come to the following conclusions about particles after doing Experiments 1.1–1.6:
* Particles in gases and liquids are in constant motion.
* Particles of different gases differ in density and move at different speeds.
* Particles move in a random, haphazard fashion.
* Particles are of different sizes.

Which experiment or experiments led to each conclusion?

1.4 Properties of different states

States of matter are never fixed and are determined by conditions of temperature and pressure, which can affect:
* the arrangement of the particles;
* the types of forces between the particles;
* the types of motion of which the particles are capable.

Density

To illustrate the variation of density as the state of matter changes, we first need to know what density is. Density is defined as the mass per unit volume:

$$\text{density} = \frac{\text{mass}}{\text{volume}}$$

From the above equation, you should be able to see that a reduction in mass or an increase in volume of a substance will result in a decrease in the density of that substance.

For a fixed mass of substance, as its state changes from solid to liquid to gas, the volume it occupies progressively increases, and hence the density decreases. Therefore, solids have the greatest density, liquids have a slightly lower density and gases are the least dense of the three phases.

We can also describe density as being a measure of the degree of compactness of a substance; the more compact it is, the greater its density. The greater the degree of attraction between the particles of a substance, the more compact its structure, and hence the greater its density. Therefore, solids have a very compact structure and are very dense. Gases, however, have minimal attraction between their particles, and so have the lowest density. What do you think the relationship between compressibility and density is?

No matter how you choose to define density, the relative densities of the three states of matter are as follows:

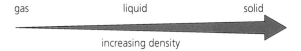

gas   liquid   solid

increasing density

Note that there is an important exception to this. Water is the only substance that is denser in its liquid state than its solid state. What evidence can you think of that supports this? Hint: If you put a block of ice in a glass of water, does it float or sink?

Energy of particles

How do the relative energies of the three states of matter compare to each other? Specifically, consider their kinetic energies, or the energy that a substance possesses due to its motion. You should realise that if you place a solid and a liquid on a flat surface, the solid will not move, the liquid will spread out, and a gas would rapidly expand to fill the space of the room or its container. We can therefore arrange the degree of motion of the three phases as follows:

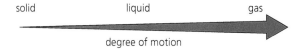

solid   liquid   gas

degree of motion

Hence their relative kinetic energies are as follows:

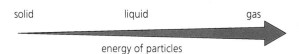

solid   liquid   gas

energy of particles

Table 1.1 compares some observable physical properties of solids, liquids and gases, and explains them based on Particle Theory. Note that we cannot see the individual particles, nor can we observe how they attract each other or the extent to which they move.

▼ **Table 1.1** A summary of the properties of the three states of matter.

Physical properties	States of matter		
	Solids	**Liquids**	**Gases**
Shape	Definite shape; many have three-dimensional lattices (they are crystalline).	Variable shape – takes the shape of the container.	No definite shape – takes the shape of the container.
Arrangement of particles	• Closely packed and regular arrangement. • Particles are strongly attracted. • Particles have restricted motion.	• Looser, irregular particle arrangement. • Particles are not as strongly attracted as in solids. • Individual particles, groups or clusters can move away from each other.	• Particles very loosely and irregularly arranged. • There is little attraction between the particles. • Particles can move about freely
Volume	Constant volume	Constant volume	Volume not constant – fills any available space.
Compressibility	Not easily compressed – pressure has no effect on volume.	Can be compressed to a small extent – pressure has a small effect on volume.	Easily compressed.
Density	High density – high level of compactness.	Medium density – medium level of compactness.	Low density – low level of compactness.
Energy of particles	Low degree of particle motion.	Higher degree of particle motion.	Highest degree of particle motion.

ORR
A&I

Experiment 1.7 Properties of different states

Procedure

In a boiling tube, heat each of the following substances, observing any changes that occur.

1 Ice

2 Liquid water

3 Butter

4 Iodine

Questions

Q1 At room temperature, both ice and iodine are solids. However, do they behave similarly when heated? If not, explain the difference.

Find out more

Theories and gases

The Kinetic Theory of Gases explains the behaviour of gases. Some of the main ideas of this Kinetic Theory are that:

- gases are composed of very small particles, called molecules, which are far apart, so most of a given volume of gas is empty space;
- the gas molecules are in rapid, random motion;
- the gas molecules collide with one another and with the sides of the container – the pressure of the gas results from the latter collisions.

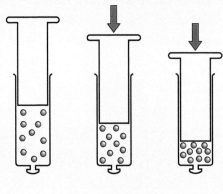

◀ **Figure 1.12** If we have a certain number of gas particles in a given space, and we increase the pressure at constant temperature, the molecules are pushed closer together; this increases the number of collisions with the sides of the container, and hence the pressure of the gas increases.

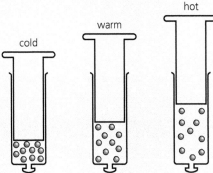

◀ **Figure 1.13** If we increase the temperature of a gas, the molecules get more energy and collide more. This would increase the pressure. To keep the pressure constant, the molecules have to move further apart, thus increasing the volume.

Q1 Use Kinetic Theory to explain what happens when (a) the pressure on a fixed mass of gas is decreased at constant temperature and (b) the temperature of a fixed mass of gas is decreased at constant pressure.

1.5 Changing states

Most substances can exist in more than one state, depending on temperature and pressure. You will be familiar with the three states of water: ice (solid), water (liquid) and water vapour (gas).

For other substances, you may be familiar with only two of the states. Dry ice, for example, is solid carbon dioxide. You may also be aware that solid gold can be melted to the liquid state – this is how gold rings are made. A change in temperature can cause a change in state, as shown in Figure 1.14.

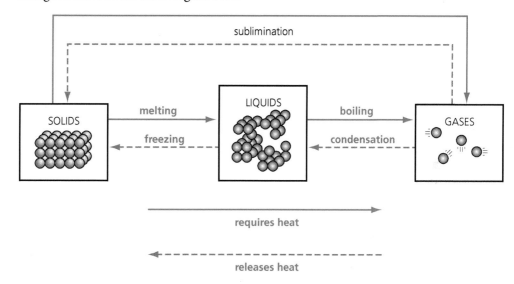

Figure 1.14
The transitions between states. Heat is supplied for melting and boiling. Heat is released during condensation and freezing. ▶

Melting

The **melting point** of a substance is the temperature at which it changes from the solid state to the liquid state.

Particles vibrate more vigorously when a solid is heated. Eventually, the vibrations become so violent that the forces of attraction no longer hold particles in position and the solid melts. The heat energy supplied during melting is used to break up the solid's structure. The temperature remains constant during melting and is called the **melting point** of the solid.

Evaporation and boiling

Evaporation is the process by which particles of a liquid leave the surface of the liquid as a vapour.

When a liquid is heated, the particles move faster, and continually collide with other particles. Occasionally, some of the particles acquire sufficient energy to break free of the surface and they escape. This process is called **evaporation**. Escaping particles take with them a lot of energy, so evaporation leads to cooling.

Boiling is the process by which a liquid is freely converted to gas or vapour at its boiling point.

Boiling is the process by which a liquid is freely converted to gas or vapour. When a pure liquid is heated, its temperature rises until the boiling point is reached. Once boiling has started, the temperature remains steady. The heat energy supplied at the boiling point goes to separate the particles from each other so that they can enter the gas state. None of the energy supplied is used to raise the temperature of the liquid any further.

Whereas evaporation occurs at the surface, boiling takes place throughout the liquid. Another point to remember is that evaporation occurs spontaneously at all temperatures, but boiling occurs at one particular temperature for a given external pressure.

The rate of evaporation depends on:

- the nature of the liquid;
- temperature;
- the amount of exposed surface.

Evaporation is greater when more surface is exposed – how can you account for this?

Find out more

Why does the boiling point change with external pressure?

Consider a liquid in a closed container (Figure 1.15). Some surface particles enter the vapour phase but cannot escape from the container. Some vapour particles re-enter the liquid state. In due course, particles leave and re-enter the liquid at the same rate. At this stage, the vapour associated with the liquid is said to be saturated.
The boiling point is the temperature at which the

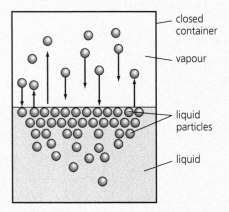

▲ **Figure 1.15**   Vaporisation in a closed container.

saturated vapour pressure of a liquid is equal to atmospheric pressure. The higher the external pressure, the higher the temperature needed for the liquid to attain the higher saturated vapour pressure.

Q1   Where would the boiling point of water be higher – at the top of a high mountain or in a pressure cooker?

▲ **Figure 1.16** Boiling.

▲ **Figure 1.17** Condensation.

▲ **Figure 1.18** Freezing.

Condensation

The change from gas to liquid can be achieved by cooling for some gases (such as water vapour), or by a combination of cooling and compression for other gases (such as propane, the liquefied gas stored in tanks, which we use for cooking). As the gas cools, the particles lose energy and move closer together, thereby increasing the attraction between them. The effect of compression is that the increased pressure pushes the particles closer together.

LPG (liquid petroleum gas) is a familiar sight around our homes. The steel containers contain liquefied propane gas stored under moderate pressure. At ordinary temperatures and pressures, propane is a gas. When it is cooled or compressed, LPG becomes a liquid. In this form, it is easily stored and transported. When you open the tap of the cylinder, the pressure is released, and some of the liquid boils and vaporises.

Freezing

As a liquid cools, its particles lose energy and move closer together. This movement increases the attraction between the particles and allows for a more regular arrangement of its particles. That is when the liquid changes to a solid. Once the liquid begins to

The **freezing point** of a substance is the temperature at which it changes from the liquid state to the solid state.

change to a solid, the temperature remains constant until all the liquid is converted. The temperature at which the solid and liquid are in equilibrium with each other at atmospheric pressure is known as the **freezing point**.

Sublimation

Sometimes a substance will change directly from a solid to a gas or from a gas to a solid, without going through the liquid state. This is called sublimation. Solid iodine and ammonium chloride are examples of substances that sublime. For example, if you heat iodine it becomes vapour. If the vapour then hits a cold surface, it sublimes back to a solid. Liquid iodine is not formed.

In summary

The effect of changing temperature on the three states of matter is summarised in Table 1.2.

▼ **Table 1.2**   The effect of heat on the three states of matter.

	Effect of heat	**Rate of diffusion**
solids	• expand slightly as temperature increases • eventually melt, often with a small increase in volume • temperature remains constant until all solid is turned to liquid	very slow
liquids	• expand slightly as temperature rises • eventually boil, with a big increase in volume • temperature remains constant during boiling, until all liquid has turned to gas	slow
gases	• expand considerably at constant pressure	rapid

Practice

Some solid stearic acid was heated until it melted. The temperature changes that occurred as the hot liquid cooled down were recorded at 2-minute intervals. The results are recorded in Table 1.3.

▼ **Table 1.3**

Time (min)	0	2	4	6	8	10	12	14	16	18
Temperature (°C)	94	79	67	61	57	56	56	56	55.7	55.3

1 Plot a graph of the data, with temperature on the vertical axis and time on the horizontal axis.

2 Use your understanding of how temperature changes as the state of a substance changes to explain the different sections of the graph.

Summary

- Solids, liquids and gases are made up of very small particles.
- These particles are not observed directly. However, we see the effect of the combined behaviour of a large number of particles.
- The inter-particle forces in solids are relatively strong. Large forces are often needed to change the shape of a solid.
- The particles of a solid vibrate about fixed points.
- The inter-particle forces in liquids are moderately strong.
- The particles in a liquid occur in clusters.
- Clusters of liquid particles slide past each other. Liquids flow.
- Liquids change their shape easily. They take the shape of their containers.
- The particles of a gas are in constant, random motion.
- The inter-particle forces in gases are very weak.
- The volume of a gas is sensitive to changes in pressure and temperature.
- In order to change a solid to a liquid or a liquid to a gas, 'extra' energy is needed to overcome inter-particle forces.

End-of-chapter questions

1 Particles are furthest apart in which of the following?
  A Liquids
  B Solids
  C Pressured liquids
  D Gases

2 Brownian motion is demonstrated when which one of the following occurs?
  A Gases and solids vibrate together.
  B Pollen grains move vigorously in water.
  C Coloured layers of liquids mix to become one colour.
  D A solid substance is put on a solid surface.

3 Which one of these has a variable shape and is not easily compressible?
  A Soap
  B Smoke
  C Water
  D Ice

4 The diagrams show the arrangement of particles in a gas (X), a liquid (Y) and a solid (Z).

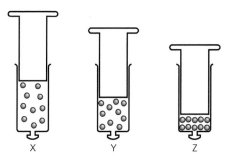

    a In which case would the inter-particle forces be weakest: (X), (Y) or (Z)?
    b If the plungers of the syringes are pushed in, in which case will it move furthest: (X), (Y) or (Z)?
    c In which case will the plunger move the least: (X), (Y) or (Z)?
    d Briefly explain your answers to parts (b) and (c).

5 A substance was cooled from the gaseous state through the liquid state to the solid state. The graph of temperature versus time for the substance as it cooled is shown below.

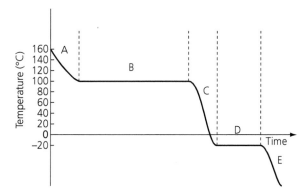

    a In which state does the substance exist at 32 °C?
    b For what temperature range is the substance a liquid?
    c Name the process occurring between B and C.
    d Name the process occurring between D and E.
    e Explain the shape of the graph between B and C.

6 Some water is placed in a beaker and kept in a closed system, as shown in the diagram.

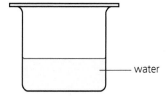

Compare the initial rate of evaporation and condensation with the rate of evaporation and condensation after 1 hour. Will evaporation eventually stop in such a system? Explain your answer.

7 Orange-coloured liquid bromine is placed in a container, as shown in the diagram.

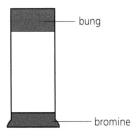

bung

bromine

Given that bromine is a volatile liquid, comment on the appearance of the system after an hour. Use Kinetic Theory to explain your observations.

8 a Define the term 'diffusion'.
   b Describe an experiment to demonstrate diffusion.

9 a What do you understand by the term 'Particle Theory'?
   b Compare the physical properties of solids and gases.

10 a What is osmosis?
   b Describe an experiment to demonstrate osmosis.
   c Draw a labelled diagram to show what happens in the experiment you described in part (b).
   d Give two examples of osmosis in everyday life.

2 Structure of the atom

In this chapter, you will study the following:

- the structure of atoms;
- the properties of electrons, protons and neutrons;
- atomic number and mass number;
- relative atomic mass;
- how to use chemical notation;
- isotopes;
- uses of radioactive isotopes.

This chapter covers
Objectives 3.1–3.7 of Section A of the CSEC Chemistry Syllabus.

In Chapter 1 we talked about particles in matter. These particles may be atoms, molecules or ions. We can think of atoms as fundamental or basic particles since molecules and ions are derived from them. However, we cannot see atoms, molecules or ions – they are too small!

2.1 The idea of atoms

Understand it better

Atoms are very small. The diameter of a typical atom is about 10^{-10} m. Table 2.1 compares the approximate size of an atom with the size of some familiar objects.

▼ **Table 2.1** The size of an atom.

Particle	Size
grain of common salt (sodium chloride)	0.0001 m
grain of talcum powder	0.000 01 m
red blood cell (visible under the microscope)	0.000 008 m
bacteria (visible under the microscope)	0.000 002 m
atom	0.000 000 0001 m

An **atom** is the smallest part of an element that can exist and still show the properties of that element.

◄ **Figure 2.1**
Grains of common salt

Q1 Compare the size of an atom with the size of a grain of salt. How much bigger is the grain of salt?

The ancient Greeks Democritus and Leucippus, who lived around 2 400 years ago, reasoned that matter was not continuous. Instead, they thought that matter was made up of indivisible bits of matter too small to be seen. They called these bits of matter 'atoms'. Throughout the years, this idea persisted, though not everyone believed in it.

Early in the 19th century, John Dalton (1766–1844), a British chemist, expanded the idea of atoms based on experiments he conducted. However, Dalton did not have the equipment to probe the structure of the atom.

Find out more

The developing story of atoms

This time line shows how scientists' ideas about atoms have changed. Scientists create theories based on the evidence they have, but these can be modified or new theories arise, as new evidence emerges.

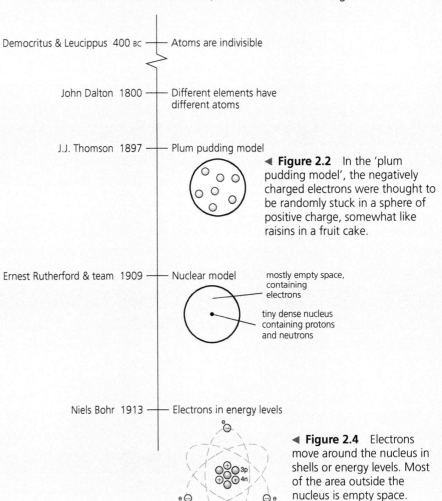

Democritus & Leucippus 400 BC — Atoms are indivisible

John Dalton 1800 — Different elements have different atoms

J.J. Thomson 1897 — Plum pudding model

◄ **Figure 2.2** In the 'plum pudding model', the negatively charged electrons were thought to be randomly stuck in a sphere of positive charge, somewhat like raisins in a fruit cake.

Ernest Rutherford & team 1909 — Nuclear model

mostly empty space, containing electrons

tiny dense nucleus containing protons and neutrons

Figure 2.3 In this model, most of the mass of the atom is concentrated at the centre or nucleus. Most of the space taken up by electrons is empty space. Rutherford called this model the 'nuclear model'. ►

Niels Bohr 1913 — Electrons in energy levels

◄ **Figure 2.4** Electrons move around the nucleus in shells or energy levels. Most of the area outside the nucleus is empty space.

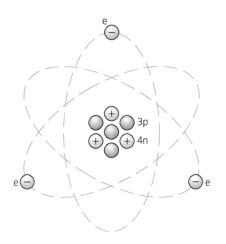

Figure 2.5 At the centre of the atom is the nucleus, which consists of its protons and neutrons. Electrons can be found moving around in shells or energy levels, while most of the area outside the nucleus is empty space. ▶

Today, scientists 'see' atoms by using a special microscope called the Scanning Tunnelling Microscope, but they still cannot see inside an atom. It has been left to people's imagination, logic and experimentation to build our current understanding of the structure of the atom. Figure 2.5 gives an idea of how scientists currently view the structure of the atom.

Atoms contain subatomic particles called electrons, protons and neutrons. The protons and neutrons make up the nucleus of the atom.

2.2 Arrangement of electrons in atoms

Electrons move around the nucleus of the atom in specific areas called shells. These shells are really different energy levels. Only electrons with the appropriate energy can occupy a given shell. The shell with the lowest energy level is closest to the nucleus. The energy level increases with each successive shell. For simplicity, the shells are represented as a series of concentric circles, as shown in Figure 2.6.

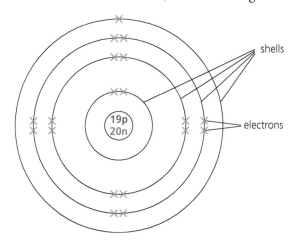

shells

electrons

◀ **Figure 2.6** The electronic configuration of a potassium atom

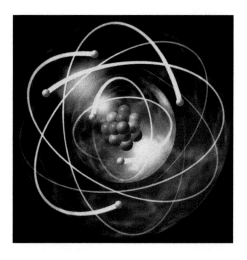

Figure 2.7
A picture of an atom ▶

Each shell can only accommodate a certain number of electrons, as shown in Table 2.2.

▼ **Table 2.2** The link between each electron shell and the maximum number of electrons it can contain

Electron shell (= n)	Maximum number of electrons(= $2n^2$)
1	2
2	8
3	18 (8, then 10)
4	32 (8, then 10, then 14)

The arrangement of electrons in an atom is called the **electronic configuration**.

If we know the number of electrons in an atom, we can work out its **electronic configuration**, because the electrons in an atom fill the shells according to the following rules:

- Electrons fill up the shells in order, beginning with the lowest energy level.
- Electrons generally fill one shell before entering the next shell (but there are some exceptions, see below).
- Each shell can accommodate the maximum number of electrons shown in Table 2.2.

The electronic configurations of helium, carbon and sodium are shown in Table 2.3.

▼ **Table 2.3** Helium, carbon and sodium atoms

Name of element	Number of electrons	Diagram	Electronic configuration
helium	2		2
carbon	6		2,4
sodium	11		2,8,1

When thinking about electronic configurations, it is important for you to note the following:

- Although the third shell can accommodate up to 18 electrons, it does this in stages.
- When the first eight electrons have entered the third shell, it becomes temporarily 'full' – additional electrons begin to fill in the fourth shell.
- When two electrons have entered the fourth shell, any additional electrons then enter and completely fill the third shell.

For example, the electronic configuration of argon (18 electrons), is 2,8,8, the configuration of potassium (19 electrons) is therefore 2,8,8,1, while that of calcium (20 electrons) is 2,8,8,2. Figure 2.8 shows the electronic configuration of calcium.

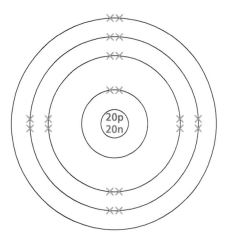

Figure 2.8
The electronic configuration of a calcium atom ▶

Practice

1 Draw diagrams showing the electronic configurations of the following atoms:
  a Nitrogen, which has 7 electrons
  b Neon, which has 10 electrons
  c Lithium, which has 3 electrons
  d Aluminium, which has 13 electrons

The valence shell

The outer occupied shell of electrons in an atom is called the **valence shell**.

The electrons in the outer shell of an atom (the **valence shell**) are the ones that take part in chemical bonding and are known as **valence electrons**. During chemical changes or reactions, these valence electrons are rearranged. The rest of the atom (i.e. the nucleus and inner shells of electrons) remains intact. We shall return to this in Chapter 4.

2.3 Protons, neutrons and electrons

Relative mass of a particle is its mass compared with the mass of a proton.

Relative charge on a particle is the charge compared with the charge on a proton.

So far, we have looked at the electrons in atoms. We now need to consider what the nucleus of the atom is made of. If you look back at Figure 2.5, you can see that the nucleus of an atom is composed of smaller particles called protons and neutrons. Protons, neutrons and electrons are sometimes called subatomic particles. We can describe subatomic particles in terms of their **relative masses** and **relative charges**.

A neutron has the same mass as a proton; therefore, they both have a relative mass of 1. The mass of an electron is far less and is considered to be negligible.

The size of the charge on the electron is the same as the size of the charge on the proton. However, the electron has a negative charge and the proton has a positive charge. The electron therefore has a relative charge of −1 and the proton has a relative charge of +1. The neutron has no charge. Table 2.4 summarises this information.

▼ **Table 2.4** Comparing protons, neutrons and electrons

Sub-atomic particle	Relative mass	Relative charge
proton	1	+1
neutron	1	0
electron	0	−1

The charge on an atom

The nucleus of an atom is composed of neutrons and protons. Remember that neutrons have no charge and protons are positive.

In each atom, the number of protons equals the number of electrons.

A particle is **neutral** when the size of its positive charge equals the size of its negative charge.

- Since the relative charge of each proton is +1, the positive charge of the nucleus is numerically equal to the number of protons present.
- Since the relative charge of each electron is −1, the negative charge of the electrons is numerically equal to the number of electrons present.
- Since number of protons equals the number of electrons, atoms are **neutral**.

Understand it better

This example is about an atom of chlorine.
- Number of protons = 17
  Positive charge = +17
- Number of electrons = 17
  Negative charge = −17
As the positive charge equals the negative charge, the atom is neutral.

2.4 Different types of atoms (elements)

Although all atoms are made of the same subatomic particles, different types of atoms have different numbers of these particles. There are two numbers that indicate the composition of an atom:

- the **nucleon number** (**mass number**) of an atom = the sum of protons and neutrons present in the nucleus
- the **proton number** (**atomic number**) of an atom = the number of protons in the nucleus

We can now work out some other pieces of information:

- the number of protons in an atom = the number of electrons
- nucleon number − proton number = number of neutrons

Understand it better

Look at these examples.

1 An atom has 10 protons and 10 neutrons. This is what we can work out:
- nucleon number = 20
- proton number = 10
- number of electrons = 10

2 An atom has nucleon number 23 and proton number 11.
number of neutrons = nucleon number – proton number
23 – 11 = 12
number of protons = proton number = 11
number of electrons = number of protons = 11

An **element** is a substance made up of only one type of atom, i.e. atoms with the same proton (atomic) number.

The proton number of an atom is like its signature; it helps us to identify it. All atoms with a proton number of 8, for example, are oxygen atoms. If the atom had a different proton number, it would be a different substance. Substances that are made up only of atoms with the same proton number are called **elements**. There are about 112 different atoms known and therefore only about 112 different elements. Hydrogen, oxygen, iron and copper are all examples of elements.

We can show the characteristics of an atom of an element by using the following chemical notation:

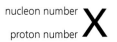

nucleon number $\quad$ **X**
proton number

where 'X' is the symbol for the element.

Table 2.5 shows the chemical notation for two elements with which you are familiar: helium and aluminium. The table also shows the number of electrons and number of neutrons, which can be calculated from the chemical notation.

▼ **Table 2.5** Numbers of subatomic particles in atoms of helium and aluminium

Name of elements	Chemical notation	Number of protons	Number of electrons	Number of neutrons
helium	$^{4}_{2}\text{He}$	2	2	2
aluminium	$^{27}_{13}\text{Al}$	13	13	14

Later on, you will see that the full chemical notation of an element or an ion includes a subscript and superscript on the right-hand side of the symbol. These represent the number of items in the element or ion (the subscript) and the charge or oxidation number (the superscript). (See Section 12.7.)

Practice

2 Construct a table like Table 2.5 and complete it for the following elements:

$^{1}_{1}$H, $^{9}_{4}$Be, $^{14}_{7}$N, $^{16}_{8}$O, $^{23}_{11}$Na, $^{34}_{16}$S, $^{40}_{18}$Ar, $^{39}_{19}$K, $^{41}_{19}$K, $^{59}_{27}$Co, $^{108}_{47}$Ag, $^{142}_{58}$Ce

2.5 Relative atomic mass of elements

The individual atoms found in substances are extremely small and have very little mass – too small to be measured by even the most accurate balance. For example, the mass of a single hydrogen atom is 1.67×10^{-27} kg. This is the same as the mass of a single proton and, by extension, the mass of a single neutron. The mass of an oxygen atom is 2.66×10^{-26} kg and the mass of a carbon atom is 2.00×10^{-26} kg.

Because of these small masses, chemists have devised a scale in which the mass of an atom is worked out by comparison with the mass of an atom of the carbon-12 isotope. Carbon-12 is assigned a mass of exactly 12.00 atomic mass units (a.m.u.). The mass of the carbon-12 isotope is therefore being used as a standard. (Isotopes are discussed later in Section 2.6.)

The **relative atomic mass**, A_r, of an element is the ratio of the average mass of one atom of the element compared to $\frac{1}{12}$ of the mass of one atom of carbon-12.

The term **relative atomic mass** of an element indicates the ratio of the mass of an atom of the element to the mass of $\frac{1}{12}$ of one atom of carbon-12. Relative atomic mass has no units. It is a ratio of two masses with the same units. Therefore, the units cancel out. We use the symbol A_r for relative atomic mass.

Understand it better

Consider three books, weighed using an ordinary balance. Book X has a mass of 200 g, book Y has a mass of 400 g and book Z has a mass of 1 200 g.

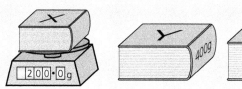

The ratio of the mass of book X : the mass of book Y is $\frac{200\text{ g}}{400\text{ g}}$ or $\frac{1}{2}$

If you make up your own scale and give Y a mass of 10 units, then what is the relative mass of X ($= X_r$) on this scale?

$$\frac{X_r \text{ units}}{10 \text{ units}} = \frac{200\text{ g}}{400\text{ g}} \text{ or } \frac{X_r}{10} = \frac{200}{400} \text{ (the units cancel out)}$$

$$X_r = \frac{200}{400} \times 10 = 5$$

Similarly:

$$Z_r = \frac{1\,200}{400} \times 10 = 30$$

We can make this more general:

$$\text{relative mass of X} = \frac{\text{mass of X (in g)}}{\text{mass of Y (in g)}} \times \frac{10}{1}$$

This can also be written as:

$$\frac{\text{mass of X (in g)}}{\frac{1}{10} \times \text{mass of Y (in g)}}$$

Note that multiplying by a fraction is the same as dividing by the reciprocal of the fraction.

The relative mass of X is the ratio of the mass of X to one-tenth the mass of Y. In this example, the mass of Y is the standard against which we compare the other masses. Note that the units cancel out.

We now need to stop considering books and look back instead to atoms. If the mass of a carbon-12 atom is assigned a value of 12, the relative mass of hydrogen on this scale can be calculated:

$$\text{relative mass of hydrogen} = \frac{\text{mass of 1 atom of hydrogen (in g)}}{\text{mass of 1 atom of carbon (in g)}} \times 12$$

or

$$= \frac{1.67 \times 10^{-27} \text{ kg}}{2.00 \times 10^{-26} \text{ kg}} \times 12$$

$$= \frac{1}{12} \times 12$$

$$= 1$$

The relative atomic mass of hydrogen is 1.

Q1 Using the mass of an atom of oxygen (given earlier), calculate the relative atomic mass of oxygen.

Q2 The mass of a magnesium atom is twice the mass of a carbon-12 atom. What is the relative atomic mass of a magnesium atom?

Because most elements have isotopes, it is more accurate to consider the average mass of the isotopes of an element when determining its relative atomic mass.

Practice

1 Calculate the A_r of copper, bromine and silver from the actual atomic masses (carbon-12 = 1.99×10^{-26} kg).

   a Copper = 1.06×10^{-25} kg

   b Bromine = 1.33×10^{-25} kg

   c Silver = 1.8×10^{-25} kg

2 Distinguish between the terms 'absolute' mass and 'relative' mass.

3 Calculate the mass (in kg) of an atom of the elements boron and chlorine from the following A_r values: B = 11 and Cl = 35. Use the mass of a carbon-12 atom as 2.00×10^{-26} kg.

2.6 Isotopes

Although all the atoms of an element *must* have the same proton number, they need not be identical. In 1919, Francis Aston (1877–1945), a British chemist and physicist, showed that some atoms of a given element may have different nucleon numbers. If atoms of a given element cannot have different numbers of protons, then this difference in nucleon number must be because the atoms have different numbers of neutrons. Aston used the term **isotopes** to describe these different atoms of the same element.

Isotopes are atoms of the same element that have different nucleon numbers (mass numbers).

Figures 2.9 and 2.10 show the subatomic particles in isotopes of hydrogen and carbon, respectively.

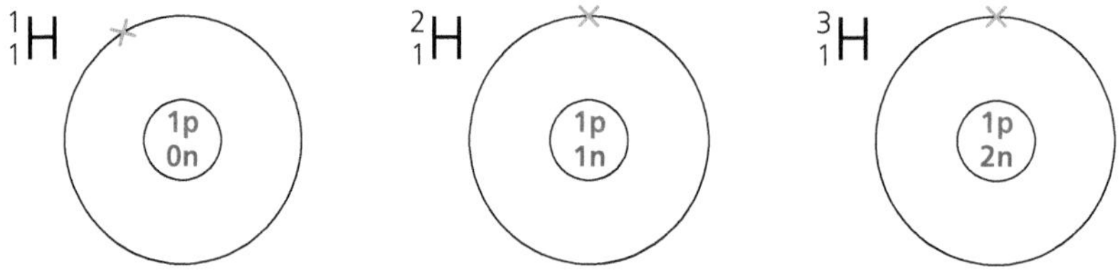

^1_1H — 1p 0n

^2_1H — 1p 1n

^3_1H — 1p 2n

Figure 2.9
Different isotopes of hydrogen ▶

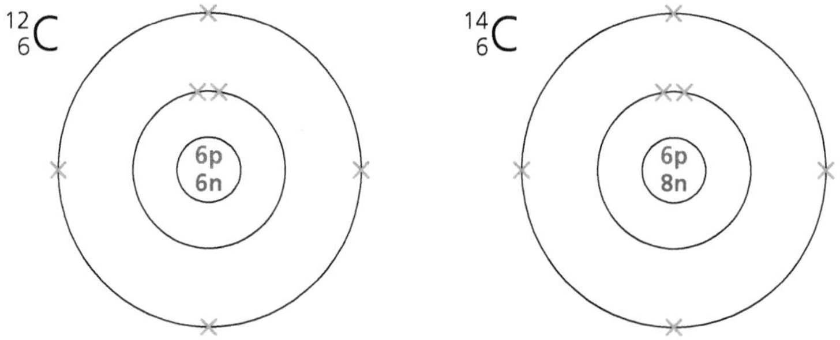

$^{12}_6C$ — 6p 6n

$^{14}_6C$ — 6p 8n

Figure 2.10
Different isotopes of carbon ▶

Isotopes are quite common, and most naturally occurring elements have more than one stable isotope. Table 2.6 shows how isotopes are alike and how they are different.

▼ **Table 2.6** About isotopes

Alike	Different
Isotopes have the same proton number and therefore the same number of protons and electrons.	Isotopes have different nucleon numbers and therefore different numbers of neutrons.
Isotopes have identical **chemical properties**.	Isotopes may have slightly different physical properties.

Chemical properties refer to how a substance changes its composition under different conditions or by reacting with other substances. Physical properties include melting point and boiling point.

Chemical notation for isotopes

The element chlorine (proton number 17), has two isotopes, with nucleon numbers 35 and 37, respectively. They may be represented in two ways, as follows:

nucleon number 35
chlorine-35

nucleon number 37
chlorine-37

Or, we can use chemical notation as described before:

$$^{35}_{17}\text{Cl} \qquad\qquad\qquad ^{37}_{17}\text{Cl}$$

Which of these two representations do you think gives more information?

2.7 A brief look at radioactivity

Isotopes may occur naturally or be artificial, i.e. synthetic. Artificial isotopes are made by bombarding atoms (or, more correctly, the nuclei of atoms) with neutrons or high-energy charged particles.

Find out more

Why do you think neutrons are suitable for bombarding atoms to make artificial isotopes?

Radioactivity is the spontaneous disintegration of unstable atomic nuclei by the emission of radiation.

Most artificial isotopes, and some naturally occurring ones, have unstable nuclei. Isotopes with unstable nuclei are **radioactive**. Some examples of radioactive isotopes are:

$$^{14}_{6}\text{C} \quad ^{24}_{11}\text{Na}^* \quad ^{235}_{92}\text{U} \quad ^{226}_{88}\text{Ra} \quad ^{60}_{27}\text{Co}^* \quad ^{35}_{16}\text{S}^* \quad ^{32}_{15}\text{P}^* \quad ^{137}_{55}\text{Cs}$$

(* denotes artificially produced isotopes)

Radioactive isotopes spontaneously eject particles and radiation from their nuclei: alpha (α) particles, beta (β) particles and gamma (γ) radiation. When radioactive isotopes eject particles and radiation, they become more stable and sometimes a different type of atom is produced.

Find out more

You know about X-rays, which are used in medicine to show up the hard parts of the body such as bones or teeth. X-rays are not produced by radioactivity.

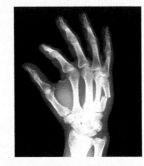

◀ **Figure 2.11**
An X-ray of a broken hand

Wilhelm Röntgen (1845–1923), a German physicist, discovered X-rays in 1895. Henri Becquerel (1852–1908), a French physicist, discovered a different type of emission. This was later called radioactivity by Marie Curie.

Q1 Find out more about Marie Curie.

Q2 How many Nobel Prizes did she win?

Q3 How many members of her family were also awarded Nobel Prizes and why?

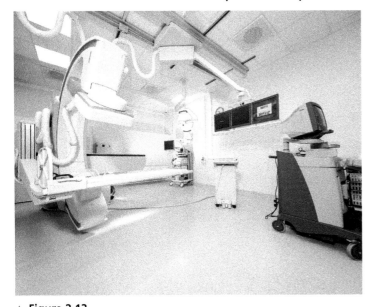

▲ **Figure 2.12** Uses of radioactivity

Uses of radioactivity

Tracers are tiny quantities of radioisotopes used to study chemical changes within living organisms.

Some forms of radiation may be dangerous to us because they can cause changes in our cells. Alpha particles, for example, can damage our DNA. Exposure to large doses of radiation can increase the risk of developing cancer. However, radiation can also be useful in many different ways.

Diagnosing and treating disease

In medicine, radioisotopes (radioactive isotopes) are used to find out what is happening inside the body. Some substances concentrate naturally in some organs of the body. Iodine, for example, accumulates in the thyroid gland in the neck and technetium-99 (an artificial isotope) accumulates in the heart and other organs. Radioactive isotopes of these elements can be used as **tracers**.

If a patient drinks a tiny amount of iodine-131, a scanner can be used to measure the amount of radioactivity in the thyroid gland. A hyperactive thyroid will accumulate more iodine than a normal thyroid. This will be detected on the scanner.

▲ **Figure 2.13**
A radiotherapy treatment room

If a radioactive isotope, such as technetium-99, is injected into a patient's bloodstream, a doctor can monitor the blood flow on a screen to see if there is reduced blood flow – a possible sign of heart disease.

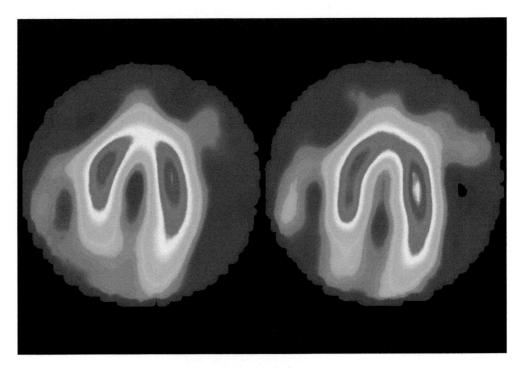

Figure 2.14
This scan shows the blood flow to the heart muscle of a patient with heart disease. The blood vessel on the left is constricted. ▶

Radiation can also be used to treat diseases. Radiation therapy, also known as radiotherapy, is used to treat cancer because it destroys cancer cells more readily than it destroys normal cells. A beam of radiation is directed at the cancerous tumour. Radioactive cobalt-60, for example, produces gamma rays that are used to kill cancerous cells.

Table 2.7 gives some examples of the uses of radioactive isotopes in medicine.

▼ **Table 2.7**   Some examples of radioactive isotopes used in medicine

Isotope	Medical uses
$^{60}_{27}\text{Co}$	treating certain types of cancer
$^{24}_{11}\text{Na}$	tracing the flow of blood and locating obstructions in the circulatory system
$^{131}_{53}\text{I}$	monitoring and treating goitre and other thyroid problems; also used in treating liver and brain tumours
$^{201}_{81}\text{Tl}$	monitoring certain heart diseases
$^{99}_{44}\text{Tc}$	monitoring certain heart diseases
^{238}Pu	providing energy for heart pacemakers

▲ **Figure 2.15**
A nuclear-reactor

▲ **Figure 2.16**
A pacemaker

Carbon-14 dating

Carbon-14 dating (also known as radiocarbon dating) is a method of finding out the age of archaeological specimens that were once living. Carbon-14 dating can be used on specimens that are no more than 50 000 years old. Bone, cloth, wood and plant fibres can be dated by this method.

Carbon-14 dating measures the levels of the isotope carbon-14 in samples of material that were once living. Each kilogram of a living organism contains a definite amount of carbon-14. The instant an organism dies, the amount of carbon-14 in it decreases due to radioactive decay. The age of a piece of wood or a dead body, for example, can be calculated by comparing its present level of carbon-14 per kilogram with that expected for the living material (or organism). The older the specimen, the smaller the amount of carbon-14 it contains.

Carbon-14 dating is used by archaeologists at the University of the West Indies to determine the age of artefacts left by the native American Indians.

In September 1991, a German couple found the body of a dead man high in the Alps, near the Austrian–Italian border. Initially, it was thought that the body was recent, perhaps that of a skier. However, carbon-14 dating showed that the body was 5 300 years old. The man was named 'Ötzi' and he has since been studied extensively.

Q1 Find out all you can about Ötzi, including the use of carbon-14 dating in determining how long ago he lived.

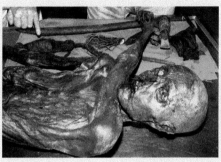

◀ **Figure 2.17**
Ötzi, on display in a museum in Bolzano, Northern Italy

Energy generation

Energy can be provided by unstable radioactive isotopes. Large radioactive atoms release energy when they are split. This is called **nuclear fission**. Fission of uranium-235 is a source of energy in some nuclear power stations.

This energy can be used to generate electricity. The use of nuclear power plants has to be monitored carefully. An accident could result in the exposure of humans and other living organisms to dangerous radiation with disastrous consequences. Besides this, the disposal of radioactive waste from nuclear power plants is a problem that has not yet been adequately solved.

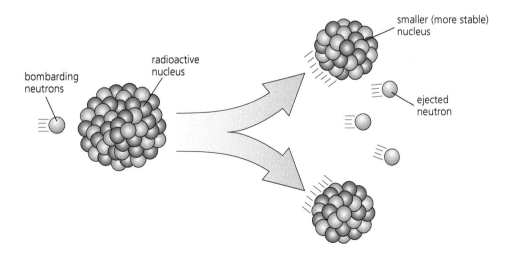

Figure 2.18
Nuclear fission ▶

The human heart has a natural pacemaker that should maintain a regular heartbeat. Sometimes this natural pacemaker is defective and has to be replaced. Some artificial pacemakers are powered by the energy produced in radioactive decay. Plutonium-238 is used in thermoelectric batteries. As the plutonium decays, the heat produced is used to generate electricity that then stimulates the heart. These batteries are expected to last for more than 25 years. As a result, the plutonium from these pacemakers must be removed and safely disposed of when the patient dies.

Summary

- All matter is made up of particles.
- An atom is the smallest particle of an element that can exist and still show the properties of that element.
- Atoms consist of:
  - a small, dense, positively charged core, called the nucleus, which contains protons (positively charged particles) and neutrons (uncharged particles);
  - negatively charged electrons, which move about in shells outside the nucleus.
- For a neutral atom, the number of protons equals the number of electrons.
- The proton number (atomic number) of an atom equals the number of protons in the nucleus of the atom.
- The nucleon number (mass number) of an atom equals the number of protons plus the number of neutrons in the nucleus.
- Isotopes are atoms with same atomic number but different mass numbers.
- Some isotopes are radioactive and may emit alpha and beta particles, and gamma radiation.
- Radioisotopes are used in medicine for diagnosing and treating disease, in carbon-14 dating to determine the age of archaeological specimens, and for producing energy.
- Radioactivity can be dangerous to humans, and radioisotopes must be handled and disposed of carefully.

End-of-chapter questions

1 The number of protons and electrons in an atom are:
  A Always the same
  B Sometimes different
  C Never the same
  D In the ratio 2:1

2 Which one of the following statements is false?

A Neutrons and protons have the same mass.

B An electron has a negative charge while a proton has a positive charge.

C The nuclear number is the sum of the protons and neutrons in an atom.

D Relative atomic mass refers to the ratio of the actual mass of an atom to that of a carbon-14 atom.

3 Element Y has three neutrons. Which one of the following elements will have similar chemical properties to Y?

A Chlorine

B Potassium

C Hydrogen

D Magnesium

4 The atoms ^{35}Cl and ^{37}Cl both have 17 protons. They will also have 17:

A Neutrons

B Isotopes

C Electrons

D Mass number

5 Different atom forms of the same element have:

A Stable and non-stable forms

B Different numbers of electrons and protons

C Different chemical properties

6 a Complete the gaps in the table below.

b Write the chemical notation of boron.

c Draw a diagram to show the electronic configuration of phosphorus.

Element	Proton number	Nucleon number	Number of protons	Number of electrons	Number of neutrons
boron	5	11			
aluminium	13				14
phosphorus		31		15	
potassium			19		20
iodine		127	53		
lead				82	126

7 Define or explain the meaning of the following terms:

a Nucleon number

b Proton number

c Gamma ray

d Radioactivity

e Isotope

f Nucleus of an atom

8   The distribution of electrons in the shells of chlorine, atomic number 17, is given in the table below. Complete this table for the other elements listed.

▼ Table 2.9

Element	Atomic number	Shell			
		1	2	3	4
chlorine	17	2	8	7	–
phosphorus					
magnesium					
calcium					
iron					

9   a   Sodium ($^{23}_{11}$Na) has a radioactive isotope with one more neutron in its atoms. Write the appropriate chemical notation for this isotope.
    b   Explain two different uses of radioactive isotopes.
    c   Explain why it is necessary to exercise caution when using radioactive isotopes.

10 a   The following terms are used in describing atoms: (i) shells, (ii) electronic configuration, (iii) neutral and (iv) chemical notation. Explain, with suitable examples, the meaning of each term listed above.
    b   What is an element? Explain why the isotopes of carbon are the same element.

3 Introducing the Periodic Table

In this chapter, you will study the following:

- how elements are classified in the modern Periodic Table;
- a brief history of the development of the Periodic Table;
- the main features of the Periodic Table;
- some patterns in groups and periods of the Periodic Table.

This chapter covers
Objective 4.1 of Section A of the CSEC Chemistry Syllabus.

In Chapter 2, you learnt that some substances are called elements and that there are over 100 known elements. Some elements occur naturally, usually combined with other elements, and others are made artificially.

(a)

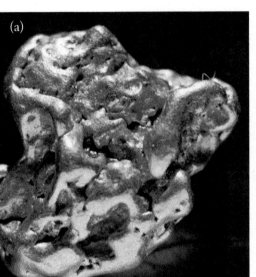

(b)

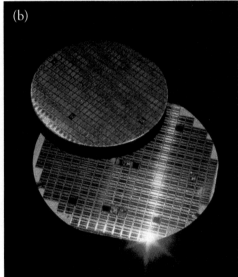

(c)

▲ **Figure 3.1** A selection of some common elements: (a) gold, (b) silicon and (c) bromine. How would you start to classify them?

Classifying anything makes study easier. You can put cricket teams into leagues, people into families, politicians into political parties, and so on. The elements form millions of compounds, and each element and each compound has its own chemical and physical properties. The modern Periodic Table is the system of classifying elements in use today, but in the past scientists have made several attempts to classify the elements.

3.1 The modern Periodic Table

The **Periodic Table** is an arrangement of the elements based on their proton (atomic) number.

The modern Periodic Table is shown in Figure 3.2. The Periodic Table contains all the known elements and arranges them on the basis of their proton (atomic) numbers into:

- horizontal rows, called periods; and
- vertical columns, called groups.

The rows are of such a length that elements with similar physical and chemical properties fall directly beneath one another.

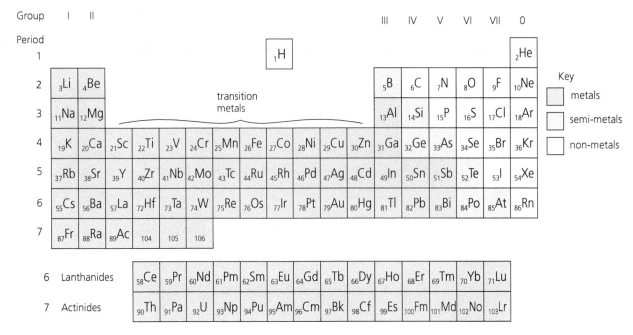

▲ **Figure 3.2**   The 'long form' of the modern Periodic Table. Chapter 6 contains more about parts of the Periodic Table.

A **period** is a horizontal row in the Periodic Table. The periods are numbered using Arabic numerals (1, 2, 3, etc.).

A **group** is a vertical column in the Periodic Table. The groups are numbered using Roman numerals (I, II, III, etc.).

If you look at the complete Periodic Table, known as the 'long form' of the Periodic Table, you will see the following:

- There are seven horizontal rows (**periods**).
- There are 18 vertical columns (**groups**).
- The lanthanides and actinides are separated from the rest of the Periodic Table and belong to Periods 6 and 7, respectively.
- Group II and Group III are separated by the transition metals.
- Hydrogen is not included in any group.

In our discussion of the Periodic Table, we will be focusing mainly on a shortened form of the Table, containing Groups I and II, and III to 0. You can see this form of the Periodic Table in Figure 3.3.

Figure 3.3   A shortened form of the modern Periodic Table ▶

Group Period	I	II			III	IV	V	VI	VII	0
1				H						He
2	Li	Be			B	C	N	O	F	Ne
3	Na	Mg			Al	SI	P	S	Cl	Ar
4	K	Ca			Ga	Ge	As	Se	Br	Kr
5	Rb	Sr			In	Sn	Sb	Te	I	Xe
6	Cs	Ba			Tl	Pb	Bi	Po	At	Rn
7	Fr	Ra								

Find out more

The origins of the Periodic Table

Several early chemists, such as Johann Döbereiner (1780–1849) and John Newlands (1837–98), made attempts to arrange the elements so as to highlight similarities among them, but these attempts were only partially successful.

Then, in 1869, Dmitri Mendeleev (1834–1907), a Russian chemist, proposed that a definite relationship existed between chemical properties of elements and their atomic masses. Mendeleev observed a recurring pattern when the elements were placed in horizontal rows in order of increasing atomic mass. With this arrangement he found that chemically similar elements could be placed in vertical columns beneath each other.

Mendeleev's classification, called the Periodic Table, has been so successful in making sense of the chemistry of the elements that it has been called 'one of the most important one-page documents ever produced'.

▲ **Figure 3.4** Mendeleev's Periodic Table, here carved 15 m high on the building in St Petersburg where he worked

H.G.H. Moseley (1887–1915) contributed to the further development of the Periodic Table by arranging the elements according to their atomic number (proton number) instead of their atomic masses.

Q1 Find out more about:
  a Dobereiner's Triads;
  b Newland's Octet Rule.

3.2 Position of elements in the Periodic Table

We can use the proton number of an element to locate it in the Periodic Table.

- As we go across a period, proton numbers increase by 1.
- If we use the proton number to write the electronic configuration of an element, we can predict the period and group to which it belongs.

The proton numbers and electronic configuration of the elements in the first three periods are shown in Table 3.1.

▼ **Table 3.1** The link between proton number and electronic configuration

	Group I	Group II	Group III	Group IV	Group V	Group VI	Group VII	Group 0
Period 1	$_1$H 1							$_2$He 2
Period 2	$_3$Li 2,1	$_4$Be 2,2	$_5$B 2,3	$_6$C 2,4	$_7$N 2,5	$_8$O 2,6	$_9$F 2,7	$_{10}$Ne 2,8
Period 3	$_{11}$Na 2,8,1	$_{12}$Mg 2,8,2	$_{13}$Al 2,8,3	$_{14}$Si 2,8,4	$_{15}$P 2,8,5	$_{16}$S 2,8,6	$_{17}$Cl 2,8,7	$_{18}$Ar 2,8,8

What patterns can you see in Table 3.1? Did you notice all of the following points?

- All the atoms in a particular period (row) have the same number of shells occupied by electrons.

- The number of electrons in the outer shell increases by 1 in consecutive elements across a particular period.

- The number of shells containing electrons is the same as the period number. For example, Period 1 elements have one shell with electrons, Period 2 elements have two shells with electrons, etc.

- When electrons enter a new shell, a new period is started.

- All the atoms in a particular group (column) have the same number of electrons in their outer shell.

- The number of inner shells of electrons increases by one in consecutive elements down a group.

- The number of electrons in the outer shell is the same as the group number. For example, Group 1 elements have one electron in their outer shell, Group 2 elements have two electrons in their outer shell, etc.

You can work out the electronic configuration of an element if you know its position in the Periodic Table. The element in Period 3 and Group VII will have three electron shells, with seven electrons in its outer shell.

The Periodic Table also gives an indication of whether a particular element is a metal or a non-metal. Look at Figure 3.5. Non-metals are found to the right of the 'staircase'. All the other elements, which are to the left of the staircase, are metals. The elements that lie along the line of the staircase show a combination of metallic and non-metallic properties, and are sometimes called **metalloids**.

Metalloids are elements that do not clearly display the properties of either metals or non-metals.

There are two groups that include metals only. These are Group I (known as the alkali metals) and Group II (known as the alkaline earth metals). There are two non-metallic groups. These are Group VII (known as the halogens) and Group 0 (known as the noble gases or the inert gases). Group III contains only metals, with the exception of boron.

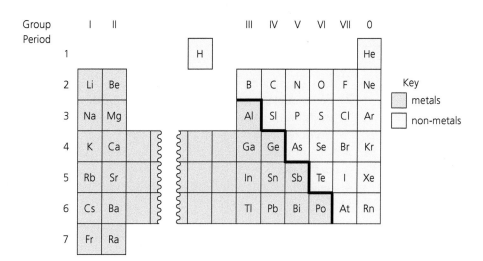

Group	I	II				III	IV	V	VI	VII	0
Period											
1				H							He
2	Li	Be				B	C	N	O	F	Ne
3	Na	Mg				Al	Si	P	S	Cl	Ar
4	K	Ca				Ga	Ge	As	Se	Br	Kr
5	Rb	Sr				In	Sn	Sb	Te	I	Xe
6	Cs	Ba				Tl	Pb	Bi	Po	At	Rn
7	Fr	Ra									

Key
■ metals
□ non-metals

Figure 3.5
Metals are on the left of the 'staircase' and non-metals are on the right. ▶

Table 3.2 summarises the differences in physical properties of metals and non-metals.

▼ **Table 3.2** Comparing metals and non-metals

Most metals	Most non-metals
solids with high melting points, giving high boiling point liquids	gases or low melting point solids, giving low boiling point liquids
shiny, reflecting light of many wavelengths	dull, reflecting light poorly or absorbing strongly
high density	low density
usually hard	usually soft
malleable, ductile, strong	often brittle, weak
good conductors of heat and electricity	insulators of heat and electricity

▲ **Figure 3.6** Copper, a representative metal

▲ **Figure 3.7** Sulphur, a representative non-metal

Experiment 3.1   Build models of elements

Procedure

1 Using commonly available materials, make models of the first twenty elements.
2 For example, you could use string or plasticine to form the nucleus and electronic shells, while you could represent the sub-atomic particles (protons, neutrons and electrons) with colour-coded styrofoam balls or marbles.
3 Use the examples of the helium, carbon and sodium atoms in Table 2.3 in Chapter 2 to practice.
4 Take some pictures of your models and display tem in your laboratory.

3.3 Patterns in the Periodic Table

When the elements are arranged in periods and groups in the Periodic Table, they show other patterns besides those in electronic configuration. We can see regular changes in both **atomic radii** and **electronegativity**.

The **atomic radius** of an atom is an indication of its size.

The **electronegativity** of an element indicates how strongly its atoms attract electrons.

Atomic radius

Figure 3.8 shows the variation in atomic radii for selected elements. Can you see the following general trends in the atomic radii?
- Atomic radii decrease across a period.
- Atomic radii increase down a group.

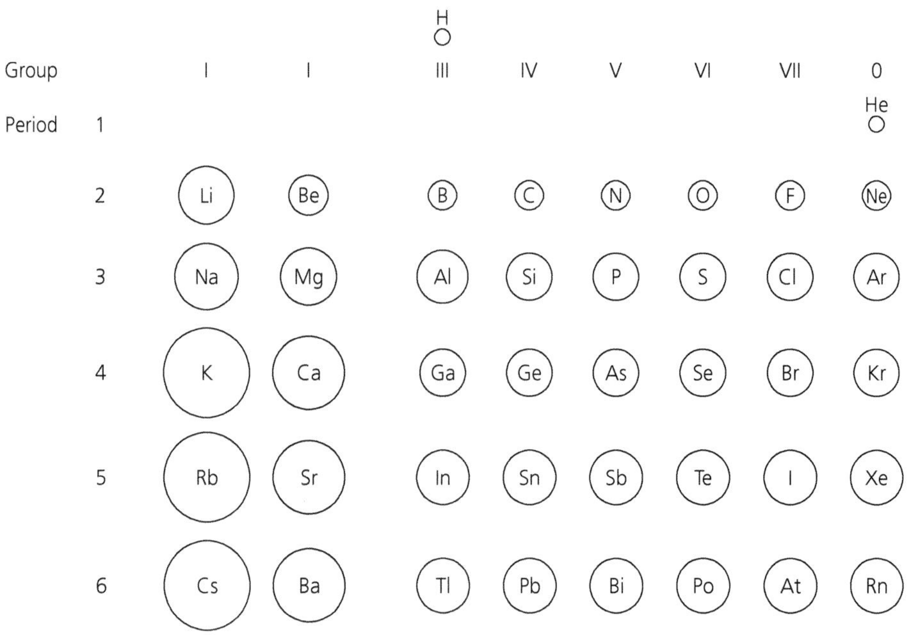

Figure 3.8
The atomic radii of selected elements ▶

To understand these trends, you will need to remember that the nucleus is positively charged and that it attracts electrons (which are negatively charged). The greater the number of shells between the outer electrons and the nucleus, the less the outer electrons feel the effect of the nucleus. As you will expect, the bigger the charge of the nucleus, the greater its pull on the outer electrons, thus making the atomic radius smaller.

As we go across a period, the nuclear charge increases but the number of electron shells remains the same. As a result, the attraction of the nucleus for the outer electrons increases and the atomic radius decreases.

On the other hand, as we go down a group, there are more electron shells and the attraction between the nucleus and the outer electrons becomes weaker. The result is that the atomic radius increases.

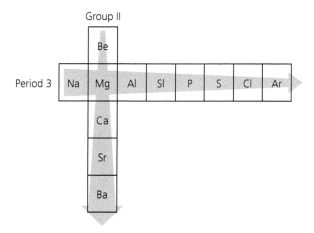

Figure 3.9
Atomic radius decreases across a period and increases down a group. ▶

Electronegativity

Some atoms attract electrons more than others. As the atomic radius decreases, the effect of the nucleus will be felt more strongly at the outside of the atom. This means that electronegativity increases across a period from left to right.

As the atomic radius increases, the effect of the nucleus will be felt less at the outside of the atom. This means that electronegativity decreases down a group.

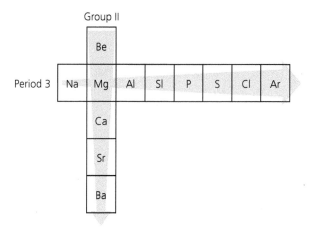

Figure 3.10
Electronegativity increases across a period and decreases down a group. ▶

The elements on the right-hand side of the Periodic Table (the non-metals) will tend to attract electrons more readily than those on the left (the metals). The noble gases are an exception to this. On the other hand, the elements on the left-hand side of the Periodic Table (the metals) will attract their outer electrons less and will show a greater tendency to lose them than those on the right (the non-metals). You will see the significance of these patterns in Chapter 6.

Practice

1 To which groups of the shortened form of the Periodic Table do the metals belong?

2 Fill in the gaps in these sentences.
    a Atomic radius _____ as a group is descended and _____ as we go across a period.
    b Electronegativity _____ as a group is descended and _____ as we go across a period.

3 Which groups and periods of the Periodic Table do the following belong to? Use the proton number to work out your answers.
    a $_{16}S$
    b $_{7}N$
    c $_{18}Ar$

Summary

- The elements are classified in a framework called the Periodic Table.
- The elements in the Periodic Table are arranged in order of increasing proton (atomic) number.
- The Periodic Table makes it easier to study the properties that elements have in common.
- The Periodic Table consists of:
  - groups – vertical columns of elements;
  - periods – horizontal rows of elements.
- All the elements in a group have the same number of electrons in their outer shell. The number of electrons in the outer shell is the same as the group number.
- All the elements in the same period have the same number of electron shells. The number of shells occupied by electrons is the same as the period number.
- In any period, there is a change from metallic elements on the left to non-metallic elements on the right.
- Metals show a greater tendency to lose their valence electrons than non-metals.
- The non-metals (with the exception of the noble gases) attract electrons more readily than metals.

End-of-chapter questions

1 A group in the Periodic Table is:
    A A horizontal row of elements
    B A vertical column of elements
    C Made up of very different elements
    D Has elements with the same number of electrons

2 Which one of the options describes some properties of metals?
  A Low melting and boiling points
  B Brittle and soft
  C Average density
  D High melting and boiling points

3 Which of the following are examples of Group VII elements?
  A Sc, Ca, K, Re
  B Li, Be, Ag, F
  C Fl, Br,Cl,At
  D He, Xe, Ca, Li

4 What happens to the atomic radii of elements across a period?
  A It increases
  B It varies
  C It grows then decreases
  D It decreases

5 What happens to the electronegativity of elements across a period?
  A It decreases
  B It fluctuates
  C It increases
  D It remains constant

---

6 Table 3.3 shows part of the Periodic Table, with the elements represented by their atomic numbers. Use this portion of the Periodic Table to answer the questions that follow.

▼ **Table 3.3**

1							2
3	4	5	6	7	8	9	10
11	12	13	14	15	16	17	18

  a Identify by proton (atomic) number:
    (i) two metals;
    (ii) two non-metals;
  b Explain the following terms:
    (i) A period
    (ii) A group
  c Describe patterns in electronic configuration:
    (i) in periods;
    (ii) in groups.

7 Explain what is meant by:
  a atomic radius;
  b electronegativity.

8 Table 3.4 is a section of the Periodic Table.

▼ **Table 3.4**

$^{7}_{3}Li$ $^{9}_{4}Be$ $^{11}_{5}B$ $^{12}_{6}C$ $^{14}_{7}N$ $^{16}_{8}O$ $^{19}_{9}F$ $^{20}_{10}Ne$

$^{23}_{11}Na$ $^{35}_{17}Cl$

$^{39}_{19}K$

a Draw diagrams to show the atomic structure of:
   (i) lithium (Li);
   (ii) potassium (K);
   (iii) fluorine (F);
   (iv) chlorine (Cl).

b Use your diagrams from part (a) to explain the following:
   (i) Lithium (Li) has a smaller atomic radius than potassium.
   (ii) Fluorine (F) is more electronegative than both lithium (Li) and chlorine (Cl).

9 Iodine is a black, crystalline, shiny solid and belongs to Group VII. Sodium is a soft, shiny solid and belongs to Group I.
   a Classify iodine and sodium as metal or non-metal.
   b List two reasons for your classification in each case.
   c Sodium belongs to Group I and Period 3 of the Periodic Table, and iodine belongs to Group VII and Period 5. Based on this information, what can you say about the atoms of sodium and iodine?

10 An outline of the Periodic Table is given below.

							E										
X												U					
	M								L	Q		V	T				
Y					S									D			

a Identify, by letter:
   (i) two elements that are in the same group;
   (ii) two elements that are in the same period;
   (iii) an element that exists as a monatomic gas and that is unreactive;
   (iv) the two elements that form non-covalent bonds;
   (v) the element with electronic configuration 2,8,2;
   (vi) the most abundant metallic element in the Earth's crust;
   (vii) the most abundant element in the Earth's atmosphere.
b Name an element that is more electronegative than element T.
c Why is element E placed separately from the rest of the Periodic Table?

4 Chemical bonding

In this chapter, you will study the following:

- the three main types of chemical bonds: ionic, covalent and metallic;
- how the atomic structure of atoms affect the types of bonds they form;
- formulae to represent ions, molecules and formula units;
- descriptions of ionic, simple molecular and giant molecular crystals;
- how the structure of sodium chloride, diamond and graphite relate to their properties and uses;
- what is meant by allotropy.

This chapter covers
Objectives 5.1–5.8 of Section A of the CSEC Chemistry Syllabus.

You now know that the individual atoms of elements are much too small for us to see. However, we can see a sheet of the element aluminium because it contains vast numbers of aluminium atoms combined together. Atoms of different types can also combine. Oxygen atoms and hydrogen atoms combine, for example, to form water. What is it that holds the atoms together?

(a) 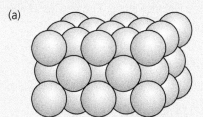   (b)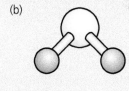

▲ **Figure 4.1** Diagrams showing the atoms in (a) aluminium and (b) water

◀ **Figure 4.2**
This glass of water contains many particles of oxygen and hydrogen atoms combined together.

4.1 What changes when elements combine?

A **compound** is a substance formed when two or more different types of elements combine chemically.

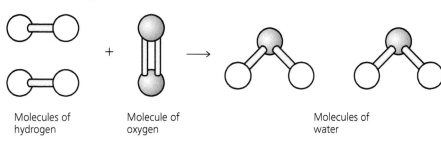

Molecules of hydrogen

Molecule of oxygen

Molecules of water

▲ **Figure 4.3** The formation of water from hydrogen and oxygen

A chemical **bond** is a force of attraction between combining atoms.

Valence electrons are the electrons in the outer electron shell of an atom.

Ions are charged particles formed when atoms lose or gain electrons.

Molecules are particles formed when atoms combine by sharing electrons.

The '**noble gases**' are found in Group 0 of the Periodic Table.

Atoms combine by forming **bonds**. When they do so, new types of particles are formed.

Here is a summary of what happens during chemical bonding.

- Atoms use their outer shell or **valence electrons** to form bonds.
- Atoms may lose, gain or share their valence electrons.
- As these changes occur, the electron configurations of the atoms change and new particles are formed. (These are **ions** or **molecules**.)
- The new configuration of each atom will be (or will appear to be) identical to that of the **noble gas** nearest to it in the Periodic Table. Noble gases have filled valence shells, which are stable electron configurations. Elements combine to acquire a stable electronic configuration.

The rest of this chapter explains all this in more detail.

Which atoms lose, gain or share electrons?

- Metal atoms containing 1, 2 or 3 valence electrons tend to lose electrons. The larger the atomic radius of the atom, the more easily it loses electrons (see Chapter 6).
- Some non-metal atoms with 5, 6 or 7 valence electrons may gain a number of electrons to fill their valence shell. The smaller the non-metal, atom the more readily it accepts electrons.
- Non-metal atoms containing 4 to 7 valence electrons may also share electrons when combining with other non-metals.

4.2 The formation of bonds

Ionic bonds are the attractive forces that hold oppositely charged ions together in electrovalent compounds.

There are three main types of chemical bonds:
- The electrovalent or **ionic bond**
- The covalent bond
- The metallic bond

4.3 The ionic bond and the transfer of electrons

Ionic bonds are formed when a metal (e.g. sodium or magnesium) reacts chemically with a non-metal (e.g. chlorine or oxygen).

'Dot and cross' diagrams are used to illustrate the formation of bonds. Only the outer shell electrons (the valence electrons) are shown for the combining atoms. Although all electrons are the same, dots (•) and crosses (×) are used to indicate which atom the electrons came from.

A **cation** is a positively charged ion. Cations are formed by metals.

An **anion** is a negatively charged ion. Anions are formed by non-metals.

We can use a '**dot and cross**' diagram to show the bonding between sodium and chlorine. Look at Figure 4.4 and note the following stages.

- Each sodium atom transfers its valence electron to the valence shell of a chlorine atom.
- As a result of this transfer, the sodium atoms are changed to positive ions (**cations**). The charge on the positive ion is equivalent to the number of electrons lost.
- Each chlorine atom is changed to a negative ion (**anion**). The charge on the negative ion is equivalent to the number of electrons gained.
- The positive ions and negative ions formed have new electronic configurations – that of the nearest noble gas in the Periodic Table.

- Attractions between the oppositely charged ions provide the binding forces that hold ionic compounds together.

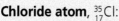

sodium atom	chloride atom	sodium ion	chloride ion
2,8,1	2,8,1	2,8	2,8,8
		EC of neon	EC of argon

▲ **Figure 4.4** Formation of sodium chloride

The smallest part of an ionic compound is known as the **formula unit**.

- The **formula unit** of sodium chloride (NaCl) consists of one sodium ion and one chloride ion.

Understand it better

Sodium atom, $^{23}_{11}$Na:
- Number of protons = 11 (11 positive charges)
- Number of electrons = 11 (11 negative charges)

A sodium atom is neutral. Remember that protons are not involved in bond formation.
The sodium atom now loses one electron:
- Number of protons = 11 (11 positive charges)
- Number of electrons = 10 (10 negative charges)

The overall charge is 1 positive charge.
A sodium cation (Na^+) is formed.

The charge on the Na^+ ion is equivalent to the number of electrons transferred.
The Na^+ ion has the same electron configuration as neon ($_{10}$Ne is 2,8).

Chloride atom, $^{35}_{17}$Cl:
- Number of protons = 17 (17 positive charges)
- Number of electrons = 17 (17 negative charges)

A chloride atom is neutral.
The chloride atom now gains one electron:
- Number of protons = 17 (17 positive charges)
- Number of electrons = 18 (18 negative charges)

The overall charge is 1 negative charge.
A chloride anion (Cl^-) is formed.

The charge on the Cl^- ion is equivalent to the number of electrons gained.
A Cl^- ion has the same electron configuration as argon ($_{18}$Ar is 2,8,8).

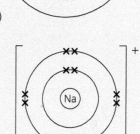

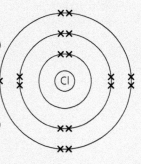

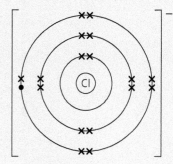

Some more examples of ionic bonding are the formation of potassium fluoride, KF (Figure 4.5); magnesium chloride, $MgCl_2$ (Figure 4.6) and lithium oxide (Li_2O, Figure 4.7).

Note that all valence electrons transferred from the metal must be accepted by the non-metal. This means that the ratio of metal to non-metal is variable. For example, a magnesium atom has two valence electrons and therefore two chlorine atoms are needed to accept the valence electrons from the magnesium atom. The formula unit of magnesium chloride ($MgCl_2$) consists of one magnesium ion and two chloride ions. (Look at Section 7.1 for more details.)

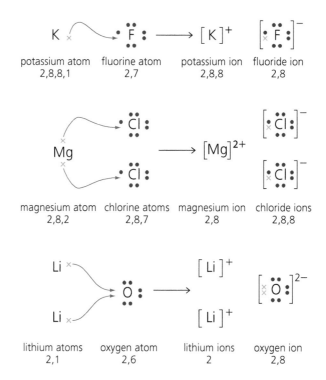

Figure 4.5
The formation of potassium fluoride – the formula unit is KF ▶

Figure 4.6
The formation of magnesium chloride – the formula unit is $MgCl_2$ ▶

Figure 4.7
The formation of lithium oxide – the formula unit is Li_2O ▶

Practice

1 Draw dot and cross diagrams to show the bonding in:
  a CaO;
  b Mg_3N_2.

2 Write the electronic configuration of the following ions:
  a Li^+
  b Mg^{2+}
  c Al^{3+}
  d Ca^{2+}
  e S^{2-}
  f N^{3-}
  g H^+

3 Which of the ions in question 2 have the same electronic configuration as that of neon?

4   In which of the following compounds does the cation and anion have exactly the same electron configuration?
    a   KCl
    b   NaF
    c $MgCl_2$
    d   CaO
    e Mg_3N_2
    f   KBr

4.4   The covalent bond

Covalent bonds are formed when atoms of non-metallic elements combine with one another. Here is a summary of covalent bonding:

*   The non-metal atoms share one or more pairs of electrons.

*   Each non-metal atom shares a number of electrons so that each atom appears to have the electronic configuration of the nearest noble gas in the Periodic Table.

*   Each shared pair of electrons constitutes a covalent bond.

*   The terms single, double, and triple covalent bonds are used to describe the sharing of one, two or three pairs of electrons, respectively, between a pair of atoms.

*   When non-metal atoms share electrons, the particles formed are called **molecules**.

*   The attraction between the nuclei of the atoms and the shared pair(s) of electrons provides the binding force which holds the atoms together.

Molecules are particles formed when two or more atoms combine by covalent bonds. The atoms in a molecule may be the same or different.

The formation of covalent bonds may be represented by using dot and cross diagrams or by a line (or dash) between the symbols for the atoms.

Covalent bonds between non-metal atoms of the same element

The fluorine molecule

Lone pairs or **non-bonded pairs** are pairs of electrons found in molecules that are not involved in the formation of simple covalent bonds.

Each isolated fluorine atom contains unpaired electrons. These isolated electrons form a pair of shared electrons in the fluorine molecule. The shared electron pair represents a single covalent bond (Figure 4.8). Each fluorine atom in a fluorine molecule contains three **lone pairs**.

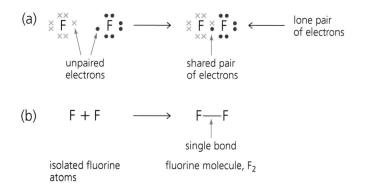

Figure 4.8
Forming a fluorine molecule, F_2 ▶

The oxygen molecule

Each isolated oxygen atom contains two unpaired electrons. These isolated electrons form two pairs of shared electrons in an oxygen molecule. The two shared electron pairs represent two covalent bonds, known as a double bond (Figure 4.9). Each oxygen atom in an oxygen molecule contains two lone pairs.

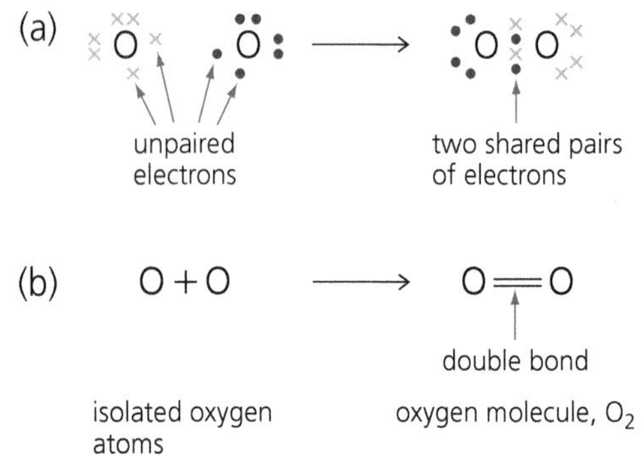

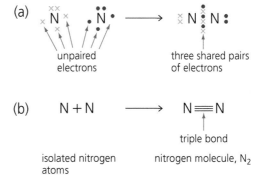

Figure 4.9
Forming an oxygen
molecule, O_2 ▶

The nitrogen molecule

Each isolated nitrogen atom contains three unpaired electrons. These isolated electrons form three pairs of shared electrons in a nitrogen molecule. The three shared electron pairs represent three covalent bonds, known as a triple bond (Figure 4.10). Each nitrogen atom in a nitrogen molecule contains one lone pair.

Figure 4.10
Forming a nitrogen
molecule ▶

Covalent bonds between different atoms

Here are a range of covalent molecules formed between different atoms.

Water (H_2O)

In water, an oxygen atom is singly bonded to two hydrogen atoms. There are two lone pairs of electrons on the oxygen atom (Figure 4.11).

Figure 4.11
Water ▶

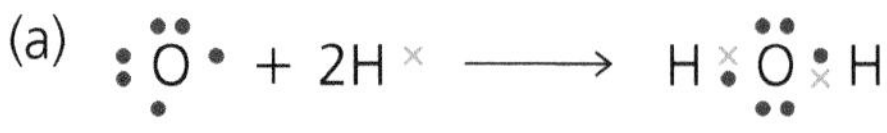

Ammonia (NH_3)

In ammonia, a nitrogen atom is singly bonded to three hydrogen atoms. There is one lone pair of electrons on the nitrogen atom (Figure 4.12).

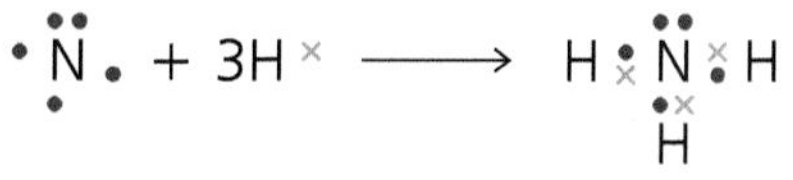

Figure 4.12
Ammonia ▶

$$\cdot \overset{\cdot\cdot}{\underset{\cdot}{N}} \cdot \ + \ 3H^\times \longrightarrow \ H \overset{\cdot\cdot}{\underset{\overset{\bullet\times}{H}}{N}} \overset{\times}{}H$$

Methane

In methane, a carbon atom is singly bonded to four hydrogen atoms. There are no lone pairs of electrons on the central (carbon) atom. All the valence electrons of carbon are used in the bonding in the methane molecule (Figure 4.13).

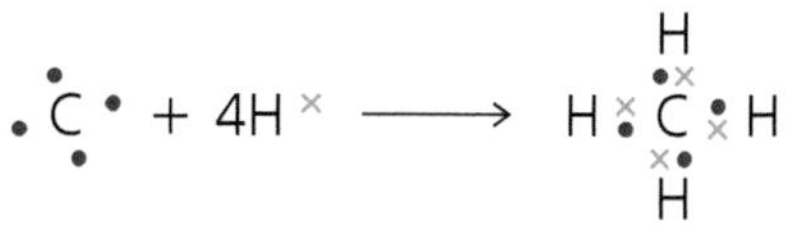

Figure 4.13
Methane ▶

$$\cdot \overset{\cdot}{\underset{\cdot}{C}} \cdot \ + \ 4H^\times \longrightarrow \ H \overset{\overset{H}{\overset{\bullet\times}{}}}{\underset{\underset{H}{\times\bullet}}{C}}H$$

Carbon dioxide

In carbon dioxide, a carbon atom is doubly bonded to two oxygen atoms. Two electron pairs are shared between each oxygen atom and the carbon atom, giving two double bonds within the molecule (Figure 4.14).

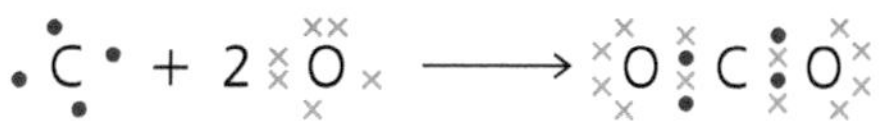

Figure 4.14
Carbon dioxide ▶

$$\cdot \overset{\cdot}{\underset{\cdot}{C}} \cdot \ + \ 2 \overset{\times\times}{\underset{\times}{O}}_\times \longrightarrow \ \overset{\times\times}{\underset{\times}{O}} \overset{\cdot}{\underset{\cdot}{C}} \overset{\times}{\underset{\times}{O}}^\times$$

Hydrogen cyanide

In hydrogen cyanide, the carbon atom forms a single bond with the hydrogen atom and a triple bond with the nitrogen atom (Figure 4.15).

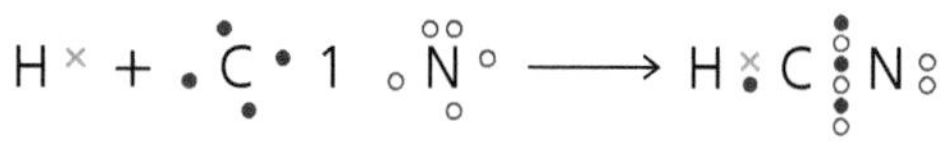

Figure 4.15
Hydrogen cyanide ▶

$$H^\times + \cdot \overset{\cdot}{\underset{\cdot}{C}} \cdot \ 1 \ {}_\circ \overset{\circ\circ}{\underset{\circ}{N}}{}^\circ \longrightarrow \ H \overset{\cdot}{\underset{\cdot}{C}} \overset{\circ}{\underset{\circ}{N}}{}_\circ$$

Practice

5 How many lone pairs of electrons are there in the following molecules?
   a CH_4
   b CF_4
   c H_2S
   d NH_3
   e HF
   f CO_2

Understand it better

Here are some quick tips for working out simple covalent bonding.

1 How many covalent bonds will each non-metal atom form?

Element	Electronic configuration	Number of electrons to add to achieve noble gas configuration	Number of covalent bonds formed by each atom
$_1H$	1	1	1
$_7N$	2.5	3	3
$_8O$	2.6	2	2

In many of the simple examples you will study, the non-metal elements will usually form the same number of bonds as the number of electrons needed to fill their valence shell.

You can test this generalisation by showing the bonding in the following covalent substances:

a F_2
b CH_4
c CO_2

2 How to work out the configuration of atoms after covalent bonding:

 Each F contains six non-bonding electrons and two shared electrons.

a For each atom, count the shared electrons and the non-bonding electrons (if any) in the valence shell.
b Consider the number of electrons in the inner shells.
c Compare the configuration with the noble gas at the end of the same period.

Coordinate covalent bonding

In all the examples shown on the previous page (Figures 4.8 to 4.15), each atom contributes one electron to the shared pairs of electrons. This is described as simple or ordinary covalent bonding.

The sharing of electrons can occur in another way. Here, both electrons in the bond are donated by one of the combining atoms. The first atom uses a pair of non-bonded electrons (a lone pair) to bond with the second atom. The second atom contributes no electrons. The bond formed is described as a **coordinate** or **dative covalent bond**.

A **coordinate** or **dative covalent bond** is formed when both electrons in the shared pair come from the same atom.

The nitrogen atom in the ammonia molecule uses its lone pair to form a coordinate bond with the H^+ ion from the acid. The ammonium ion, NH_4^+, is formed. The ammonium ion has a positive charge and is electrostatically attracted to the Cl^- ion from the acid.

(a)

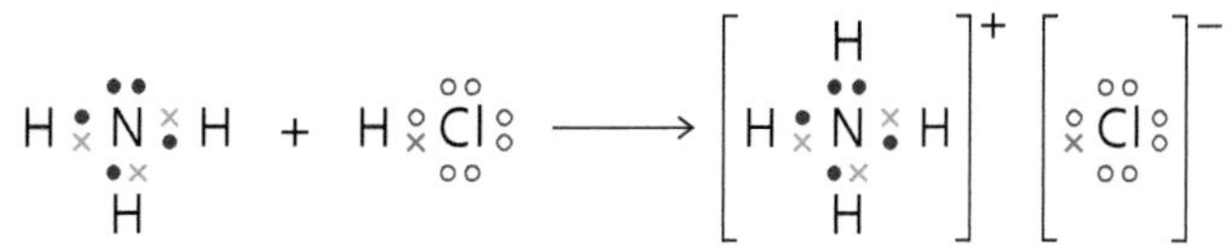

$$H \overset{\cdot\cdot}{\underset{\cdot\times}{\cdot N \times}} H \ + \ H \overset{\circ\circ}{\underset{\circ\circ}{\circ Cl \circ}} \longrightarrow \left[H \overset{\cdot\cdot}{\underset{\cdot\times}{\times N \cdot}} H \right]^{+} \left[\overset{\circ\circ}{\underset{\circ\circ}{\times Cl \circ}} \right]^{-}$$

Figure 4.16
Forming a coordinate
bond in the ammonium
ion ▶

(b) NH_3 +   HCl   ⟶ NH_4Cl

ammonia   hydrogen   ammonium
             chloride   chloride

One of the lone pairs in a water molecule can also form a coordinate covalent bond to a hydrogen ion which has no electrons. The product is the oxonium (or hydroxonium) ion. It is this ion that is characteristic of acids.

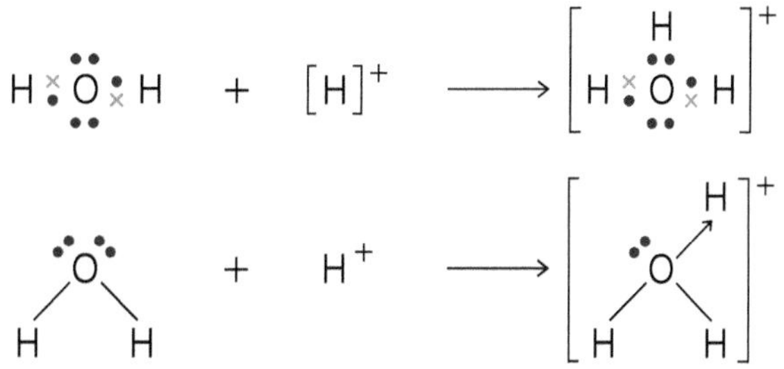

(a)

$$H \overset{\cdot\cdot}{\underset{\cdot\cdot}{\times O \cdot}} H \ + \ [H]^{+} \longrightarrow \left[H \overset{\cdot\cdot}{\underset{\cdot\cdot}{\times O \cdot}} H \right]^{+}$$

(b)

Figure 4.17
The coordinate bond in
the oxonium ion ▶

a water molecule   a hydrogen   the oxonium ion
with two lone pairs   cation with
                        no electrons

Practice

6 Write electron dot and cross diagrams for the following compounds:
  a CO_2
  b $SiCl_4$
  c BF_3
  d H_2O_2
  e CS_2
  f OCl_2
  Which of these molecules can form coordinate bonds?

4.5 Non-polar and polar covalent molecules

When a covalent bond is formed between two different types of atom, the atoms attract the electron pair to different extents. Atoms of fluorine, oxygen, nitrogen and chlorine attract electrons more strongly than atoms of other elements, and are therefore strongly electronegative (see Section 3.3). In molecules such as hydrogen chloride, the electron pair of the bond is pulled closer to the more electronegative chlorine atom. As a result of this, the chlorine atom develops a tiny negative charge (written as δ−), whereas the hydrogen atom develops a tiny positive charge (written as δ+). The hydrogen chloride molecule has a slight separation of charge within it and is described as a **polar molecule**.

A **polar molecule** is a
covalent molecule that
has areas with different
charges due to unequal
sharing of electrons.

$$\overset{\delta+}{H} \text{---} \overset{\delta-}{Cl}$$

Understand it better

Look back at the information on variation in electronegativity of elements in the Periodic Table (see Chapter 3). Locate the elements listed in Table 4.1 in the Periodic Table and explain the trend in electronegativity shown.

▼ **Table 4.1** Electronegativities of H, C, Cl and F

Element	Electronegativity
hydrogen	2.1
carbon	2.5
chlorine	3.0
fluorine	4.0

Q1 Calculate the difference in electronegativity between the pair of atoms in the following molecules: HCl, HF, CH_4.

Q2 Which molecule is the most polar?

By contrast, a molecule of hydrogen (H_2) or chlorine (Cl_2) is non-polar, since the atoms are identical and have the same attraction for the bonding electron pair.

Examples of highly polar molecules include hydrogen fluoride (HF), water (H_2O), and ammonia (NH_3) (Figure 4.18). It is important to note that all polar molecules are electrically neutral overall, since the sum of the small positive and negative charges equals zero.

Polar covalent compounds have higher melting and boiling points, and react differently with water than do non-polar molecules.

Figure 4.18
Water and ammonia are polar covalent molecules. ▶

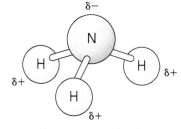

the water molecule   the ammonia molecule

ORR

Experiment 4.1 Testing liquids for polarity

Apparatus and materials

- Water
- Propanone
- Ethanol
- Methylbenzene
- Trichloroethane
- A burette
- A plastic rod (e.g. the barrel of a ball-point pen or a comb)

Procedure

1 Fill the burette with the liquid.

2 Open the tap and allow the liquid to flow out in a steady stream.

3 Bring the plastic rod near the stream of liquid (but without touching it) and observe what happens.

4 Now rub the plastic rod on your trousers or skirt.

5 Bring it close to the stream of liquid again and observe what happens.

Figure 4.19 illustrates what happens when the liquid used is water.

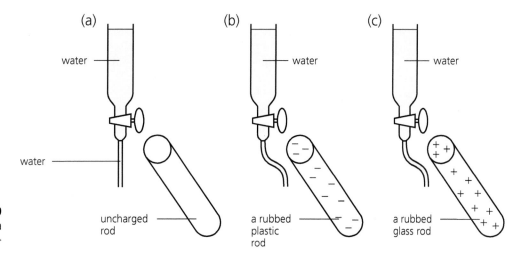

Figure 4.19
Deflecting water with a charged rod ▶

Explanation

The plastic rod acquires a negative charge when it is rubbed. Since water is attracted to the negatively charged rod, it is reasonable to infer that water is polar and has partial positive charges.

A stream of water will also be attracted to a Perspex rod rubbed on fur. In this case, the rubbed Perspex rod acquires a positive charge, showing that water also has partial negative charges.

4.6 Metallic bonding

Metal atoms do not bond together by ionic bonds or by covalent bonds. In a metal, the valence electrons leave the atoms, which then form positive ions. The electrons lost by a particular atom do not necessarily remain associated with the resulting ion. In fact, the electrons are mobile and they flow through the spaces between the positive ions.

Metals can be viewed as orderly arrangements of positive ions held together in a 'sea' of freely moving electrons.

This type of bonding, known as metallic bonding, is used to explain how the particles in metallic elements such as iron, aluminium and copper, are held together.

The metallic bond is therefore the strong electrostatic force of attraction that exists between the stationary positive ions and the mobile electrons (the electron 'sea') in the metal.

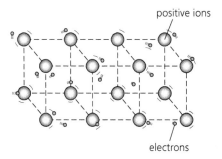

Figure 4.20
A lattice of positive ions (cations) in a sea of electrons ▶

The properties of metals are summarised in Table 4.2.

▼ **Table 4.2** Some properties of metals

Property of metals	Comment
Metals are usually hard solids with high melting points (exceptions are mercury, the only liquid metal, and Group I metals, which are soft enough to be cut with a knife).	Metals contain strong electrostatic bonding; positive ions are held together by a 'sea of electrons'.
Metals conduct electricity.	The mobile electrons are responsible for conducting the current (not ions as in electrolytes).
Metallic conductors have certain features that are different from the giant ionic conductors.	There is no separation of ions as in ionic compounds.
Metals conduct in the solid as well as the liquid state. Electrical conductivity decreases with increasing temperature.	Mobile electrons are present in both the solid and liquid state.
Metals are good conductors of heat in both the solid and the liquid state. If two ends of a metal rod are at different temperatures, there is an energy flow from the hot end to the cold end until the whole bar is at the same temperature. Examples of metals with high thermal conductivity are silver, copper, gold, aluminium, tungsten, zinc and iron.	The particles at the warmer end of the metal rod have more kinetic energy than those at the colder end. Since the particles are closely packed together in the metal structure, they pass on energy by collisions.

Metals are malleable (can be easily beaten into shape) and ductile (can be drawn into a wire).	When metals are beaten, the layers of atoms slide easily over one another. The layers of metal ions can move relative to one another without losing cohesion because of the free electrons between the planes.
Metals are shiny when clean. **Note:** Some metal surfaces tarnish, for example, aluminium reacts with oxygen to form a thin layer of oxide; silver objects tarnish when they form silver sulphide on their surface.	Light falling on any surface is either absorbed or reflected. Most metals reflect light of all wavelengths and they are therefore shiny.

As you will recall, metals contain positive ions (cations) in a 'sea of electrons' and the bonding is electrostatic in nature. The positive ions in a metal are arranged in a regular repeating pattern and are capable of only restricted motion (vibration) when cold.

It is the free-moving electrons (the sea of electrons) that account for many of the properties of metals.

4.7 Structure and properties

Bonding helps us to understand how different substances are formed. In this section, we will see that the properties of elements and compounds depend on the type of particles formed in bonding and the forces between the particles.

> The **structure** of a solid specifies the arrangement of the particles and the types of bonds that keep the particles in fixed positions within the solid.

In solids, where, as you know, the particles occupy fixed positions, the properties of the solid also depend on the arrangement of the particles, that is the **structure** of the solid.

We can use our understanding of chemical bonding in elements and compounds to explain their properties. Here are some examples:

- Graphite, which is a form of the element carbon, is slippery.
- Copper(II) sulphate crystals are brittle.
- Candle wax melts easily.
- Copper conducts electricity well.
- Sodium chloride (common salt) dissolves well in water but does not dissolve in oil.

Ionic, simple molecular, and giant molecular crystals

A crystal is a solid substance, the structure of which has an ordered arrangement of atoms, molecules or ions. We can classify the structure of such solids as either (i) an ionic crystal, (ii) a simple molecular crystal or (iii) a giant molecular crystal, depending upon the type of bonding present (ionic or covalent), as well as the arrangement of the atoms or molecules in the compound. Solids that do not have a crystalline structure are called amorphous.

4.8 Structure and properties of ionic compounds

Ionic crystals

A feature of ionic compounds is that they form structures that have a well-defined crystal lattice. This is facilitated by the oppositely charged ions (cations and anions), which arrange themselves in an ordered and regular way. Sodium chloride (NaCl) is probably the most common example of an ionic compound that you will come across. The ions are arranged in such a way that each Na^+ is surrounded by six Cl^-, and each Cl^- is surrounded by six Na^+.

▶ **Figure 4.21**
What is the shape of these sodium chloride crystals?

▶ **Figure 4.22**
Table salt is sodium chloride.

Properties of ionic compounds

The **heat of fusion** is the amount of energy that must be absorbed per unit mass of the sample, to melt the sample without a change in temperature

Sodium chloride is a good example of an ionic compound. Here is a summary of the properties of ionic compounds:

- They are hard, brittle crystalline solids.
- They have high melting points and **heats of fusion**.
- They conduct electricity well when molten or when dissolved in water.
- They do not conduct electricity in the solid state.
- (Most) dissolve readily in water.
- They react readily with each other in solution.

Structure of ionic compounds

A **lattice** is a regular arrangement of points in two or three dimensions. A crystal lattice is the arrangement of points on which atoms, molecules or ions are centred in a crystal.

Ionic solids are formed by ionic bonding – the particles in the solids are ions. These solids are crystals because of their structure. The regular structure of a crystal tells you that the ions are arranged in an orderly manner (see Figure 4.21). We call this orderly arrangement a crystal **lattice**.

If you look at crystals of salt in Figure 4.21, you will see that they have a regular shape. This is because the sodium chloride structure is a regular, repeating arrangement of ions. Each sodium ion is surrounded by six chloride ions as its nearest neighbours. Each chloride ion, in turn, has six sodium ions as its nearest neighbours. This arrangement of ions results in a closely packed three-dimensional structure. Two models of sodium chloride crystal structures are shown in Figure 4.23.

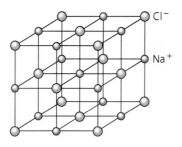
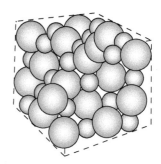

Figure 4.23
The sodium chloride crystal lattice ▶

Ionic compounds are said to consist of 'a giant structure of ions'. Table 4.3 explains how the properties of ionic compounds depend on the type of particles they contain (that is, ions) and also on their structure.

▼ **Table 4.3**   Linking the properties of ionic compounds to their structure

Properties of ionic compounds	Comment
crystalline solid	due to regular arrangement of ions, resulting from strong attractions between opposite charges
conduct electricity when molten	on melting, ions are set free: these ions move to an oppositely charged electrode when a voltage is applied
high melting points, high boiling points, high heats of fusion, high heats of vaporisation	these high values indicate that the ions are held together strongly – therefore lots of energy is needed to separate the ions

4.9   Structure and properties of simple covalent (molecular) substances

Properties of simple covalent substances

Most substances that contain covalent bonds are also described as simple molecular substances, because they consist of separate molecules. Many covalent substances show the properties outlined in Table 4.4.

▼ **Table 4.4**   Some typical properties of covalent substances

Properties of molecular substances	Comment
some are liquids or gases at room temperature	they consist of small molecules with weak attractive forces between them
low melting and boiling temperatures, low heats of fusion and vaporisation	again due to weak intermolecular forces
some are soluble in water, while some are also soluble in non-polar organic solvents such as methylbenzene	the non-polar molecular substances dissolve in non-polar solvents; polar substances dissolve in both types of solvent
do not conduct electricity when molten	this is due to the absence of ions (**Note:** A few react with water to produce ions and the resulting 'solution' conducts electricity.)

Intermolecular forces are forces between individual molecules (such as van der Waals forces). They are weak forces.

Intramolecular forces are forces within molecules (covalent bonds). They hold the atoms together in the molecule. These are strong forces.

Covalent substances tend to be liquids or gases at room temperature because the forces between their particles (the molecules) are weak. These intermolecular forces must not be confused with intramolecular forces. 'Inter-' means between; 'intra-' means within.

- **Intermolecular forces** are discussed below.
- **Intramolecular forces** are the chemical bonds we have discussed so far in this chapter.

Simple molecular crystals

In simple molecular compounds, the number and arrangement of each different type of atom is fixed in the molecule. The bond between any two atoms is a covalent one. Therefore, when we write carbon dioxide as CO_2, it means that for every molecule of carbon dioxide present, there is one atom of carbon separately bonded to two oxygen atoms. Similarly in molecular iodine, I_2, there are two atoms of iodine bonded to each other.

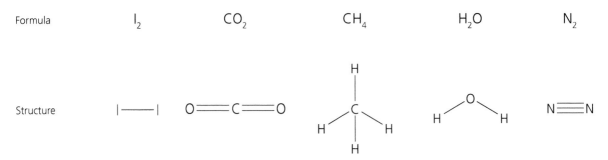

▲ **Figure 4.24**   Examples of common simple molecular structures

In a crystal of iodine, the I_2 molecules pack together in layers, which we describe as a herring-bone pattern. In this pattern, the iodine molecules are further away (about 1.25 times) from the molecules above and below them, than they are from molecules in their own layer.

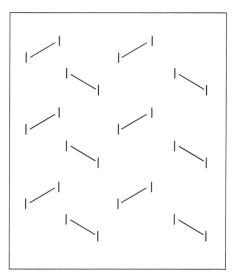

▲ **Figure 4.25** The arrangement of iodine molecules in a crystal of iodine

▲ **Figure 4.26** A sample of iodine crystals

4.10 Intermolecular forces

We will discuss two types of intermolecular forces:

- Van der Waals forces
- Hydrogen bonds

These forces are neither ionic nor covalent.

Van der Waals forces

The uneven distribution of electronic charge leads to positive and negative charges within molecules. These are called **dipoles**.

All covalent molecules, whether polar or non-polar, develop temporary or instantaneous **dipoles**. This results from the uneven movement of all the electrons within the molecules. Van der Waals forces are the weak attraction between oppositely charged ends of molecules with temporary dipoles.

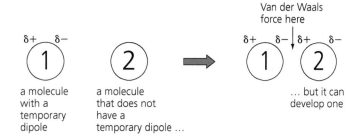

▲ **Figure 4.27** Van der Waals forces. When molecule 2 develops a temporary dipole then it will be attracted to molecule 1.

Hydrogen bonds

The **hydrogen bond** is the weak attraction between an electron-deficient hydrogen atom in one molecule and an electron-rich electronegative atom in another molecule.

The **hydrogen bond** is the weak attraction between the slightly positive hydrogen atom in one polar molecule and the slightly electronegative atom in another polar molecule of the same type or of a different type.

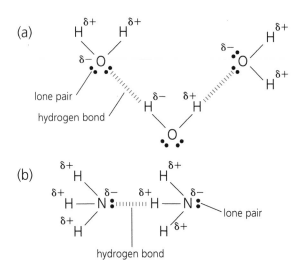

(a)

lone pair

hydrogen bond

(b)

lone pair

hydrogen bond

▲ **Figure 4.28**   Hydrogen bonding in (a) water and (b) ammonia

4.11   Structure of simple covalent solids

Solids containing discrete covalent molecules generally have lower melting and boiling points than giant ionic solids, because the covalent molecules have weak intermolecular forces. Such solids are referred to as **simple covalent solids**. Compare the melting points of covalent molecules such as hydrogen chloride (melting point −115 °C) and of ionic compounds such as sodium chloride (melting point 801 °C).

Examples of such simple molecular solids are the elements sulphur (S_8) (see Figure 4.29), phosphorus (P_4) and iodine (I_2), the compounds solid carbon dioxide (dry ice) and solid water (ice). The sublimation of iodine (see Section 1.5) is evidence for the presence of weak intermolecular forces in such solids.

Figure 4.30 shows the arrangements of the molecules in crystals of solid ice.

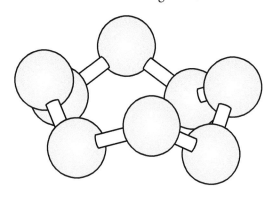

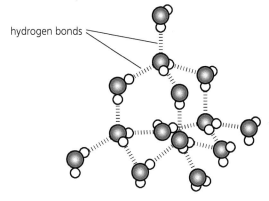

hydrogen bonds

▶ **Figure 4.29**
Sulphur exists as S_8 molecules.

▲ **Figure 4.30**   The molecular structure of ice. Discrete molecules of water are held together by hydrogen bonds.

Allotropes are different forms of the same element existing in the same physical state.

Sometimes the particles of the same element can be arranged in different ways, giving rise to different forms. For example, the elements sulphur and phosphorus have more than one crystalline form. When an element can exist in different forms in the same physical state, it is said to exhibit allotropy. The different forms of the element are known as **allotropes**.

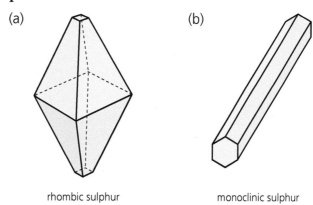

▲ **Figure 4.31** These two allotropes of sulphur form differently shaped crystals. (a) Rhombic sulphur is also known as alpha sulphur. (b) Monoclinic sulphur is also known as beta sulphur.

4.12 Giant covalent solids

A few covalent solids (such as diamond, graphite and silica) have high melting and boiling points as well as high heats of fusion and vaporisation. Like simple covalent solids, these solids are also formed by covalent bonding, but they do not form individual molecules. Instead, they exist as macromolecules in which very strong covalent bonds extend in three dimensions ('macro-' means 'large'.) Such compounds are referred to as giant covalent solids. They are usually insoluble in both polar and non-polar solvents.

Giant molecular crystals

Just as with simple molecular compounds, compounds with a giant molecular structure also have covalent bonds that join the atoms together. However, unlike simple molecular structures, the number of atoms joined to each other in giant molecular structures is not definite, with the size of the molecule theoretically continuing indefinitely for the size of the sample. Even though the size of the molecule is undefined, the arrangement of the atoms follows such a clearly defined and ordered structure, that we say that they have a repeat unit. This repeat unit represents what the expanded molecule looks like.

Some examples of giant molecular structures that you would have encountered at home include the following:

▲ **Figure 4.32** The Hope Diamond at the Smithsonian Museum of Natural History

- Polymers (see Sections 4.1–4.3 for more information).
- Silicon dioxide: The major component of beach sand and the glass in your windows is silicon dioxide. Silicon dioxide has an empirical formula of SiO_2.
- Diamond: Diamond is composed of covalently bonded carbon atoms. Each carbon atom is bonded to four other carbon atoms in a tetrahedral formation. This arrangement is repeated throughout the entire diamond lattice in all directions.

  Pure diamonds are colourless; it is the presence of trace amounts of minerals that produce diamonds of various colours, such as blue, red, green and yellow diamonds.

- Graphite: Just like diamond, graphite is also made up of covalently bonded carbon atoms. However, the atoms are arranged differently in graphite – the covalent bonds exist only in two dimensions. Graphite's structure comprises hexagonal layers of carbon atoms, and each carbon atom is bonded to only three other carbon atoms. As you will see in Section 5.7, this difference in structure between graphite and diamond means that they have very different physical properties, which hence impacts upon their uses.

▲ **Figure 4.33** Sample of graphite crystals.

Experiment 4.2   Make models of ionic and covalent structures

Procedure

1 Collect suitable materials that you can use to make models. For example, you could use plasticine, Styrofoam balls of varying sizes, plastic straws or similar type materials.

2 Refer to diagrams in the text and make models to represent the structures of NaCl, graphite and diamond. Take pictures of your models.

It is important that we are able to classify compounds in this way. As you will see in the next chapter, each of these structural classes has its own set of physical properties. These properties distinguish each class from the other classes. This is especially important when we discover or create a new compound, as its physical properties will provide good clues as to what type of bonding holds it together and the shape of its structure.

▼ **Table 4.5**   Comparing diamond and graphite

Diamond	Graphite
In diamond, each atom is bonded to four others.	In graphite, the atoms are arranged in flat six-membered rings.
Diamond is one of the hardest substances known. It is widely used in cutting and drilling.	The lubricating properties of graphite can be explained in terms of the ability of the flat sheets to slide past each other.
Due to the absence of free (uncombined) electrons, diamond does not conduct electricity.	Graphite conducts electricity because of the presence of free electrons in its structure.

▲ **Figure 4.34** Diamond

▲ **Figure 4.35** Graphite

Silica is the main component in sand. It is a very hard solid, though not as hard as diamond. (Diamond is the hardest naturally occurring substance.) The structure of silica is similar to diamond.

4.13 Differences in physical properties of ionic and molecular solids

Consider the nature of the bonding in an ionic compound: the ionic bonds extend in all directions throughout the solid. These electrostatic forces of attraction create a crystal lattice that is tightly held together. Think of the common ionic compounds that you would have encountered before, such as NaCl (table salt) and $MgSO_4$ (Epsom salts). They are both solids at room temperature and have a high melting point.

Some ionic compounds, such as aluminum oxide (Al_2O_3; melting point = 2 072 °C), have such high melting points that they are used as refractory materials. Refractory materials are substances that are chemically and physically stable at high temperatures, and so can be used as linings for furnaces and incinerators.

In contrast, there are those substances that have a simple molecular structure. For example, most substances with a single molecular structure, except iodine, are either liquids or gases at room temperature. These substances are liquids at room temperature because the forces that hold the separate molecules together in simple covalent structures are much weaker than those in ionic compounds. In fact, we can describe these simple molecular substances as having a structure analogous to a brick wall held together by weak mortar. In this example, while the individual bricks (the molecules) are difficult to destroy (because of the strong covalent bonds), it is fairly easy to separate individual bricks from each other in the wall, because of the weak intermolecular forces of attraction holding the structure together.

Figure 4.36
Structure of a simple
molecular solid. A and
B represent the atoms
of a molecule bonded
together covalently. ▶

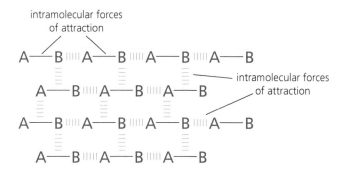

intramolecular forces
of attraction

intramolecular forces
of attraction

The result of having these weak intermolecular forces is that only a small amount of energy is needed to overcome the force of attraction between molecules in a sample, and so room temperature often provides enough energy for these substances to be in a liquid or gaseous phase. Simple molecular compounds that are solid at room temperature have low melting points. In some cases, for example, iodine, gentle warming causes sublimation, which is when the solid is transformed directly to a gas, without passing through the intermediate liquid state.

▲ **Figure 4.37** Iodine crystals at room temperature

▲ **Figure 4.38** Iodine sublimes to purple vapours when gentle warmed.

There are other differences between ionic and molecular solids, all of which are due to the differences in structure and intermolecular forces of attraction between the two types of substances.

If you go to the beach and accidentally get the sea water in your mouth, you would have noticed that it tastes salty, even though you cannot see any particles of NaCl. The saltiness of sea water is due mainly to the presence of NaCl, which is highly soluble in water. This solubility is a feature of ionic compounds, but what makes ionic solids so water soluble? Even though a water molecule is electrically neutral overall, its hydrogen atoms have a slight positive charge and its oxygen atom carries a slight negative charge. The charged ions of an ionic solid are easily surrounded by the oppositely charged centres of the water molecules, which overcome the electrostatic attractions of the ionic solid in a process called solvation.

Both Na^+ and Cl^- ions are absolutely necessary for us to lead healthy lives. The easiest way of getting these ions is by including salt in your diet. Approximately 60% of the human body is water, and this provides the required medium for transporting NaCl around the body.

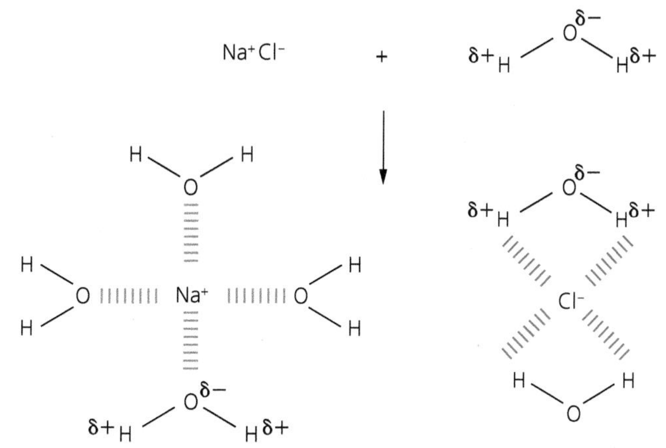

$$Na^+Cl^- \quad + \quad \delta+_H \diagup O^{\delta-} \diagdown H^{\delta+}$$

▲ Figure 4.39 Solvation of a typical ionic solid by water

Organic solvents (such as hexane and toluene) do not have this charge-separation that water has, and we say that they are non-polar or hydrophobic (water-fearing) solvents. Since the separation of ions is necessary for ionic solids to dissolve, these organic solvents cannot facilitate solvation, and so we have ionic compounds that have low or no solubility.

Conversely, simple molecular solids are insoluble in water but soluble in organic solvents. This is because, in terms of solubility, 'like attracts like'. This means that solvents that have only a small charge separation will attract and dissolve those solids that also have only a small charge separation. Some molecular compounds appear to dissolve in water, but this is because they chemically react with it, and are not simply dissolving in the water (which is a strictly physical process).

Let us now consider another property of ionic solids. You may have been told at some point that you should not turn electrical appliances, electrical plugs and light switches on or off when your hands are wet, because of the danger of electrocution. Pure water itself does not conduct electricity, but water that you get from your taps has dissolved ions in them. These ions act as vehicles for transporting electrical charge, and help to explain why aqueous solutions of ionic solids conduct electricity. Like any vehicle, for it to be of use, it must be able to move. In the rigid crystal lattice of ionic solids, the ions are confined to their positions and cannot move, and so ionic solids do not conduct electricity. However, when molten (liquid), the ions are free to move, which means that in a liquid state, ionic compounds are good conductors of electricity.

Contrast that to simple molecular solids, which do not conduct electricity whether dissolved in organic solvents or in their liquid state. This is because there are no ions or electrons that can carry the electrical charges.

Experiment 4.3 Differences in the physical properties of ionic and simple molecular solids

Procedure

1 Gently warm 0.5 g of iodine crystals in a boiling tube.

2 Now gently heat 0.5 g of NaCl.

3 Record your observations and account for any differences you observed.

4 Prepare the following mixtures:

    a 0.5 g I_2 + 10 ml hexane

    b 0.5 g I_2 + 10 ml water

    c 0.5 g NaCl + 10 ml hexane

    d 0. 5 g NaCl + 10 ml water

5 Comment on the relative solubilities of I_2 and NaCl in hexane and water respectively.

6 Using a conductivity meter, measure the conductivity of each of the mixtures.

7 Explain your observations in steps 3, 5 and 6, making particular reference to the structure and bonding present in iodine and sodium chloride.

4.14 Properties and uses of sodium chloride, diamond and graphite

We have already discussed the properties of ionic compounds, using sodium chloride as an example. The uses of these compounds are related to their properties. In this section, we will focus on the uses of diamond and graphite.

You have already been introduced to the fact that although both diamond and graphite have a giant molecular structure, the actual arrangement of their atoms is significantly different, which in turn impacts upon their properties and uses.

Think about a use for graphite and diamond. As an example, the 'lead' in the pencils that you write and draw with is really graphite mixed with some clay (the amount of clay added determines the hardness of the pencil giving rise to the classifications of 2B, 2HB, HB, and so on). Diamonds are precious stones that are often used in rings and other items of jewellery. Think about the properties that diamond and graphite must have in order to be used in the way that they are. If you drop a pencil in water, or if water falls on a diamond ring, neither dissolves, unlike NaCl. This represents the first important property of diamond and graphite: they are both insoluble in water. This should not be surprising if you think about what makes a substance soluble in water (see section 5.6). To be water soluble, a substance must either have ions that can be surrounded by the water molecules (diamond and graphite are covalently bonded structures) or react with the water (carbon does not chemically react with water).

We can account for the other properties of graphite and diamond by examining their structures:

- Melting point: Both diamond and graphite have very high melting points ($> 3\,500°C$), which are much higher than most ionic compounds. Remember that to melt a substance, you need to supply enough energy to overcome the attractive forces that hold the individual molecules or ions together. However, we can regard both diamond and graphite as having thousands of atoms joined together in one very large molecule. The extended network of covalent bonds present in both structures requires a large amount of energy to be separated, which explains why they have such high melting points. Graphite's ability to maintain its structure at high temperatures explains its use as a refractory material.

- Conductivity: In order for a substance to conduct electricity, it must have mobile electrons or ions that are capable of carrying current. In graphite, each carbon atom has only three bonds, and since carbon has four valence electrons, there is a free electron per carbon atom. In a similar way to metallic bonding, these free electrons carry electric charge, and we can use graphite in electrodes in circuits and batteries. If you have ever seen the inside of a normal AA alkaline battery, the black central part is made of graphite. Conversely, diamond has no free electrons (or ions), and so does not conduct electricity.

- Hardness: The tetrahedral covalently bonded arrangement of carbon atoms extends throughout the entire diamond lattice, giving it a very rigid structure. While a very popular use for diamond is in making jewellery, it is its property as the hardest substance known allows it to be used as an important industrial material in abrasion, cutting metals and drills for mining. In contrast, graphite is a much softer material, because the covalent bonding network extends in only two dimensions and not three. Even though the hexagonal layers of carbon atoms are held together by strong covalent bonds, the bonds between these hexagonal layers are much weaker (similar to those present in simple molecular solids).

- Lubricating power: In graphite, the individual hexagonal layers can easily slide over each other (they are not attracted to each other by strong bonds), which means that graphite is an excellent lubricant, either alone or suspended in a liquid. Bicycle chains commonly use a graphite-based lubricant, because unlike oil, it is not sticky. The ability of the individual layers to slide over and separate from each other is also why graphite can be used in pencils, because it can leave a marks.

(a) (b)

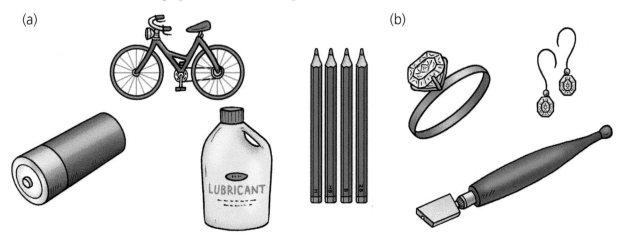

▲ **Figure 4.40** Uses of (a) graphite and (b) diamond

4.15 Allotropy

In the previous section, you learnt that both diamond and graphite are composed of only carbon atoms. The two substances have different structures of the same element. We call this phenomenon allotropy. Allotropy occurs where different forms of the same element of the same phase (solid, liquid or gas) exist. We call these forms allotropes.

Carbon has allotropes other than graphite and diamond, one of the most common being Buckminsterfullerene, which is a covalent compound with 60 carbon atoms and which resembles a football. The C60 molecule is very stable, and it can withstand high temperatures and pressures. The exposed surface of the structure can selectively react with elements and still keep its spherical shape. It can hold atoms and small molecules within its structure. The scientists who discovered the 'buckyball' in 1985; Sir Harold Kroto, Robert Curl and Richard Smalley; thought that its structure resembled the geodesic domes proposed by Buckminster Fuller, and hence borrowed his name for their new discovery. Small amounts of this allotrope are found in soot and have also been detected in outer space. It is also mentioned in some popular video games, science fiction movies and novels. Do you know the names of any?

▲ **Figure 4.41** Structural model of a buckyball or Buckministerfullerine. The carbon atoms are represented by the black spheres, which are bonded to each other via covalent bonds (white sticks).

Many common elements, besides carbon, exhibit allotropy. Some examples include the following:

- Sulphur: Orthorhombic and monoclinic sulphur, among others
- Phosphorus: Red, white, black and violet phosphorus

▲ **Figure 4.42** Allotropes of phosphorus. Notice the difference in their colour.

Practice

7 As you have seen, substances that exist as solids at room temperature may have different types of bonding and may also have different structures. Solids are usually classified into the following types:
- Giant ionic solids
- Simple covalent solids
- Giant covalent solids
- Giant metallic solids

Complete Table 4.6 for the solids shown.

▼ **Table 4.6**

Substance	Structure	Type of bonding	Appearance
copper			
sulphur			
sand			
salt			
graphite			

Summary

▼ Table 4.7

Type of bonding	Occurs between…	Involves…	Particles formed	Characterstics of bond	Properties of substances formed	Structure of solid state
Ionic	Metals and non-metals	Loss of electrons by metals and gain of electrons by non-metals.	Ions: cations (positive) and anions (negative)	Strong electrostatic attraction between oppositely charged particles.	Hard, crystalline solids High melting points, etc.	Giant ionic: e.g. sodium chloride
Covalent	Non-metal atoms	Sharing of electrons. May be single, double or triple bonds. Simple covalent bonds – one electron from each atom. Co-ordinate covalent bonds – both electrons from same atom.	Discrete molecules	Strong bond within molecule between nucleus and shared electrons. May be polar if atoms involved have different electronegativity. Weak intermolecular forces.	Usually liquids and gases or low melting point solids.	Usually simple molecular: e.g. solid carbon dioxide, ice
			A few form macromolecules		Some are hard, with high melting points.	Some are giant molecular: e.g. diamond, silica
Metallic	Atoms of the same metal	Atoms losing electrons which become mobile	Positive ions and mobile electrons.	Electrostatic bond between positive ions and electrons.	Hard, shiny, electrical; thermal conductors	Giant metallic

End-of-chapter questions

1 Which of the following describes water?
  A A polar liquid
  B A giant molecular structure
  C A mixture of hydrogen and oxygen
  D A compound of hydrogen and oxygen

2 All of the following, except one, have a giant molecular structure. Which is the odd one out?
  A Diamond
  B Silicon dioxide
  C Copper
  D Graphite

3 Which of the following groups are good conductors of electricity?
  A All ionic substances
  B Graphite, copper, iron
  C Graphite, diamond, sodium chloride
  D Water, hydrogen, carbon

4 Which one of the following describes the properties of metals?
  A They are solids at room temperature and have low melting points.
  B They conduct electricity only in a liquid form.
  C They can be easily made into different shapes.
  D Their electrical conductivity increases with increasing temperature.

5 Ionic bonding occurs between:
  A Atoms of the same metal
  B Metals and non-metals
  C Giant molecules
  D Different forms of the same element

---

6 The compound water (H_2O) is formed between ^1_1H and $^{16}_8O$.
  a Use dot and cross diagrams (valence shell only) to show the bonding in a molecule of water.
  b Name the type of bonding in a molecule of water.
  c On your diagram for part (a), label:
    (i) a bonding pair of electrons;
    (ii) a lone pair of electrons.
  d What is coordinate or dative bonding?
  e Use diagrams to show how a molecule of water can form the hydroxonium ion (H_3O^+) by dative bonding.

7 Table 4.8 shows some of the properties of five substances R, S, T, U and V.

▼ **Table 4.8**

Substance	Electrical conductivity when solid	Electrical conductivity when molten	Soluble in water?	Melting point (°C)
R	poor	good	yes	700
S	poor	poor	no	−65
T	good	good	no	2 000
U	poor	poor	no	−120
V	poor	poor	no	1 750

Select from Table 4.8:

a two substances that are solids at room temperature;

b a metal;

c a molecular substance;

d an ionic compound;

e a giant molecular substance.

8 Explain the following:

a Chlorine and fluorine can form the molecule ClF by covalent bonding but ionic bonding is not found in this compound;

b A solid substance X melts at 1 290 °C but it does not conduct electricity when solid or in the molten state

c Diamond and graphite are both pure forms of carbon, but graphite conducts electricity and diamond does not.

9 Compare electrical conduction of sodium chloride and the metal silver.

10 Explain the following:

a The melting point of water is much lower than the melting point of sodium chloride.

b A molecule of chlorine is non-polar but a molecule of hydrogen chloride is polar.

c Covalent bonding exists in both diamond and sulphur, but the melting point of diamond is much higher than that of sulphur.

11 When sodium bonds with fluorine to form sodium fluoride, ions are formed, but when fluorine bonds with fluorine, molecules are formed.

a Distinguish between ions and molecules, using the examples above.

b Describe how ionic bonding differs from metallic bonding.

(Proton numbers: Na = 11; F = 9)

5 Mixtures and separations

In this chapter, you will study the following:

- the differences between pure substances and mixtures;
- different types of solutions;
- suspensions and colloids, and their properties;
- how temperature affects the solubility of solids in water;
- separation techniques and their application;
- the extraction of sucrose from sugar cane.

This chapter covers
Objectives 2.1–2.6 of Section A of the CSEC Chemistry Syllabus.

The meaning of the word 'pure' in everyday life may be quite different from its meaning in Chemistry. We may, for example, refer to water that has been chlorinated as pure. When we say 'pure' in chemical terms we mean that a given material contains only one substance – either a single compound or a single element. In Chemistry, pure water contains only the compound water and is made up of water molecules only. Pure iron contains only the element iron and is made up of iron atoms only.

◄ **Figure 5.1**
Which type of water is pure?

5.1　How can we tell when a substance is pure?

Some of the characteristics of pure substances are shown in Table 5.1.

▼ **Table 5.1** Some characteristics of pure substances

Characteristic	Example
definite and constant composition	in water, the ratio of hydrogen to oxygen is always the same and water is always represented by the formula H_2O
definite and constant physical properties, such as fixed melting and boiling points, under a given set of conditions	pure water, for example, melts/freezes at 0 °C (273 K) and boils at 100 °C (373 K) at a pressure of 1 atmosphere
distinct chemical properties	water reacts vigorously with the metal sodium
show only a single spot when analysed by chromatography	a pigment containing only one dye shows a single spot (discussed later in this chapter)

The following two experiments can be used to show whether or not a substance is pure.

Experiment 5.1　To determine the melting point of a solid

Procedure

1　Insert a fine capillary tube, open end down, into the finely crushed solid to be tested.

2　Turn the capillary tube over and tap it gently on a hard surface to ensure that all materials move towards the closed end and are evenly packed.

3　Add more material until the capillary tube contains not less 5 mm of compacted solid at the bottom (closed) end.

4　Set up the apparatus as shown in Figure 5.2.

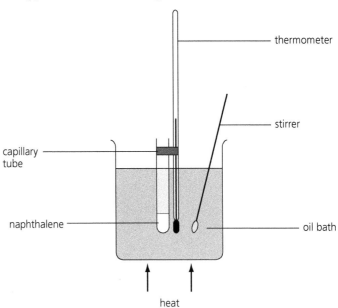

Figure 5.2
The apparatus for determining the melting point of a solid ▶

5 Gently heat the oil bath, while stirring, and note the temperature, T_s, at which melting just begins and the temperature, T_f, at which melting is completed.

6 Take the mean of the readings T_s and T_f as the melting point of the solid. For pure solids, T_s and T_f should not differ by more than 1 °C.

Experiment 5.2   To determine the boiling point of a pure liquid

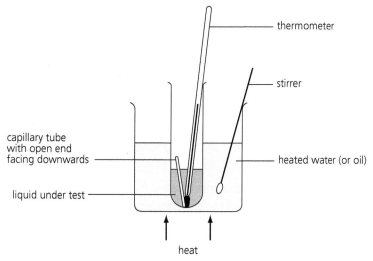

Figure 5.3
The apparatus for determining the boiling point of a liquid ▶

Procedure

1 See Figure 5.3 for how to arrange the apparatus.

2 Place 1 or 2 cm³ of the liquid under test in a test-tube.

3 Place a fine capillary tube (one end sealed), open end downwards, into the liquid sample.

4 Place the test-tube in the water bath and gently heat, while stirring, until steady streams of bubbles issue from the open end of the capillary tube.

5 Allow the temperature of the bath to decrease slowly until bubbling from the capillary tube just ceases. Note this temperature.

6 This temperature is the boiling point of the liquid.

Question

Q1   Why is the boiling temperature taken as the temperature when bubbling just ceases?

Impurities in solids result in the solid melting over a range of temperatures. Impurities can cause a lowering of the melting temperature.

The presence of dissolved solids causes the boiling temperature to increase.
An increase in external pressure leads to an increase in boiling temperature.

Practice

A pure substance is either a single element or a single compound.
1 **a** Define (i) element and (ii) compound.
   b List the differences between a compound and an element.

5.2   Mixtures

A mixture is made up of two or more substances not chemically combined together.

Examples of mixtures are:
- mixtures of elements such as copper and zinc, as in brass;
- mixtures of elements and compounds, as in air;
- mixtures of several compounds, such as in crude oil.

Mixtures melt and boil over a range of temperatures rather than at fixed points. This is in keeping with their composition. The properties of the mixture are a combination of the properties of the different substances it contains. Also, as you shall see later, the components of a mixture can be separated relatively easily by using physical methods, such as filtration and distillation.

There are two broad categories of mixtures.

- **Homogeneous mixtures** have the same composition throughout. Examples are a **solution** of sodium chloride (common salt) in water and brass, which is an **alloy** of copper and zinc.

- **Heterogeneous mixtures** have a composition that is not uniform. Examples are a mixture of sand and water; and milk, which is a mixture of butter fat and water.

Often the different components of a heterogeneous mixture can be readily distinguished by the naked eye. However, in some cases, such as blood and milk, the mixture appears uniform, unless you use a microscope.

A **solution** is a homogeneous mixture containing one or more solutes dissolved in a solvent.

An **alloy** is a solution of a metal and another solid element (usually another metal).

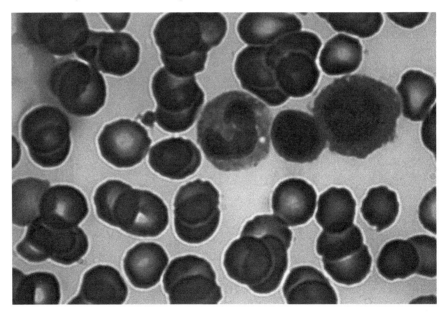

Figure 5.4
Blood is a heterogeneous mixture, but you need to use a microscope to see the different components. ▶

Homogeneous mixtures

Solutions

A solution is a stable homogenous mixture in which one substance (the solute) is dissolved in another substance (the solvent).

Solutions in which water is the solvent are called **aqueous solutions**.

Solvents other than water are called **non-aqueous solvents**.

Homogeneous mixtures are solutions. Any solution has two components: the solute and the solvent. In most of the solutions you will deal with, the solvent is water and the solute is a solid. However, a wide range of solutions exist. The solute and solvent may be solid, liquid or gas, as shown in Table 5.2. The solvent is considered to be the component that is present in the larger quantity. Note that all gaseous mixtures are solutions.

▼ **Table 5.2** Examples of different types of solutions

Solute	Solvent	Examples
gas	liquid	oxygen in water carbon dioxide in fizzy drinks
liquid	liquid	alcoholic drinks (alcohol in water) gasoline
solid	liquid	sugar in water iodine in ethanol
solid	solid	alloys, e.g. brass, bronze, coinage metals

The distinguishing features of solutions are:

* the solute and solvent are thoroughly mixed – all parts of the solution have the same chemical composition, chemical properties and physical properties;

* the solute and solvent do not separate when the solution is allowed to stand;

* the particles of the solute are not visible, even under an optical microscope;

* the solution may be coloured, but is usually transparent if the solvent is a liquid;

* the solute may, in many cases, be separated from the solvent by purely physical means.

▲ **Figure 5.5**
The world famous Angostura Bitters from Trinidad and Tobago is a unique solution.

Find out more

Water dissolves many solutes, and is called the 'universal solvent'. However, water does not dissolve all solutes. Non-aqueous solvents are needed to dissolve things such as paint and nail polish. Examples of non-aqueous solvents are propanone (also known as acetone), ethanol (often just known as alcohol), 1,1,1-trichloroethane and methylbenzene.

Q1 Which non-aqueous solvents are used to dissolve the following:
   a Grass   **d** Oil
   b Paint   **e** Chewing gum
   c Nail polish

Q2 What does the term 'like dissolves like' mean?

Q3 **a** What is 'dry cleaning'?
     b What solvent or solvents are used in dry cleaning?

More about solutions

Solubility

A solute is considered soluble if it dissolves readily in a given solvent. Sparingly soluble substances dissolve only to a small extent in a given solvent. Often when we think of a solid as insoluble, it is really sparingly soluble. Sugar and sodium chloride are readily soluble in water, but solid iodine is only sparingly soluble in water. Solid iodine dissolves easily in ethanol, but solid sodium chloride is insoluble in ethanol.

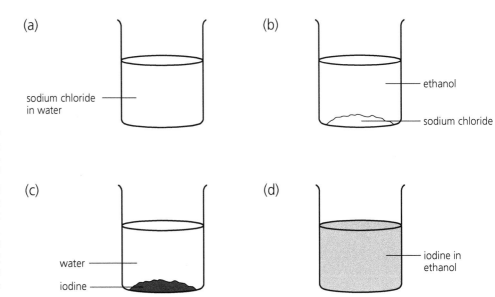

Figure 5.6
Solubility depends on both the nature of the solute and the nature of the solvent. (a) Sodium chloride dissolves in water, but does not dissolve in ethanol (b). (c) Iodine dissolves slightly in water, but completely in ethanol (d). ▶

The extent to which a solute dissolves in a particular solvent depends on:

- the nature of the solute and the solvent;
- the temperature;
- the pressure.

Ionic and polar covalent substances often dissolve in polar solvents, such as water. Non-polar covalent substances usually dissolve in non-polar solvents, such as hexane, ether and 1,1,1-trichloroethane.

It is generally the case that more solid dissolves as the temperature increases. However, the solubility of gases in liquids generally decreases as the temperature increases.

Pressure is significant for gases dissolved in liquids. Most gases increase their solubility as pressure increases.

The **solubility** of a solute is the number of grams of it that dissolve in 100 g of solvent at a given temperature and pressure.

To define **solubility** then, we must specify certain conditions, such as the solvent and the temperature. Graphs that show how solubility varies with temperature are known as solubility curves (see Experiment 5.3).

The following procedures will increase the rate at which a solid dissolves in a solvent (how quickly it dissolves):

- Crushing increases the surface of the solute exposed to the solvent.
- Stirring brings more solvent in contact with solute.
- Heating increases the movement of the solute particles, causing more mixing.

Concentration of solutions

Sometimes we can describe the **concentration** of a solution in general terms, while at other times we need to be more precise. Here are some general descriptions of solutions.

The **concentration** of a solution is the amount of solute that is dissolved in a fixed volume of solution.

- A dilute solution contains a small quantity of solute dissolved in the solvent.
- A concentrated solution contains relatively large quantities of solute dissolved in the solvent.
- A saturated solution contains as much solute as the solvent can possibly dissolve at a particular temperature and pressure in the presence of undissolved solute. A solution that is saturated at one temperature may not be saturated at another temperature.
- A supersaturated solution contains more solute than the solvent can normally dissolve at a given temperature and pressure. Supersaturated solutions are unstable. Disturbances, for example, shaking, stirring or the addition of crystals (a process known as seeding), cause a supersaturated solution to throw out excess solute and become saturated.

More precise descriptions of solutions are given in Chapter 9.

Figure 5.7
Copper(II) sulphate solutions of different concentrations. Which is the most dilute and which is the most concentrated? ▶

 ORR
M&M
A&I

Experiment 5.3   Determining the solubility of potassium nitrate at different temperatures

Apparatus and materials

- boiling tube
- burette
- thermometer
- potassium nitrate crystals

Theory

A saturated solution at any temperature can be prepared as follows:

- Dissolve the solute at a temperature that is 10–20 °C higher than that required.
- When no more solute goes into solution, cool the solution to the desired temperature. Excess solute will crystallise (i.e. fall out of solution), leaving a saturated solution of potassium nitrate.

Procedure

1 Using the burette, put 10 cm³ of water into the boiling tube.

2 Add 15 g of potassium nitrate crystals to the water and heat gently, until all the crystals dissolve.

3 Cool the solution until crystallisation just begins. Record the temperature of crystallisation, T_1, and the volume of solvent, V_1.

4 Add 2 cm³ of water to the above mixture. Heat again, while stirring to dissolve the crystals. Cool the resulting solution until crystallisation begins. Again record the temperature of crystallisation, T_2, and the new volume of solvent, V_2.

5 Add further 2 cm³ portions of water, repeating the above process and recording the temperatures, T_3, T_4, T_5, etc. and the corresponding total volume of solvent V_3, V_4, V_5, etc.

6 Use the following relationship to determine the solubility at each temperature:

solubility in grams of solute per 100 g of water $= \dfrac{15 \times 100}{V_s}$,

where V_s is the volume of solvent

Results

In one such experiment, the following results were obtained:

▼ **Table 5.3**

Temperature (°C)	74	66	58	54	49	45	42	40
Solubility of KNO₃ (g per 100 g of water)	150	125	107	94	83.5	75	68	62.5

1 Plot a graph of solubility versus temperature.

2 Use your graph to:
    a predict the solubility of potassium nitrate at 100 °C and at room temperature;
    b calculate the mass of crystals that fall out of a saturated solution of potassium nitrate in 100 g of water when it is cooled from 60 °C to 40 °C.

Practice

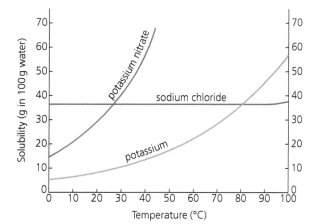

Figure 5.8
Standard solubility curves for some salts. Note that solubility varies with temperature. ▶

2 Which salt is the most soluble at 30 °C?

3 For which salt does the solubility not increase significantly as the temperature increases?

4 At which temperature is the solubility of sodium chloride and potassium nitrate the same?

5 How many grams of undissolved solid will be present in a mixture of 100 g potassium nitrate and 100 cm³ water at 20 °C?

Heterogeneous mixtures

Heterogeneous mixtures include suspensions and colloids. In these mixtures, the components do not dissolve in each other; rather one substance is initially dispersed in the other. Then, in the case of a suspension, settling will take place on standing whereas in the colloid, it remains suspended.

Suspensions

A suspension is a heterogenous mixture in which solid or liquid particles are suspended in a solid, liquid or gas.

Figure 5.9 ▶
The Amazon River brings fertile alluvial soil deposits to the coastal areas of Guyana in its muddy water. The soil is suspended in the water.

The following are examples of suspensions:

- Mud in water
- Powdered chalk in water
- Some medicines separate out into layers on standing (they may carry a label that reads 'shake well before use').

The distinguishing features of a suspension are:

- the components separate out when the suspension is allowed to stand (some separate out more rapidly than others);
- the suspended particles are larger than those of solutions and colloids, and are visible to the naked eye (that is, without the use of a microscope);
- the particles are not individual atoms or molecules, but are believed to be clusters of them.

Colloids

A colloid is a heterogenous mixture in which particles of one component are dispersed in another component.

Colloids can be thought of as being intermediate between true solutions and suspensions:

- The particles of one component do not dissolve in the other, which makes a colloid similar to a suspension.
- The particles do not separate out on standing, which makes them similar to a solution.
- Emulsions, foams and sols are types of colloids.
- An emulsion is formed when drops of one liquid are dispersed in another liquid in which it is not soluble, for example, emulsion paint is formed from polymer particles dispersed in water.
- A foam consists of a gas dispersed in a liquid, for example, shaving cream and fire extinguishing foam.
- A sol consists of very small solid particles dispersed in a liquid, for example, blood or pigment-based paints.

Examples of colloids are a starch/water mixture, toothpaste, smoke, whipped cream, soap suds and milk.

The distinguishing features of colloids are:

- the components do not separate out on standing;
- the components are not separated by simple filtration;
- the dispersed particles are intermediate in size between those of a solution and those of a suspension;
- the particles are clusters of molecules or atoms that are big enough to scatter a beam of light but too small to settle.

▲ **Figure 5.10** Is this a solution or a suspension? Discuss with your classmates and give reasons for your answers

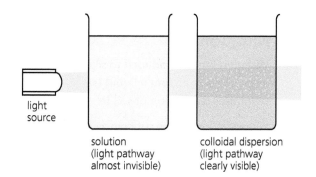

Figure 5.11
Passing light through a
solution and a colloid ▶

solution
(light pathway
almost invisible)

colloidal dispersion
(light pathway
clearly visible)

Comparing solutions, suspensions and colloids

Table 5.4 compares the properties of the three types of mixtures.

▼ **Table 5.4**   A comparison of solutions, suspensions and colloids (Note: 1 nm = 10^{-9} m)

	Comparative size of particles (nm)	Able to pass through filter paper?	Settles out?	Exhibits Brownian motion?	Appearance
solutions	0.1–2	yes	no	yes	transparent (if solvent is liquid)
colloids	2–1 000	yes	not usually	yes	not transparent; may appear homogeneous
suspensions	⩾1 000	no	yes	no	not transparent; heterogeneous

5.3   Separating mixtures

At home, we separate mixtures all the time. For instance we strain fruit juice and filter tap water to remove impurities. Chemists also separate mixtures, for instance removing the oxygen from the air so that it can be used in hospitals. Or, chemists may have prepared a compound in the laboratory and then need to purify it (remove the contaminants). A range of methods of separating mixtures is available.

In this section of this chapter, you will study:
- some methods of separating mixtures in the laboratory;
- the principle on which each method of separation is based;
- examples of mixtures that can be separated by each method;
- the apparatus used for each method of separating mixtures.

How can mixtures be separated?

You will remember that the components of a mixture are not chemically combined. They can therefore be separated by physical methods. This is quite different from separating the elements in a compound (e.g. getting sodium and chlorine from sodium chloride), where chemical processes (such as electrolysis) are necessary.

The components of a mixture also retain their properties. The physical methods used in separating mixtures depend on the differences in physical properties of the components of the mixture. To use a simple example, a mixture of iron and sulphur can be separated by using a magnet since only one of the components of the mixture (the iron) is attracted to the magnet.

Methods of separating mixtures
Filtration

Filtration is a method used to separate suspended solids from a liquid (such as soil from a suspension of soil in water).

Filtration is based on differences in particle size of the components. The filter paper acts as a selective physical barrier that allows the liquid to pass through it, but does not allow the solid to pass through (Figure 5.12). Filtration can be used if you want to keep either the liquid or the solid (or even both) after the separation.

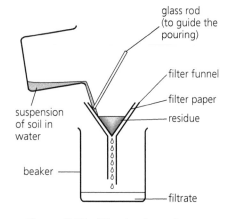

▲ **Figure 5.12** Filtration is used to separate a solid from a suspension.

▲ **Figure 5.13**
Water filtration at a water processing plant in Willemstad Curacao, Caribbean Sea

The same principle is used in purifying drinking water. Filter beds are used by water authorities to remove suspended particles from water. Motor cars also use filters to remove solid dirt particles from oil in the engine.

Sublimation

Sublimation is used to separate a solid that sublimes from a mixture of solids. You will recall that sublimation is the direct conversion of a solid to a vapour on heating, and a vapour to a solid on cooling.

A mixture of sodium chloride and ammonium chloride can be separated by the process of sublimation (Figure 5.14). Ammonium chloride sublimes but sodium chloride does not. When a mixture of the two compounds is gently heated, the ammonium chloride collects on the base of the test-tube whereas sodium chloride, which is unaffected by the gentle heat, remains behind.

Figure 5.14
Sublimation as a technique for separating solids ▶

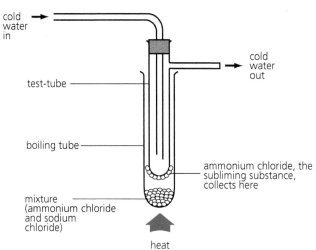

One other substance that sublimes is iodine.

Use of solvents

▲ **Figure 5.15**
In Grenada, nutmeg and oleoresins are extracted using organic solvents. These are then used to manufacture flavourings and perfumes.

Solids may also be separated from one another by the use of solvents. In this method, a solvent is chosen that dissolves one of the components of the mixture but not the other.

For example, if you wanted to remove copper(II) sulphate from a mixture of sand and copper(II) sulphate, you could add water. Water dissolves the copper(II) sulphate, but not the sand. To retrieve both the copper(II) sulphate and the sand, a combination of other methods of separation needs to be used (Figure 5.16).

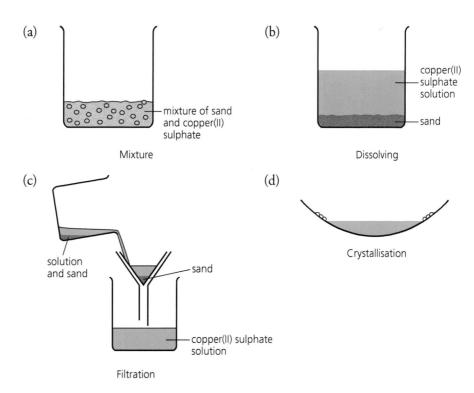

Figure 5.16
Obtaining sand and copper(II) sulphate from a mixture ▶

The stages in this separation are as follows.

1 Solution: Water is added to dissolve the copper(II) sulphate.

2 Filtration: The sand is retrieved by filtration.

3 Crystallisation: The filtrate in this case is a solution of copper(II) sulphate in water. It is then heated in an evaporating dish to concentrate the solution until it becomes saturated. The saturated solution is allowed to cool. As it does so, crystals separate out. The slower the solution cools, the bigger the crystals of solid that are formed.

Find out more

Sodium chloride (common salt) was at one time a scarce commodity though it was in much demand, mainly for preserving meat and fish. Climatic conditions in certain Caribbean islands made them suitable for producing salt by solar evaporation of sea water in natural salt ponds. Anguilla and St. Martin, for example, are dry and flat. They have a history of salt production, exporting salt to many parts of the world, including Europe and North America, as well as to Trinidad and Tobago for use in the oil industry. The salt industry was an important source of revenue and employment for Anguillans from as early as the 18th and 19th centuries, and was a traditional industry there until recently. Bonaire and the Netherlands Antilles still produce salt in a modern industry.

▲ **Figure 5.17**
Salt from salt ponds in St Marten. What process is used to extract the salt?

According to one early account of the process of salt production in St. Martin, the salt ponds were no more than 3 feet (1 m) deep. The pond water needed to weigh 10 Dutch pounds per gallon before it began to make salt. It required a temperature of 90–92 °F (around 33 °C) in the shade to 108–110 °F (around 43 °C) in direct sun. The surface of the pond became still and like a mirror. It then crusted over with a thin cake of salt, which thickened. As the day wore on, the water cooled and the crust sank heaping up and forming crystals. Workers then waded into the ponds and picked up the cakes of salt, rinsing them and piling them into flat-bottomed wooden boats ('flats'), from which they were loaded into barrels. It was a common sight to see heaps of salt along the sides of the natural salt ponds.

Questions

Q1 Using the following conversion factors, calculate the density of the salt water of weight 10 Dutch pounds per gallon in $g\ cm^{-3}$. Compare the density of this water with the density of fresh water ($1.0\ g\ cm^{-3}$) and the average density of sea water ($1.02\ g\ cm^{-3}$).
(1 Dutch pound = 494 g; 1 UK gallon = 4.5 dm^3.)

Q2 Comment on the density of the water needed for salt formation.

Q3 Explain the process by which salt is formed.

Q4 Use the internet to find out more about:
a salt production in Bonaire today;
b the use of salt in the oil industry.

Simple distillation

Simple distillation is used to obtain a pure solvent from a solution. An example would be the purification of water from sea water. The method depends on the fact that the solvent vaporises at a much lower temperature than the solute.

The processes involved in simple distillation are summarised as follows:

$$\text{solution} \xrightarrow[\text{heating}]{\text{evaporation of solvent}} \text{pure solvent (as vapour)} \xrightarrow[\text{cooling}]{\text{condensation}} \text{pure solvent (as liquid)}$$

The solid remains in the distillation flask and is called the residue. The distilled solvent is the desired product and is called the distillate.

Look at Figure 5.18. Note the following details concerning the apparatus used in simple distillation:

- Condensation and boiling take place in different parts of the apparatus.

- The thermometer is in contact with vapour – in this position, the thermometer registers the temperature of the pure vapour.

- It is advisable to add anti-bumping granules (or broken porcelain) to the flask to achieve steady boiling.

- Water to cool the vapour enters the condenser from the end closer to the receiver, and leaves from the end close to the distilling flask.

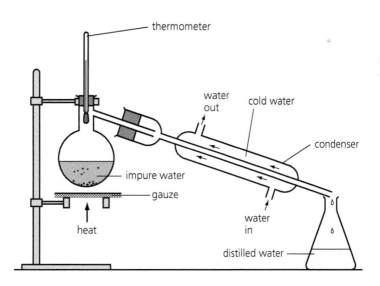

Figure 5.18
The usual laboratory apparatus for simple distillation ▶

Fractional distillation

Pairs of liquids that mix well together, that is, they dissolve readily in one another, are said to be **miscible**. An example is ethanol and water.

Fractional distillation is used to separate **miscible** liquids where the components of the liquid mixture have boiling points that are close together. A mixture of ethanol (boiling point = 78 °C) and water (boiling point = 100 °C), for example, is poorly separated by simple distillation so fractional distillation should be used.

Figure 5.19
Gasoline is produced by fractional distillation of crude oil or petroleum at this refinery in Trinidad and Tobago. ▶

The apparatus used in fractional distillation is shown in Figure 5.20.

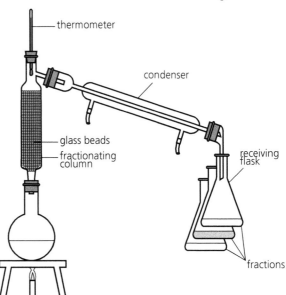

Figure 5.20
A laboratory fractional distillation apparatus ▶

Note the following about the apparatus used in fractional distillation:

- The vapour passes through a column of glass beads (called a fractionating column) before going through the condenser.
- The temperature of the fractionating column decreases as you go from bottom to top.
- The thermometer is in contact with the vapour as it enters the condenser.

A simple explanation of the process, using the example of ethanol and water, is as follows:

- As the liquid mixture is heated, the components vaporise.
- Ethanol has the lower boiling point and is therefore more volatile.
- The vapour contains more of the more volatile component (ethanol).
- The mixture of vapours passes up the fractionating column.
- The vapour condenses and vaporises in the column many times as it rises.
- The more volatile component (the ethanol) eventually rises to the top of the column

as vapour while the less volatile component (the water) condenses and falls down the column.

- The more volatile component (the ethanol) passes through the condenser and is collected as liquid in the receiving flask – it is referred to as the first fraction.
- The next fraction is the next most volatile component – in mixtures of many liquids, several fractions are collected.

Fractional distillation is also used on an industrial scale to separate the components of liquid air as well as in the refining of crude oil. These processes are further described in Section 17.13.

Using a separating funnel

Pairs of liquids that do not mix, but separate into distinct layers, are said to be **immiscible**. An example is cooking oil and water.

A separating funnel is used to separate liquids that are **immiscible**. The method is based on the fact that immiscible liquids do not mix but form two distinct layers, with the less dense liquid on top and the more dense liquid below.

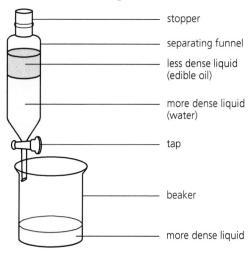

Figure 5.21
Arrangement of apparatus using a separating funnel. ▶

Note the following:
- The mixture of immiscible liquids is placed in the funnel.
- The components are left to separate.
- The denser liquid is withdrawn through the tap into one container.
- The tap is closed just before the second layer gets to it.
- The container is substituted and a little liquid is allowed to run out until only the upper layer begins to flow – this liquid (which will contain a little of both layers) is discarded.
- The container is changed again and the upper liquid is run out and collected.

Using a separating funnel is an important part of solvent extraction, which is discussed next.

Solvent extraction

Solvent extraction is used to separate a component from a mixture by using two solvents. Solvent extraction depends on:
- the desired component being more soluble in one solvent than in the other;
- the two solvents being immiscible.

Experiment 5.4 will help you to understand the processes involved in solvent extraction.

Experiment 5.4   To compare the solubility of iodine in two solvents

Procedure

1 Add a spatula of iodine crystals to 25 cm³ of water in a separating funnel.

2 Cover the funnel and shake the mixture until no more iodine dissolves.

3 Observe the contents of the separating funnel carefully.

4 Add 25 cm³ of 1,1,1-trichloroethane to the separating funnel.

5 Cover the funnel. Rock it gently to and fro to shake the mixture. Remove the cover from time to time to release the pressure in the funnel, and replace it.

6 Observe the mixture carefully.

Explanation

- The water and the 1,1,1-trichloroethane are immiscible.
- The water is more dense than the 1,1,1-trichloroethane and forms the bottom layer.
- The iodine is more soluble in the 1,1,1-trichloroethane than in the water and therefore more of it dissolves in the top layer.
- You can tell by the colours of the layers where the iodine is dissolved.

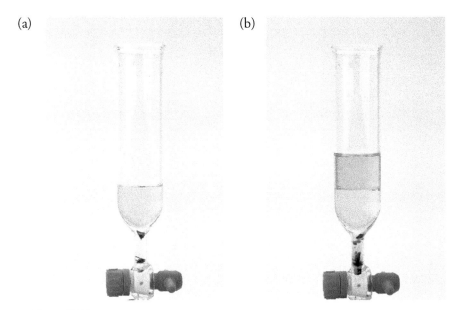

(a)   (b)

▲ **Figure 5.22**
(a) Iodine is not very soluble in water.   (b) The iodine moves into the 1,1,1-trichloroethane layer.

Solvent extraction is a good method to extract organic substances from aqueous solutions. An example is the removal of caffeine from tea or coffee. Caffeine is a stimulant that affects the nervous system. It is an organic compound that is more soluble in dichloromethane than it is in water. Also, dichloromethane and water are immiscible.

ORR
M&M
A&I

Experiment 5.5 Removing caffeine by solvent extraction (Teacher demonstration)

⚠ This experiment involves the use of organic solvents, so it should be carried out in a fume cupboard or a well-ventilated room.

Procedure

1 Boil either 5 g of instant coffee or 2 g of tea with 200 cm³ of distilled water for 15 minutes. The caffeine and other solids will dissolve in the water.

2 Filter the liquid to remove any undissolved solids.

3 Add 5 cm³ of 10% lead(II) ethanoate solution to the filtrate and boil for a further 15 minutes, then filter. The lead ethanoate precipitates some of the components of the solution.

4 Place the cooled filtrate in a separating funnel.

5 Add 25 cm³ of dichloromethane and shake thoroughly. More of the caffeine dissolves in the dichloromethane than in the water. Allow the layers to separate, then remove and save the lower dichloromethane layer.

6 Repeat step 5, but this time add 25 cm³ dichloromethane to the upper aqueous layer. Shake as before and combine the lower layer with the first dichloromethane extract.

7 Remove the dichloromethane by simple distillation using a hot water bath to obtain a crude sample of caffeine.

The caffeine is purified as follows:

8 Dissolve the crude caffeine in 15 cm³ of hot methylbenzene, then add 10 cm³ of cool, high boiling point petroleum spirits. Observe that white crystals are produced.

9 Filter the crystals and dry them between two pieces of filter paper. The white crystals are caffeine.

Paper chromatography

Paper chromatography is a technique used to separate, purify or identify substances. This method involves the use of a stationary phase and a mobile phase (these terms are explained later).

The mixture to be separated is placed on a strip of filter paper (or chromatography paper) and a solvent is allowed to move through it. The separation is based on the differences in the rates of movement of the different components in the mixture along the paper.

To separate substances by paper chromatography, a fine capillary tube is used to place a spot of the solution containing the mixture to be separated near the edge of the piece of chromatography paper (Figure 5.23). This point of application is called the origin or baseline.

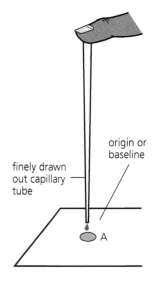

origin or baseline

finely drawn out capillary tube

A

▲ **Figure 5.23** Preparing for a chromatographic separation

The sequence of the three stages shown in Figure 5.24 illustrates the chromatographic separation of ink that contains three components. Because of the absorbent nature of paper, the solvent moves against gravity and carries the ink dyes along with it. The ink dyes are well separated if they move at different rates.

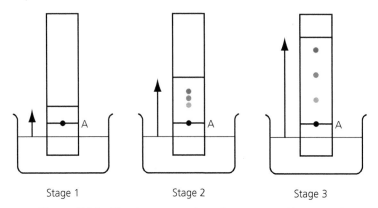

Stage 1   Stage 2   Stage 3

▲ **Figure 5.24**   The three stages in a chromatographic separation

Step 1: The ink dye is spotted and allowed to dry. The original spot is identified as A. The solvent begins to move up the paper by capillary action.

Step 2: Solvent moves up the paper taking different components along at different rates.

Step 3: The separation of the mixture is complete. The different components string out along the paper like runners in a race.

Note the following about the process:

• The water molecules attached to the chromatography paper form the stationary phase and the solvent moving through the paper is the mobile phase.

• The stationary phase will tend to stop the components of the mixture from moving, while the mobile phase will carry them along the paper.

• The components of the mixture will be held to the stationary phase and will dissolve in the mobile phase to different extents.

• The components of the mixture will therefore move along the paper at different rates.

Each substance can be recognised by a value known as an R_f value, which is a constant for a given type of chromatography paper and a given solvent system.

$$R_f = \frac{\text{distance travelled by a component}}{\text{distance travelled by solvent front}}$$

The solvent can either move upwards against gravity in ascending chromatography, or downwards with gravity in descending chromatography.

Note that colourless substances can be separated by chromatography but the chromatogram must be treated with a reagent (called a locating agent), which reacts with the spots to give colours.

The **solvent front** is the distance travelled by the solvent from the origin.

▲ **Figure 5.25**
Chromatography can be used to help solve crimes in drug testing and in sports.

Understand it better

Paper chromatography is often used to identify substances in a mixture, as in the following example.

A scientist wanted to identify the components of a mixture of amino acids (Sample U). She had pure samples of three amino acids (Samples G, L and T, respectively). She dissolved each sample in dilute acid and spotted them on a strip of chromatography paper. She placed the strip into a tank containing a solvent, covered it and allowed the chromatogram to develop until the **solvent front** was just below the top of the paper. She sprayed the paper with ninhydrin solution (a locating agent for amino acids) and allowed it to dry in an oven.

Her chromatogram is shown in Figure 5.26.

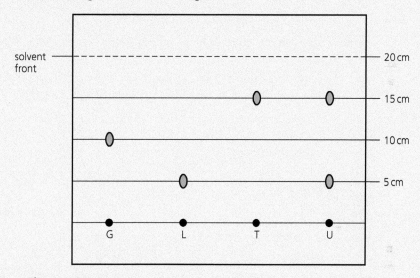

▲ **Figure 5.26** The scientist's chromatogram

Q1 Which amino acids appear to be in the mixture?

Q2 Calculate the R_f values for G, L and T in this system.

Recrystallisation

Solids, especially organic solids, are often impure when first prepared. One method of cleaning solids is called recrystallisation. This is the process by which crystals of solid material are produced, usually from a saturated solution. The steps in the process are shown in Figure 5.27.

If crystals do not appear in stage (c), either seed the solution with a pure crystal or scratch the side of the flask with a clean glass rod. Seeding is the addition of a crystal (usually of the solute) to encourage crystallisation. Note that the collected crystals may need to be washed with some of the solvent before finally removing them from the filter paper.

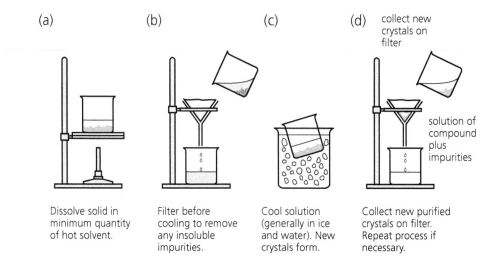

(a) (b) (c) (d) collect new crystals on filter

solution of compound plus impurities

Figure 5.27
The stages in recrystallisation ▶

Dissolve solid in minimum quantity of hot solvent.

Filter before cooling to remove any insoluble impurities.

Cool solution (generally in ice and water). New crystals form.

Collect new purified crystals on filter. Repeat process if necessary.

5.4 The extraction of sucrose from cane sugar

The methods of separation discussed above can be applied to industrial processes such as the extraction of sucrose from the juice of cane sugar.

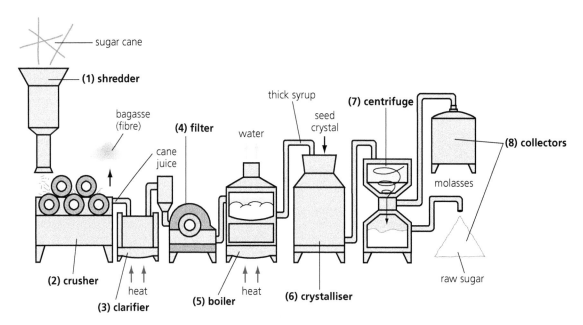

▲ **Figure 5.28**   A flow diagram showing the stages in the extraction of sucrose

1   Shredder: Contains rotary knives for cutting cane into small pieces.

2   Crusher: Juice is extracted as water is sprayed on the cane and rollers apply pressure to crush the cane. The cane fibre, known as 'bagasse', is later burnt to supply fuel for the boilers.

3   Clarifier: cane juice is acidic and is neutralised by the addition of calcium hydroxide. Heating allows for the precipitation of impurities in the form of insoluble calcium salts.

▲ **Figure 5.29** Sugar is produced at the Portvale Sugar Factory in Barbados using modern methods. The Frank Hutson Sugar Museum housed on this compound shows the history of sugar production in the 18th and 19th centuries.

4 Filter: The precipitated solid materials are removed from the cane juice. The clarified juice passes to the first of several boilers.

5 Boiler: The juice is evaporated under reduced pressure to prevent charring. Several boilers, each at a more reduced pressure than the previous one, are used.

6 Crystalliser: The thick syrup is supersaturated. Pure sugar crystals are added to cause crystallization of the liquid.

7 Centrifuge: Here the mixture from the crystalliser is spun at high speeds to separate it into molasses and raw sugar crystals.

8 Collectors: Containers for collecting the separated molasses and sugar.

Summary

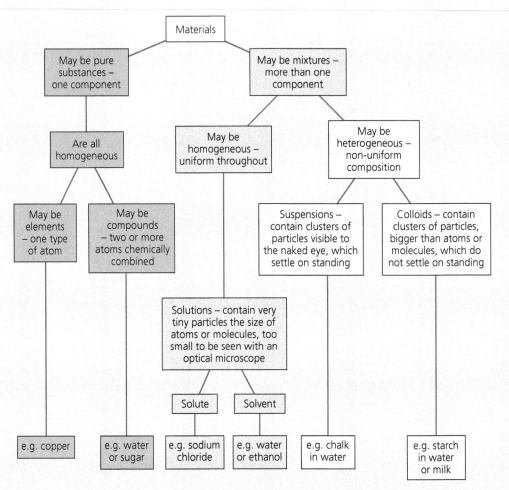

Method of separation	Basis of separation	Used to separate
filtration	difference in particle size	insoluble solid from liquid, e.g. soil from water
use of solvents, filtration and crystallisation	differences in solubility	two solids, e.g. sodium chloride and sand
sublimation	one solid sublimes and the other does not	two solids, e.g. ammonium chloride from sodium chloride
simple distillation	widely differing boiling points	solvent from solution, e.g. water from copper sulphate solution
fractional distillation	small differences in boiling points	miscible liquids, e.g. ethanol and water

Summary *continued*

Method of separation	Basis of separation	Used to separate
separating funnel	immiscible liquids forming separate layers	immiscible liquids, e.g. oil and water
chromatography	difference in rates of migration of different solutes in a solvent due to differences in solubility	a wide range of mixtures including coloured solutions, e.g. pigments in ink or grass
solvent extraction	difference in solubility in different solvents that are immiscible	organic substances from aqueous solutions, e.g. caffeine from tea or coffee

End-of-chapter questions

1 Which characteristic does a pure substance show?
 A It has no distinct chemical properties.
 B It has a range of melting and boiling points.
 C It has a constant composition.
 D It is heterogeneous.

2 All of the following are examples of heterogeneous mixtures except one. Which is the odd one out?
 A Colloids, emulsions, suspensions
 B Milk, shaving foam, blood
 C Milk, jello, alcohol
 D Muddy water, sols, whipped cream

3 Which of the following is the best method to use in a desalination plant?
 A Sublimation
 B Filtration
 C Fractional distillation
 D Solvent extraction

4 Which one of these processes occurs in the extraction of sucrose from cane sugar?
 A Separate collectors hold molasses and sugar.
 B The cane juice is made acidic in the clarifier.
 C Bagasse is the name for the cane fibres before the juice is extracted.
 D In the filtering process, solids are added to the cane juice.

5 What is a compound?
 A It is made up of two or more substances chemically combined together
 B It is a heterogeneous substance.
 C It is made up of a solvent and a solute
 D A suspension or a colloid

6 a Classify the following as element, compound or mixture:
   (i) Paint
   (ii) Ethanol
   (iii) Soil
   (iv) Water
   (v) Copper sulphate
   (vi) Aluminium
   (vii) Air
   (viii) Juice
   (ix) Milk
   (x) Nitrogen

 b Draw up a table to compare elements, compounds and mixtures. Consider the following properties:
   (i) Composition
   (ii) Physical properties such as melting and boiling points
   (iii) Separation

7 Copy and complete Table 5.5, which is about separating mixtures.

▼ **Table 5.5**

Mixture to be separated	Example	Method of separation
solid from liquid: (a) suspension	chalk in water	
solid from liquid: (b) solution		evaporation or crystallisation
solid from solid	ammonium chloride and sodium chloride	
	sodium chloride and sugar	
solvent from solution	water from copper sulphate solution	
liquid from liquid: (a) miscible liquids		
liquid from liquid: (b) immiscible liquids		
mixture of coloured substances		

8 The apparatus shown below can be used to obtain pure water from sea water.

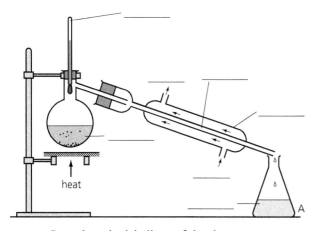

heat

a Complete the labelling of the diagram.
b What is the purpose of the water in the condenser?
c Name one test you can do to determine that the liquid in A is pure.
d Draw and label a diagram of the apparatus you could use from only materials available around the home to separate water from sea water.

9 a Describe a test you could use to decide whether:
   (i) a liquid is pure;
   (ii) a solid is pure.
b How does the presence of impurities affect the results of the tests described in part (a)?

10 a Define solubility.
b State three factors that affect the solubility of a solid in a liquid.
c State three factors that will affect how quickly a given ammount of copper sulphate dissolves in a fixed volume of water.
d Will stirring a solution increase the solubility of a solute in a given solvent, all other factors remaining the same? Explain your answer.

11 Each of two rival companies has marketed a 'new red food colouring'. A cook suspects that the 'new' food colouring is identical to an older food colouring. Describe, in detail, a procedure she can use to test her suspicion.

12 An organic solvent L has a density of 0.9 g cm^{-3} and is immiscible in water.
a Describe, with the aid of diagram(s), how you would separate solvent L from a mixture of water and solvent L.
b After separation, solvent L is cloudy.
   (i) Suggest a reason for this cloudiness.
   (ii) Outline a method for getting rid of the cloudiness.
c An organic solute X is moderately soluble in water, but more soluble in liquid Y. Describe how you would use liquid Y to remove the solute X from a solution of it in water.

Multiple-choice questions for Chapters 1–5

1 All of the following are mixtures EXCEPT:
  A Air
  B Sea water
  C Water
  D Milk

To answer Questions 2 and 3, choose the term that BEST describes the process.
A Evaporation
B Boiling
C Diffusion
D Osmosis

2 A puddle of water in the road dries up after a few hours on a hot day.

3 Cucumber slices become soft and leak water after salt and pepper are added.

4 An atom has 31 electrons, 39 neutrons and 31 protons. It follows that its mass (nucleon) number is:
  A 31
  B 70
  C 62
  D 39

5 To which period of the Periodic Table does the element sulphur ($_{16}^{32}$S) belong?
  A 6
  B 3
  C 2
  D 1

6 All of the following elements are in the same group of the Periodic Table EXCEPT:
  A $_3$Li
  B $_{11}$Na
  C $_{19}$K
  D $_9$F

7 Look at the atomic structure shown below:

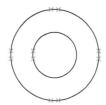

The element with this atomic structure is in:
  A Group II and Period 2
  B Group 0 and Period 1
  C Group II and Period 8
  D Group 0 and Period 2

8 The pair of ions with the same electron configuration as neon ($_{10}^{20}$Ne) is:
  A Li^+ and Cl^-
  B Mg^{2+} and Cl^-
  C Mg^{2+} and F^-
  D Li^+ and F^-

9 Which of the following pairs of elements can combine to form an ionic compound? (Note: you will need to use your Periodic Table to answer this question.)
  A Hydrogen and carbon
  B Aluminium and magnesium
  C Silicon and chlorine
  D Magnesium and sulphur

10 What is the correct formula for the substance represented in the diagram below?

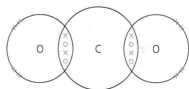

  A CO
  B CO_2
  C C_2O
  D C_2O_2

11 Which of the following diagrams represents a mixture of elements? (Note: each ball represents an atom.)

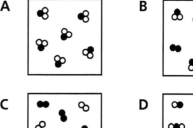

12 Sea water, when compared to pure water, will have:

A A lower boiling point and a lower freezing point

B A lower boiling point and a higher freezing point

C A higher boiling point and a higher freezing point

D A higher boiling point and a lower freezing point

For Questions 13–15 use the following information:

In the following chromatogram, Y and Z are dyes of unknown composition and R, S, T and U are pure pigments.

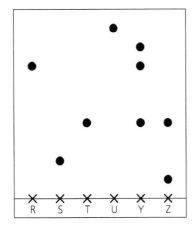

13 Which pigments are NOT components of dye Y?

A R and U

B S and U

C S and T

D T and U

14 Both dyes Z and Y contain:

A R

B S

C T

D U

15 The pigment that is most soluble in the solvent used is:

A R

B S

C T

D U

16 Water and ethanol cannot be used as the two solvents in a solvent extraction separation because alcohol and water:

A Are both good solvents

B Have different boiling points

C Are both colourless

D Are miscible

17 Which of the following atoms is an isotope of $^{14}_{7}X$?

A $^{15}_{7}X$

B $^{14}_{8}X$

C $^{7}_{3}X$

D $^{28}_{14}X$

18 Which of the following particles is a negative ion?

Particle	No. of protons	No. of electrons	No. of neutrons
R	13	10	14
S	17	17	20
T	17	18	18
U	14	14	14

A R

B S

C T

D U

19 Atoms are neutral. This means that:

A They have no charged particles

B They have the same number of neutrons and protons

C Their valence shells have the same number of electrons as an inert gas

D The amount of positive charge in the atom equals the amount of negative charge

20 The mass (nucleon) number of an atom equals:

A The number of neutrons it contains

B The number of protons it contains

C The number of protons plus the number of electrons

D The number of protons plus the number of neutrons

6 Groups II and VII, and Period 3

In this chapter, you will study the following:

- the changes in the physical and chemical properties in metals as you move down Group II;
- the changes in the physical and chemical properties in elements as you move down Group VII;
- the changes in the physical and chemical properties as you move across Period 3;
- how to predict the properties of an element based on properties of nearby elements.

This chapter covers
Objectives 4.2–4.5 of Section A of the CSEC Chemistry Syllabus.

In Chapter 3, you saw that when all the known elements are arranged in order of increasing atomic (proton) numbers, elements of similar properties recur at regular intervals. This recurrence is referred to as 'periodicity'.

6.1 Group I and Group II elements

All the elements in Groups I and II are metals. The metals in Group II are the alkaline earth metals and the metals in Group I are known as the alkali metals.

The metals in Group II show a variation in physical and chemical properties as the group is descended. To help you to understand this part of the chapter, you should review the following ideas in Chapters 3 and 4 of this book.

- Metals are solids with giant crystal lattices.
- Metals usually have high melting and boiling points.
- Metals have large atomic radii.
- Metals tend to have high densities.
- Metals tend to lose electrons easily.
- Metals react chemically with non-metals to form ionic compounds.

1 **H** Hydrogen 1																	2 **He** Helium 4
3 **Li** Lithium 7	4 **Be** Beryllium 9											5 **B** Boron 11	6 **C** Carbon 12	7 **N** Nitrogen 14	8 **O** Oxygen 16	9 **F** Fluorine 19	10 **Ne** Neon 20
11 **Na** Sodium 23	12 **Mg** Magne-sium 24											13 **Al** Aluminium 27	14 **Si** Silicon 28	15 **P** Phos-phorus 32	16 **S** Sulphur 32	17 **Cl** Chlorine 35	18 **Ar** Argon 40
19 **K** Potassium 39	20 **Ca** Calcium 40	21 **Sc** Scandium 45	22 **Ti** Titanium 48	23 **V** Vanadium 51	24 **Cr** Chromium 52	25 **Mn** Manganese 55	26 **Fe** Iron 56	27 **Co** Cobalt 59	28 **Ni** Nickel 58	29 **Cu** Copper 63	30 **Zn** Zinc 64	31 **Ga** Gallium 69	32 **Ge** Germ-anium 75	33 **As** Arsenic 75	34 **Se** Selenium 80	35 **Br** Bromine 79	36 **Kr** Krypton 84
37 **Rb** Rubidium 85	38 **Sr** Strontium 88	39 **Y** Yttrium 89	40 **Zr** Zirconium 90	41 **Nb** Niobium 93	42 **Mo** Molyb-denum 98	43 **Tc** Technetium 114	44 **Ru** Ruthenium 102	45 **Rh** Rhodium 103	46 **Pd** Palladium 106	47 **Ag** Silver 107	48 **Cd** Cadmium 114	49 **In** Indium 115	50 **Sn** Tin 120	51 **Sb** Antimony 121	52 **Te** Tellerium 130	53 **I** Iodine 127	54 **Xe** Xenon 132
55 **Cs** Caesium 133	56 **Ba** Barium 138	57–71	72 **Hf** Hafnium 180	73 **Ta** Tantalum 181	74 **W** Tungsten 184	75 **Re** Rhenium 187	76 **Os** Osmium 192	77 **Ir** Iridium 193	78 **Pt** Platinum 195	79 **Au** Gold 197	80 **Hg** Mercury 202	81 **Ti** Thallium 205	82 **Pb** Lead 208	83 **Bi** Bismuth 209	84 **Po** Polonium 209	85 **At** Astatine 210	86 **Rn** Radon 222
87 **Fe** Francium 223	88 **Ra** Radium 226	89–103	104 **Unq** Unnil-quadium 260	105 **Unp** Unnil-pentium 262	106 **Unh** Unnil-hexium 263	107 **Uns** Unnil-septium 262	108 **Uno** Unilocntium 265	109 **Une** Unnilen-nium 266									

57 **La** Lanth-anum 168	58 **Ce** Cerium 140	59 **Pr** Praseo-dymium 141	60 **Nd** Neody-mium 142	61 **Pm** Prome-thium 145	62 **Sm** Samarium 152	63 **Eu** Europium 153	64 **Gd** Gadolinium 158	65 **Tb** Erbium 159	66 **Dy** Erbium 164	67 **Ho** Holmium 165	68 **Er** Erbium 168	69 **Tm** Thulium 169	70 **Yb** Ytterbium 174	71 **Lu** Lutetium 175
89 **Ac** Actinium 257	90 **Th** Thorium 232	91 **Pa** Protac-tinium 257	92 **U** Uranium 238	93 **Np** Neptunium 237	94 **Pu** Plutonium 244	95 **Am** Americium 243	96 **Cm** Curium 247	97 **Bk** Berkelium 247	98 **Cf** Californium 251	99 **Es** Einsteinium 254	100 **Fm** Fermium 257	101 **Md** Mende-levium 258	102 **No** Nobelium 255	103 **Lr** Lawre-ncium 209

▲ **Figure 6.1** The Periodic Table of elements

Group II – the alkaline earth metals

Trends in physical properties of Group II elements

The names and symbols of the Group II metals and their electronic configurations are shown in Table 6.1.

▼ **Table 6.1** Group II elements

Name of element	Symbol	Electronic configuration
beryllium	Be	2,2
magnesium	Mg	2,8,2
calcium	Ca	2,8,8,2
strontium	Sr	2,8,18,8,2
barium	Ba	2,8,18,18,8,2

Practice

1 What type of bond do you expect to be formed when magnesium combines with bromine? Explain your answer.

2 Which of the ions Sr^{2+} or Mg^{2+} is formed more readily from the corresponding atoms? Explain your answer.

Some trends in the physical properties of Group II elements are given in Table 6.2.

▲ **Figure 6.2** Group II elements, beryllium, magnesium and calcium

▼ **Table 6.2** Trends in the physical properties of Group II elements

Element	Atomic radius (pm)	Ionisation energy (kJ mol⁻¹)	Melting point (°C)	Boiling point (°C)	Relative density	Electronegativity
magnesium	160	747	650	1 110	1.74	1.2
calcium	197	596	850	1 487	1.54	1.0
strontium	215	558	768	1 350	2.62	1.0
barium	217	512	714	1 640	3.51	0.9

Explaining the trends in physical properties down Group II

An analysis of the data in Table 6.2 shows many trends. These are discussed below.

1 Atomic radius increases down Group II

As the number of shells containing electrons increases from magnesium to barium, the radius of the atom increases. Shielding of valence electrons by the electrons in filled inner shells also contributes to increasing atomic radius down the group. (See Chapter 3.)

2 Ionisation energy decreases down Group II

Metal atoms ionise by losing their valence electrons to form cations.

Ease of ionisation indicates how easily an atom forms ions.

Ionisation energies reflect the **ease of ionisation** of metals. The lower the ionisation energy value, the more easily the metal ionises.

Larger metal atoms lose their valence electrons more easily than smaller ones. This is because:

- the valence electrons are further from the nucleus;
- the valence electrons are shielded by the repulsion of electrons in filled inner shells.

A look at the pattern of electron configurations of the Group II metals (see Table 6.1) helps to reinforce the increases in atomic radii and shielding as the group is descended.

3 Group II elements have relatively high melting and boiling points

Group II metals have giant metallic structures. This is reflected by their high melting and boiling points.

4 There is a general increase in density from calcium to barium

Your study of atoms in the Periodic Table showed that the numbers of protons and neutrons in the nuclei of the atoms increase significantly from period to period as the group is descended. This means that the masses of the atoms increase down the group. This contributes to their increasing density, because:

$$\text{density} = \frac{\text{mass}}{\text{volume}}$$

5 Electronegativity values are generally low and show a slight decrease down the group

As atomic radius increases down the group, the nucleus of the atom attracts electrons less readily (see Section 3.3).

In conclusion ...

From these trends, we can conclude that metallic character increases down Group II. This means that barium is more metallic than strontium which, in turn, is more metallic than calcium, and so on. This trend of increased metallic character down the group is general for all groups.

Trends in chemical properties of Group II elements

All Group II elements have two electrons in their outer shells, with their inner shells having a noble gas electron configuration (see Table 6.1). When Group II elements combine with non-metals, they lose (transfer) these two electrons to form doubly charged positive ions.

This is also true for their reactions with water and dilute non-oxidising acids.

The more easily a metal ionises, the more reactive it is. Therefore, the Group II metals will become more reactive as the group is descended. This increase in reactivity parallels the increase in metallic character and greater ease of ionisation as the group is descended.

Table 6.3 summarises the chemical reactions of magnesium, calcium and barium.

▼ **Table 6.3** Reaction of Group II metals with oxygen, water and dilute acids

	Magnesium	**Calcium**	**Barium**
Reaction with oxygen or air	reacts but only if heated	reacts if heated in air	reacts rapidly at room temperature
Reaction with water	a slow reaction; hydrogen is evolved; insoluble magnesium hydroxide is formed	brisk reaction; hydrogen is evolved; soluble calcium hydroxide is formed	very rapid reaction; hydrogen is evolved; soluble barium hydroxide is formed
Reaction with dilute acids, e.g.HCl(aq)	rapid reaction; hydrogen is evolved; a salt, e.g. $MgCl_2$, is formed	very rapid reaction; hydrogen is evolved; a salt is formed	violent reaction; hydrogen is evolved; a salt is formed

Note that the reaction conditions and the vigour of the reactions mentioned in Table 6.3 give a clue to the reactivity of the metals in the group. Here are some examples:

- Magnesium and calcium react with oxygen or air only if heated, whereas barium reacts rapidly at room temperature.

- Magnesium only reacts very slowly with cold water, whereas calcium and barium react vigorously with cold water. Hot water or steam is needed for a more vigorous reaction with magnesium.

- The reaction with dilute hydrochloric acid intensifies as the group is descended. (Why does this not hold for dilute sulphuric acid?)

You should note that the reaction of the metals with dilute acid is generally more vigorous than their reactions with water.

Why do metals react more vigorously with dilute acids than with water?

When a metal such as magnesium reacts with water, steam, or with acid, the reaction may be represented by the ionic equation:

$$Mg(s) + 2H^+(aq) \rightarrow Mg^{2+}(aq) + H_2(g)$$

Water is ionised to only a very small extent and therefore contains only a small number of hydrogen ions (H^+) compared to acids. The higher the concentration of H^+ ions, the more rapid the evolution of hydrogen gas.

In all of the chemical reactions described here, the Group II element transfers its two valence electrons to form divalent cations. Group II elements are therefore reducing agents (see Section 12.7). Since the ease of ionisation of the Group II elements increases down the group, the reducing property of the elements also increases down the group.

Practice

3 Use the data in Table 6.3 to predict how strontium will react:
    a with oxygen;
    b with water;
    c with dilute hydrochloric acid.

4 Give balanced chemical equations for the reactions you mention in Question 3.

Solubility of some Group II compounds

The Group II metals react with oxygen, water and dilute acids to form ionic compounds. However, note that the metallic oxides formed from the reaction of Group II metals with oxygen are not all soluble in water. Solubility increases as the group is descended.

- The strong ionic bonding in magnesium oxide resists the dissolving properties of the solvent, water.
- Calcium oxide (also known as 'quick lime') behaves in a unique way with water, eventually producing an aqueous suspension of calcium hydroxide (lime water), which has weakly alkaline properties.
- The solubility of the sulphates, formed by the reaction of the metals with dilute sulphuric acid, decreases down the group (see Section 11.4). The formation of insoluble sulphates interferes with the reaction of the metals with the acid.

Practice

5 Select the term from the pair of words that are <u>underlined</u> to correctly complete the sentences.

a For a metal, ease of ionisation indicates how easily a metal atom <u>gains /</u> <u>loses</u> electrons.

b The <u>smaller / larger</u> the metal atom, the more easily it ionises.

c The <u>smaller / larger</u> the atomic radius of a metal atom, the more reactive the metal atom.

d The more reactive the metal, the <u>lesser / greater</u> is its ease of ionisation.

e The <u>closer / further</u> the valence electrons are from the nucleus of the metal, the greater the ease of ionisation.

f The <u>closer / further</u> the valence electrons are from the nucleus of the metal, the more reactive the metal.

Group I – the alkali metals

▲ **Figure 6.3** Group I elements, sodium, potassium and lithium

Like typical metals, the elements of Group I are solids at room temperature and have low electronegativity. However, they do not show all of the typical physical properties of metals. Here are some of the differences in physical properties:

- They are soft and can be easily cut with a knife.
- They have relatively low densities.

Sodium and potassium can float on oil but copper sinks. What can you work out about the densities of these three metals from this?

Groups I and II show similar reactions and trends in both physical and chemical properties. You can see how well you understand the principles you studied about Group II by applying them to Group I.

Practice

6 Use the Periodic Table to find the atomic numbers of the Group I metals and complete a table like Table 6.1 for Group I.

7 Table 6.4 shows some of the physical properties for Group I metals.

▼ **Table 6.4** Trends in physical properties of the Group I metals

Group I metal	Physical state at room temperature	Density at 25 °C (g cm^3)	Melting point (°C)	Boiling point (°C)	Ionisation energy (kJ mol^1)
lithium	solid	0.53	179.1	1340	519
sodium	solid	0.97	97.5	585	494
potassium	solid	0.86	63.5	775	418
rubidium	solid	1.53	39.0	690	403
caesium	solid	1.88	28.5	670	374

a How do these trends in density, melting point and ionisation energy compare with those shown for Group II metals?

b Compare the ionisation energies for Group I and Group II metals. State, with a reason, whether the Group I metals are more or less reactive than the Group II metals.

c Predict the change in atomic radius down Group I.

d How do you expect this change to affect:
  (i) the ease of ionisation;
  (ii) the reactivity of Group I metals as the group is descended?

8 Write equations for the reactions of sodium with water, dilute acid and oxygen.

9 Why should students not attempt these reactions in the laboratory? Why is sodium metal stored in oil?

10 What will happen if (i) red litmus and (ii) blue litmus are added to the solution left after reacting the alkali metals with water?

11 How will the reactivity of sodium differ with water, ethanoic acid and dilute hydrochloric acid?

12 How will the reactivity of rubidium compare with that of sodium?

Find out more

Are transition metals similar to Group I and Group II metals?
You will notice that some familiar metals, such as zinc, copper and iron, are not included in Group I or Group II. This is because they are transition metals and these are not included within one of the main groups of the Periodic Table (see Chapter 3). The transition metals occur in a block between Groups II and III.

▲ **Figure 6.4** Transition metals iron, copper and zinc

Figure 6.5 gives the symbols and atomic numbers of the transition elements scandium (Sc) to zinc (Zn).

Group II

Group III

| Sc 21 | Ti 22 | V 23 | Cr 24 | Mn 25 | Fe 26 | Co 27 | Ni 28 | Cu 29 | Zn 30 |

◄ **Figure 6.5**
The transition elements from scandium to zinc

Like the Group II metals, the transition metals show typical physical properties of metals, for example:
- they are hard solids with high density;
- they have high melting and boiling points.

Some of these properties are shown in Table 6.5.

▼ **Table 6.5** Selected physical properties of some transition elements

Element and symbol	Atomic radius (nm)	Relative density	Boiling point (°C)	Hardness relative to diamond
vanadium, V	0.135	6.1	3 400	
chromium, Cr	0.129	7.2	2 480	0.9
manganese, Mn	0.137	7.2	2 097	0.5
cobalt, Co	0.125	8.7	2 900	0.5
nickel, Ni	0.125	8.9	2 722	0.5

Chemically, the transition metals are less reactive than the Group I and Group II metals, as shown by the examples in Table 6.6. Can you remember how Group I and Group II metals would react in comparison?

▼ **Table 6.6** Selected chemical properties of iron and copper

	Iron	Copper
Reaction with water	no reaction; rusting occurs if air (or oxygen) is present	no reaction
Reaction with dilute acid	slow reaction; hydrogen is evolved	no reaction
Reaction with chlorine gas	a vigorous reaction when heated; $FeCl_3$ formed	a vigorous reaction when heated; $CuCl_2$ formed

Transition metals also differ from the Group I and Group II metals in the following ways:

- They form ions with different charges and show variable oxidation states (see Section 12.7).

- Aqueous solutions of their compounds are often coloured, whereas Group I and Group II elements form only colourless ions in aqueous solution.

- Transition metals and their compounds are frequently good catalysts – for example, iron is used in the synthesis of ammonia and vanadium(V) oxide is used in the manufacture of sulphuric acid (see Chapter 25).

6.2 The Group VII elements

In this section, we continue our study of trends in the Periodic Table by looking at some non-metals. We will look at the elements in Group VII, which are known as the halogens.

▲ **Figure 6.5**
What are the advantages and disadvantages of using these halogen-containing items?

We will:

- examine the change in physical properties of the halogens as Group VII is descended;
- consider the ease of ionisation of these non-metals and use this pattern of ionisation energies to predict the relative strength of the elements as oxidising agents.

The halogens – fluorine, chlorine, bromine, iodine and astatine – are non-metallic elements found in Group VII of the Periodic Table. These non-metals are highly reactive elements, as can be predicted from their position to the far right of the Periodic Table (see Chapter 3).

▼ **Table 6.7** The halogens: their symbols, molecular formulae, atomic numbers and electron configurations

Element	Symbol	Molecular formula	Atomic number	Electronic configuration
fluorine	F	F_2	9	2,7
chlorine	Cl	Cl_2	17	2,8,7
bromine	Br	Br_2	35	2,8,18,7
iodine	I	I_2	53	2,8,18,18,7
astatine	At	At_2	85	2,8,18,32,18,7

Our study will focus mainly on the elements chlorine, bromine and iodine. However, the trends or patterns shown by these elements can be extended to include the other members of Group VII – fluorine and astatine.

Physical properties of the halogens

As you would expect, the atoms of the halogens each contain seven valence electrons. They then share the single (unpaired) valence electron to form diatomic covalent molecules in which the atoms are held by strong single covalent bonds. Weak forces then exist between the molecules. Weak intermolecular forces are typical of substances with simple molecular structures (see Section 4.9).

The halogens therefore:
- have low melting and boiling points;
- are more soluble in non-polar solvents than in water (a polar solvent);
- are non-electrolytes;
- exist as gases, volatile liquids or soft solids at room temperature, depending on the relative strengths of the intermolecular forces.

Appearance of the halogens

All the halogens are coloured, the colour becoming more intense as Group VII is descended.

The main physical properties of the halogens, chlorine to iodine, are summarised in Table 6.8.

▼ **Table 6.8** Some physical properties of the halogens

Element	Atomic radius (nm)	Physical state at room temperature and atmospheric pressure	Colour		Melting point (°C)	Boiling point (°C)	Electro-negativity	Solubility in water (mol dm⁻³)
chlorine	0.099	gas	yellow green		−113	−35	3.0	1.5
bromine	0.114	liquid	deep red with deep red vapour		−0.07	59	2.8	4.2
iodine	0.133	solid	black with a sheen; gives purple vapour on sublimation		114	184	2.5	2×10^{-2}

You can observe several trends as Group VII is descended. These trends and their explanations are shown in the Table 6.9.

▼ **Table 6.9** Trends in Group VII

Trends as the group is descended	Explanation
atomic radius increases	the number of shells occupied by electrons increases
electronegativity decreases	as the atoms get larger, the ability to attract electrons decreases
melting and boiling points increase	as the size of the halogen molecules increases, the strength of the intermolecular forces also increases
physical state changes from gas to liquid to solid	as intermolecular forces increase, molecules are pulled closer together

Practice

13 Given that room temperature could range from 28–30 °C on a hot day in the tropics, justify the physical states shown for the elements chlorine, bromine and iodine in Table 6.8.

14 If there is a spill of chlorine gas on the third floor of an industrial building, which workers are more at risk, those on the first floor or those on the fifth floor? Explain your answer.

15 Bromine is a volatile liquid whereas iodine is a solid. Explain this difference in terms of intermolecular forces.

Like other highly reactive elements, the halogens are not found naturally in their elemental states but can be prepared from naturally occurring compounds of the elements.

Chemical reactivity of the elements

Fluorine, the first element of the group, is the most reactive non-metal known and its chemistry can be complicated by the fact that it will oxidise many substances, including water. Astatine, the last element in the group, is radioactive, which makes it unsafe and therefore not suitable for experiments in a school laboratory. Astatine is also extremely rare: it is estimated that there is less than 25 g of astatine in the Earth's crust!

Unlike metals, the reactivity of non-metals depends on the ease with which they gain electrons, that is, their electronegativity. The halogens are highly electronegative and are therefore very reactive. Smaller atoms gain electrons more readily than larger ones and therefore the reactivity decreases as the group is descended.

Practice

16 Select the term from the pair of words that are <u>underlined</u> to correctly complete the sentences.

  a For a non-metal, ease of ionisation indicates how easily a non-metal atom <u>gains / loses</u> electrons.

  b The <u>smaller / larger</u> the non-metal atom, the more easily it ionises.

  c The <u>smaller / larger</u> the atomic radius of a non-metal atom, the more electronegative it is.

  d The more electronegative the non-metal atom, the <u>lesser / greater</u> its ease of ionisation.

  e The more reactive the non-metal, the <u>lesser / greater</u> is its electronegativity.

  f The <u>closer / further</u> the valence electrons are from the nucleus of the non-metal, the greater the ease of ionisation.

  g The <u>closer / further</u> the valence electrons are from the nucleus of the non-metal, the more reactive the non-metal.

The halogens as oxidising agents

The halogens readily accept electrons from the substances with which they react. As you will see in Section 12.7, substances that accept electrons are oxidising agents. The ability of the halogens to attract electrons decreases down the group (see Table 6.9). This means that their ease of ionisation decreases, so we can deduce that their oxidising strength also decreases down the group. It follows that, of the three halogens we are looking at, chlorine is a more powerful oxidising agent than both bromine and iodine, and bromine is a more powerful oxidising agent than iodine.

When the halogens act as oxidising agents, the coloured halogen molecules are changed to colourless halide ions. These colour changes help us to recognise when the halogens are acting as oxidising agents. The relative strength of the halogens as oxidising agents is reflected in displacement reactions. A colour change indicates that the added halogen has displaced another halogen from a solution of its halide ions.

$$X_2 \quad + \quad 2KY \quad \longrightarrow \quad 2KX \quad + \quad Y_2$$

halogen X	salt of halogen Y	salt of halogen X	halogen Y
(coloured)	(colourless)	(colourless)	(coloured)

This reaction indicates that halogen X is more likely to form ions (attract electrons) than halogen Y. X is a more powerful oxidising agent than Y.

ORR

M&M

A&I

Experiment 6.1 Displacement reactions of halogens

The aim of this experiment is to test the oxidising power of the halogens.

Procedure

1 Use chlorine gas or an aqueous solution of chlorine. Bubble the gas or add drops of the solution into:
   a distilled water;
   b an aqueous solution of potassium iodide;
   c an aqueous solution of potassium bromide.

2 Using an aqueous solution of bromine, add drops to:
   a distilled water;
   b an aqueous solution of potassium chloride;
   c an aqueous solution of potassium iodide.

3 Using an aqueous solution of iodine, add drops to:
   a distilled water;
   b an aqueous solution of potassium chloride;
   c an aqueous solution of potassium bromide.

4 Record any colour changes that occur in any of the above mixtures in a suitable table.

Questions

Q1 What is the purpose of the distilled water in the experiments?

Q2 In which tests did a reaction occur?

Q3 What test would you carry out to identify iodine produced in a reaction?

Q4 Which halogens, if any, did (a) chlorine (b) bromine and (c) iodine displace?

Q5 Which halide ion(s), if any, did (a) chlorine (b) bromine and (c) iodine oxidise?

Q6 Use the results from tests 3 and 4 to arrange chlorine, bromine and iodine in order of increasing oxidising power.

The halogens behave as oxidants not only with aqueous solutions of salts containing halide ions but also when they react with metals and non-metals. Some chemical reactions of halogens are recorded in Table 6.10.

▼ **Table 6.10** Some chemical reactions of the halogens.

Name of halogen	Reactions of aqueous solutions of halogens with salts of halogens	Reaction of halogens (gas or vapour) with hot iron wire	Reaction with other elements
chlorine	displaces bromine from bromides and iodine from iodides	a brisk reaction; iron(III) chloride is formed	reacts with most metals and non-metals
bromine	displaces iodine from iodides	a less vigorous reaction; iron(III) bromide is formed	reacts with most metals and non-metals, but less vigorously
iodine	will not displace either chlorine or bromine from their salts	a slow reaction; iron(II) iodide is formed	reacts slowly with most metals

From Table 6.10 it can be seen that:

- chemical reactivity decreases regularly from chlorine to iodine;
- the more reactive halogen displaces the less reactive halogen from a solution of its salts;
- oxidising power decreases down the group.

Practice

17 Down Group VII there is a gradation in physical state and intensity of colour of the elements at room temperature.

    a State the physical state and colour of the elements fluorine to iodine inclusive.

    b Explain the change in physical state in terms of the bonding in the elements.

    c Predict with a reason what the physical state of astatine will be.

18 Fluorine is the first member of Group VII and astatine is the last. Use the relative positions of these two elements in the group to answer the following questions:

    a Which of the elements will convert chloride ions to chlorine gas? Explain your answer.

    b Which of the elements produces an anion that is readily oxidised by chlorine gas? Explain your answer and write a balanced chemical equation for the reaction.

    c Will fluorine be a more or less powerful oxidising agent then astatine? Explain your answer based on the relative sizes of the molecules.

19 Write balanced chemical equations for each of the following reactions:

   a chlorine + iron → iron(III) chloride

   b bromine + hydrogen → hydrogen bromide

20 Complete and write balanced equations for each of the following reactions:

   a bromine + potassium iodide →

   b chlorine + sodium iodide →

6.3 Period 3 elements

Introduction

We continue our study of trends within the Periodic Table by looking at the elements in Period 3. In this section, we will study the gradual changes from metals to non-metals.

See Chapter 3 for an introduction to the Periodic Table.

Groups I and II of the Periodic Table contain only metals, and Group VII contains only non-metals. However, the elements of Period 3 range from metals on the left to non-metals on the right. Table 6.11 shows some properties of the elements of Period 3.

The pattern of electron configurations in Table 6.11 is typical of a period of elements. All the elements of a given period have the same number of fully occupied inner shells. However, as the atomic number of the element increases by one, so does the number of electrons in the valence shell of the atom. Metals tend to have one, two or three valence electrons, whereas non-metals have four or more. The change in the number of valence electrons supports the variation from metal to non-metal across the period (see Chapter 3).

Changes in physical properties

As expected, the change from metals to non-metals across Period 3 will be reflected in changes in the physical properties of the elements. These trends in physical properties are summarised in Table 6.12.

▼ **Table 6.11**   The elements of Period 3, sodium to argon

	Sodium	Magnesium	Aluminium	Silicon	Phosphorus	Sulphur	Chlorine	Argon
Atomic number	11	12	13	14	15	16	17	18
Electron configuration	2,8,1	2,8,2	2,8,3	2,8,4	2,8,5	2,8,6	2,8,7	2,8,8
Brief description	A soft metal of low density. Sources are rock salt, salt mines and sea water. Obtained by electrolysis of molten common salt.	Silvery grey metal. Important in plant and animal life.	Silvery grey metal. The most abundant metal. Sources are bauxite and alumino-silicate rocks.	A non-metal (sometimes described as a metalloid). Has the diamond structure.	Non-metallic solid. Widely distributed in the Earth's crust. An essential constituent of cell protoplasm, the nervous system and bone.	Yellow solid. Occurs near volcanoes, hot springs, in natural gas and petroleum.	A green-yellow gas. Prepared by electrolysis of concentrated brine.	An inert gas extracted from the atmosphere, especially during production of liquid air or the manufacture of ammonia. Used where an inert atmosphere is required.

▼ **Table 6.12** Physical properties of the elements of Period 3 from sodium to argon

	Sodium	Magnesium	Aluminium	Silicon	Phosphorus	Sulphur	Chlorine	Argon
Metal or non-metal?	metal	metal	metal	semi-metal	non-metal	non-metal	non-metal	non-metal
Atomic radius (nm)	190	145	118	111	98	88	79	71
Electronegativity	0.9	1.2	1.5	1.8	2.1	2.5	3.0	0
Melting point (°C)	97	649	660	1 410	44 (white)	123	−113	−189
Boiling point (°C)	882	1 090	2 467	2 355	280	444	−35	−186

Practice

21 Use melting and boiling point values to predict the physical state of each of the Period 3 elements at room temperature.

Look at the data in Table 6.12, and note the following as we go across a period:

- Atomic radius decreases – the tendency to lose electrons will therefore decrease and therefore metallic character decreases.

- Electronegativity increases – the tendency of the atoms of the elements to gain electrons increases and therefore non-metallic character increases.

- Melting and boiling points increase up to a maximum value and then decrease.

Practice

Patterns in physical properties become more obvious when represented in the form of line graphs rather than in a table.

22 Plot graphs of atomic number on the horizontal axis against the following physical properties given for the elements of Period 3 in Table 6.13:
   a Atomic radius
   b Melting point
   c Electronegativity
   (Keep these graphs as part of your notes on periodicity.)

23 Use your graphs to help you complete the following statements about the elements in Period 3:
   a The metals in Period 3 have _____ electronegativity than the non-metals.
   b The atomic radii of elements _____ from left to right across Period 3.
   c A metal atom in Group II has a _____ atomic radius than the Group I metal atom. The Group II metal is therefore _____ reactive than the Group I metal.

d A non-metal atom in Group VI has a _____ atomic radius than the element in Group VII. The Group I non-metal is _____ reactive than the Group VII non-metal.

e Silicon has the highest melting point because it has a _____ atomic structure, unlike the elements phosphorus and sulphur, which follow it in the period and which have _____ structures.

f Sodium, magnesium and aluminium have higher melting points than phosphorus and sulphur because they have _____ structures with strong _____ bonds between the cations and _____ electrons.

Other physical properties

Another important physical property is that of electrical conduction. The general pattern is that metals are good conductors of electricity and the greater the number of mobile electrons in their structure the more pronounced the conductivity. There are obvious problems with the use of a highly reactive metal such as sodium as a conductor. Can you think what these problems are? The conducting ability of aluminium is well known. Other typical metallic properties, such as malleability and ductility, are also more applicable to aluminium and magnesium than to sodium.

By contrast, non-metals are not electrical conductors because of the absence of mobile electrons. Do you know of a non-metal, not in Period 3, that conducts electricity?

Certain other non-metals, such as silicon, are considered to be semi-conductors.

Trends in chemical properties

We can see the variation from metallic to non-metallic character across Period 3 by looking at the reactions of the elements in the period with oxygen, chlorine and water, and the nature of the products formed (see Table 6.13).

Note the following about the reactions shown in Table 6.13:

- In their reactions with non-metals such as oxygen and chlorine, the metals transfer electrons and form ionic compounds, whereas the non-metals form covalent compounds by sharing electrons.

- The oxides of metals are basic and the oxides of non-metals are acidic.

- As metallic character decreases across the period, the reaction with water is less vigorous. Metals react with water or steam to form metallic oxides or hydroxides and to release hydrogen gas. Non-metals do not react with water.

- Aluminium oxide and aluminium chloride show some characteristics of both metallic and non-metallic compounds.

▼ **Table 6.13** Reactions of Period 3 elements with oxygen, chlorine and water

Element	Reactions with oxygen	Reactions with chlorine	Reactions with water
Sodium	when heated, sodium forms an oxide: $4Na(s) + O_2(g) \rightarrow 2Na_2O(s)$ ionic compound; basic oxide	$2Na(s) + Cl_2(g) \xrightarrow{heat} 2NaCl(s)$ ionic compound; crystalline solid	this is a vigorous to violent reaction; hydrogen is liberated; the resultant solution is alkaline: $2Na(s) + 2H_2O(l) \rightarrow 2NaOH(aq) + H_2(g)$
magnesium	magnesium burns with a brilliant flame, producing magnesium oxide: $2Mg(s) + O_2(g) \xrightarrow{heat} 2MgO(s)$ ionic compound; basic oxide	$Mg(s) + Cl_2(g) \xrightarrow{heat} MgCl_2(s)$ ionic compound; crystalline solid	reacts slowly with hot water to form the metallic oxide and hydrogen gas: $Mg(s) + H_2O(l) \rightarrow MgO(s) + H_2(g)$
aluminium	a thin oxide coating forms: $4Al(s) + 3O_2(g) \rightarrow 2Al_2O_3(s)$ shows both ionic and covalent character; amphoteric oxide	$2Al(s) + 3Cl_2(g) \xrightarrow{heat} 2AlCl_3(s)$ shows both ionic and covalent character; crystalline solid	reacts with steam to form aluminium oxide and hydrogen gas: $2Al(s) + 3H_2O(g) \rightarrow Al_2O_3(s) + 3H_2(g)$
silicon	no reaction	powdered silicon combines with chlorine: $Si(s) + 2Cl_2(g) \rightarrow SiCl_4(s)$ covalent compound; colourless liquid	no reaction
phosphorus	phosphorus forms two oxides: P_4O_6 and P_4O_{10} $P_4(s) + 3O_2(g) \xrightarrow{heat} P_4O_6(s)$ covalent compound; acidic oxide	phosphorus ignites in chlorine, burning with a pale yellow flame: $P_4(s) + 6Cl_2(g) \xrightarrow{heat} 4PCl_3(l)$ covalent compound; colourless liquid	no reaction
sulphur	sulphur forms two oxides: SO_2 and SO_3 $S(s) + O_2(g) \xrightarrow{heat} SO_2(g)$ covalent compound; acidic oxide	chlorine passed over molten sulphur yields covalent chlorides slowly $S_8(s) + 4Cl_2(g) \rightarrow 4S_2Cl_2(l)$	no reaction
chlorine	chlorine forms a variety of covalent oxides	not applicable	dissolves in and reacts with water to form a mixture of acids: $Cl_2(g) + H_2O(l) \rightarrow HCl(aq) + HOCl(aq)$

Practice

▼ **Table 6.14** Properties of chlorides and oxides of Period 3 elements

	Properties of chlorides			Properties of oxides		
	Bonding structure	Melting point (°C)	Electrical conductivity when molten	Bonding structure	State	Boiling point (°C)
R	giant ionic	808	good	giant ionic	solid	sublimes at 1 275
S	simple molecular	–70	non-conducting	simple molecular	gas	–10
T	giant ionic	715	good	giant ionic	solid	3 600
U	ionic with covalent character	183	poor	ionic with covalent character	solid	2 980

Table 6.14 shows some of the properties of the chlorides and oxides of some of the elements of Period 3. Use the letters R, S, T and U in your answers.

24 Use the information in Table 6.14 to answer the following questions:

   a Identify one metal in Table 6.14. Give reasons for your answer.

   b Identify one non-metal in Table 6.14. Give reasons for your answer.

   c Of the elements shown, which is likely to be the most metallic?

   d Of the elements shown, which is likely to be the most non-metallic?

   e Arrange the elements as they might occur from left to right in Period 3.

   f Which element is likely to have the smallest atomic radius?

   g Which element is likely to have the highest electronegativity?

The following exercise tests whether you can apply what you have learnt about Periods 2 and 3.

25 Give explanations for the following statements. Use information from Chapter 3 if in any doubt.

   a Why does the atomic radius of the Period 2 elements decrease across the period from left to right?

   b Across Period 2, 'metallic character decreases and non-metallic character increases'. Give two underlying reasons for this change.

   c How would you expect reducing power and oxidising power to change across Period 2?

   d Which of the elements in Period 2 are engaged in:
      (i) metallic bonding;
      (ii) ionic (electrovalent) bonding;
      (iii) covalent bonding?

Summary

Group II elements (magnesium to barium)

- Are all metals.
- Have two electrons in their valence shell, with all the inner shells being filled.
- All elements show a common and fixed oxidation of +2 in their compounds.
- All elements form compounds that are ionic.
- Show gradual changes in physical properties as the group is descended:
  - Densities increase
  - Melting and boiling points generally decrease (metallic bonding becomes weaker)
  - Electronegativities decrease
- Show gradual changes in chemical properties as the group is descended: elements are increasingly more reactive due to increasing ease of ionisation.

The Group I elements show similar trends, but they have an oxidation state of +1 in their compounds and are more reactive than the Group II elements.

Transition metals

- Show similar physical properties, but are less reactive than Group I and Group II metals.
- Have variable oxidation states in their compounds.
- Their compounds act as catalysts.
- They form coloured compounds.
- They form complex compounds.

Group VII elements

- Group VII elements, the halogens, have similar electronic configurations and, generally, have similar chemical properties.
- The halogens accept electrons easily – they are highly electronegative.
- The more electronegative the halogen, the more reactive it is.
- Because of their comparatively small atomic size and high electronegativity, the halogens are powerful oxidising agents.
- Fluorine is the most powerful oxidising agent and iodine is the least powerful oxidising agent in Group VII.
- The halogens oxidise hydrogen, other non-metals and most metals.
- Reactivity decreases down Group VII. An element high in the group displaces one lower down the group from its compounds.

Period 3

- The following trends are observed when going across Period 3, sodium to argon:
  - Atomic radii decrease.
  - The elements change from metals (e.g. sodium and magnesium) to non-metals (e.g. sulphur and chlorine), i.e. there is a decrease in metallic character.

- Whereas the oxides of the elements on the left of Period 3 are basic, the oxides of the elements on the right are acidic (although aluminium oxide is amphoteric – having both basic and acidic properties).
- The oxides and chlorides of the metals in Period 3 are ionic compounds with high melting and boiling points.
- The oxides and chlorides of the non-metals in Period 3 are covalent compounds.

End-of-chapter questions

1 As we descend Group II from beryllium to barium, which of the following is true of the elements?
  A Ease of ionisation increases
  B Electronegativity increases
  C Reactivity decreases
  D Atomic radius decreases

2 All of the following statements are true for both Group II and Group VII elements, EXCEPT:
  A Reactivity increases as we descend the group
  B All the members of the group form electrovalent compounds
  C The more easily the elements form ions, the more reactive they are
  D More reactive elements in the group will displace less reactive elements from their compounds

3 The electronegativity of an element refers to:
  A Size of its negative charge
  B Its ability to attract electrons
  C The ease with which it forms ions
  D The ease with which it reacts with negative ions

4 Some of the properties of four elements in Period 3 of the Periodic Table are shown in the table below:

Element	Physical state	Nature of oxide	Reaction with water	Electrical conductor
R	gas	acidic	no reaction	no
S	solid	acidic	no reaction	no
T	solid	basic	violent	yes
U	solid	amphoteric	reacts if activated	yes

The correct order for the elements in Period 3, from left to right, is most likely to be:
  A U, T, R, S
  B R, S, U, T
  C R, S, T, U
  D T, U, S, R

5 Which two of the following statements are correct?

I   Fluorine gains electrons more readily than chlorine.

II Fluorine loses electrons more readily than chlorine.

III Fluorine is a more powerful oxidising agent than chlorine.

IV Fluorine is a more powerful reducing agent than chlorine.

A I and III

B I and IV

C II and III

D II and IV

6 Which of the following elements is NOT a member of Group II of the Periodic Table?

A Calcium

B Barium

C Sodium

D Magnesium

7 Which of the following lists represents the atomic numbers of five members of the same period of the Periodic Table of elements?

I   2, 10, 18, 36, 54

II 9, 17, 35, 53, 85

III 8, 9, 10, 11, 12

IV 13, 14, 15, 16, 17

A I and II

B III and IV

C I only

D IV only

The following information is provided for Questions 8–11.

Six elements in the same period of the Periodic Table have the physical properties shown in the table below.

Element	Physical state	Melting point (°C)	Type of compounds formed
U	solid	181	forms ionic compounds
V	gas	−219	forms covalent or ionic compounds
W	gas	−249	does not form compounds
X	gas	−220	forms covalent or ionic compounds
Y	solid	2 130	forms ionic compounds
Z	solid	1 280	forms ionic compounds

8 The most likely order of the elements in the period, from left to right, is:

A W, V, X, Y, Z, U

B U, Z, Y, V, X, W

C Y, Z, U, V, X, W

D W, X, V, U, Z, Y

9 The compound formed between X and V is most likely to be:

  A Ionic

  B Covalent

  C A solid with a high melting point

  D Ionic with some covalent character

10 Z and Y will most likely:

  A Form ions that migrate to the anode during electrolysis

  B Form oxides that dissolve in water to form acids

  C Have high electronegativity

  D Have giant structures

11 The element most likely to be a semi-metal is:

  A Y

  B W

  C U

  D X

---

12 Use your knowledge of the Group I and II elements and transition elements to answer the following questions:

  a Which of these elements show fixed oxidation states in their compounds?

  b Which type of element is likely to have the highest melting and boiling points? Explain your answer.

  c Explain whether Group I or Group II elements will be more reactive with oxygen.

  d An oxide of this type of element is used to increase the rate of release of oxygen from hydrogen peroxide.

      (i) Which type of element can form this oxide?

      (ii) What is the term used for the action of this oxide?

  e Indicate the less reactive element in the following pairs of elements:

      (i) Ca and Sr

      (ii) K and Na

      (iii) Mg and Ca

      (iv) Li and Na

13 The element radium (Ra), atomic number 88, is in the same group as calcium. Predict the following, and in each case explain how you arrived at your answer:

  a The number of electrons in the valence shell of radium

  b How the element is likely to react with

      (i) water and (ii) oxygen

  c Whether the melting point of radium will be higher or lower than that of calcium

14 Write the chemical formulae for all the compounds formed:

  a when sodium reacts with water;

  b when magnesium reacts with steam;

  c when barium reacts with dilute hydrochloric acid;

  d when calcium reacts with oxygen.

15 Astatine is a member of Group VII. Little is known of its properties because it is unstable and radioactive. Based on your knowledge of chlorine, bromine and iodine, predict:
   a the formula of astatine molecules;
   b the likely colour and physical state, at room temperature and atmospheric pressure, of astatine;
   c the reaction, if any, of astatine with iron;
   d the reaction, if any, of astatine with aqueous potassium iodide;
   e the reaction, if any, of astatine with aqueous sodium hydroxide.

16 Identify the halogen that:
   a gives a blue-black colour with starch;
   b sublimes to a purple vapour when heated;
   c gives a red-brown vapour at room temperature;
   d is used to disinfect water;
   e is radioactive;
   f will oxidise chloride, bromide and iodide ions.

17 Use the elements of Period 3 to explain clearly the meaning of each of the following terms:
   a Periodicity
   b Electronegativity
   c Ease of ionisation
   d Reducing property
   e Oxidising ability

18 Illustrate the variation in chemical properties across Period 3, by considering the reaction of the elements with (a) water and (b) oxygen.

19 Metal oxides are usually basic or amphoteric, whereas non-metal oxides are usually acidic or neutral. Use the chemical formulae of the Period 3 elements and their compounds in your answers to the following questions.
   a Which of the four types of oxides described in the introduction to the question is an acid anhydride?
   b **(i)** Write the chemical formulae for two acid anhydrides formed by elements of Period 3.
   (ii) Write a balanced chemical equation for the reaction of one of these acid anhydrides with water.
   c **(i)** Which element in Period 3 forms an amphoteric oxide?
   (ii) Write the chemical formula for this oxide.
   (iii) Write balanced chemical equations, including state symbols, for the reaction of this oxide with dilute hydrochloric acid and with dilute sodium hydroxide.
   d Which Period 3 element will form an oxide with:
   (i) the strongest basic property;
   (ii) the strongest acidic property?
   (iii) Write balanced chemical equations, including state symbols, for the reactions of the basic oxide with water and with dilute sulphuric acid.

7 Writing formulae and equations

▲ **Figure 7.1** How many water molecules does this glass contain?

▲ **Figure 7.2** How many molecules of oxygen are in this jar?

In this chapter, you will study the following:

- how to interpret and write correctly the formulae of covalent and ionic compounds;
- how to write and balance chemical equations;
- how to write ionic equations;
- the usefulness and limitations of chemical equations.

This chapter covers
Objective 6.5 of Section A of the CSEC Chemistry Syllabus.

Just as we write sentences, our names or our addresses in a particular way, chemists follow set conventions in writing formulae and equations. We begin our quantitative studies by learning how to write formulae and equations – as these are the chemist's way of describing substances and their chemical reactions.

- If we want to make soap in the laboratory, how do we know what volume and what concentration of reactants to use?
- How does an analytical chemist calculate the concentration of nitrate or phosphate ions in drinking water to ensure that it does not exceed acceptable levels?
- Can chemists tell how many particles are in a certain mass or volume of a substance?

7.1 Molecular formulae

The **molecular formula** of a compound gives the actual numbers of the different types of atoms in one molecule of a covalent compound or the ratio of ions present in one formula unit of an ionic compound.

The **molecular formula** of a compound gives:

- the number of atoms of each element present in one molecule for covalent compounds;
- the ratio of the ions present in one formula unit for ionic compounds (because the term 'molecule' cannot be applied to ionic compounds).

Understand it better

a The molecular formula of glucose is $C_6H_{12}O_6$. This means that 1 molecule of glucose contains:
- 6 atoms of carbon;
- 12 atoms of hydrogen;
- 6 atoms of oxygen.

b The molecular formula of ammonium sulphate is $(NH_4)_2SO_4$. The ratio of ions is:
- two NH_4^+ ions;
- one SO_4^{2-} ion.

Q1 How many atoms of each type do the following contain?
    a C_4H_8 **b** $Ca(NO_3)_2$ **c** $Al_2(SO_4)_3$

Q2 Which two of the compounds in Question 1 are ionic compounds?

Q3 What is the ratio of cations to anions in each of the ionic compounds?

Note: Chemical formulae are really worked out by experiment and you will learn how this is done in Chapter 8. Here, however, we will give you some simple tools to help you write the molecular formula of a compound correctly, using other types of information. You need to write formulae correctly in order to write equations.

Working out numbers and ratios in compounds

In Chapter 6, we worked out the ratio of atoms in molecules or formula units by using the number of outer electrons of elements (their 'valence electrons') to form bonds so that they achieve a stable electronic configuration. We can also work out the formulae by using charges or oxidation numbers.

Using charges

This method only works for ionic compounds.

A **polyatomic ion** (sometimes called a radical) is a charged group of atoms that often occur together in compounds. Examples are sulphate (SO_4^{2-}), carbonate (CO_3^{2-}), nitrate (NO_3^-) and ammonium (NH_4^+).

Tables 7.1 and 7.2 contain lists of the charges carried by common anions and cations (including **polyatomic ions**). Remember that in one formula unit of an ionic compound, the total positive charge must equal the total negative charge.

▼ **Table 7.1** The charges on some common negative ions (anions)

1− ions	2− ions	3− ions
fluoride, F^-	sulphate, SO_4^{2-}	phosphate, PO_4^{3-}
chloride, Cl^-	sulphite, SO_3^{2-}	nitride, N^{3-}
bromide, Br^-	carbonate, CO_3^{2-}	
iodide, I^-	oxide, O^{2-}	
hydroxide, OH^-	sulphide, S^{2-}	
manganate(VII), MnO_4^-	chromate(VI), CrO_4^{2-}	
hydrogen carbonate, HCO_3^-	dichromate(VI), $Cr_2O_7^{2-}$	
ethanoate, CH_3COO^-	ethanedioate, $C_2O_4^{2-}$	
methanoate, $HCOO^-$		

▼ **Table 7.2** The charges on some common positive ions (cations)

1^+ ions	2^+ ions	3^+ ions
metals of Group I: Li^+, Na^+, K^+ hydrogen, H^+ ammonium, NH_4^+ silver, Ag^+ copper(I), Cu^+	metals of Group II: Mg^{2+}, Ca^{2+}, Ba^{2+} lead(II), Pb^{2+} iron(II), Fe^{2+} zinc, Zn^{2+} tin(II), Sn^{2+} copper(II), Cu^{2+}	aluminium, Al^{3+} iron(III), Fe^{3+} chromium(III), Cr^{3+}

Writing formulae using the charges on ions in ionic compounds

Step 1: Write the name of the compound.
Step 2: Write down the symbols of the elements (or polyatomic groups) and their charges.
Step 3: Since the total negative charge must equal the total positive charge, balance out the charges by adjusting the numbers of ions as necessary. Do not change the charge on any of the ions.
Step 4: Write the formula using the numbers of each ion as subscript.

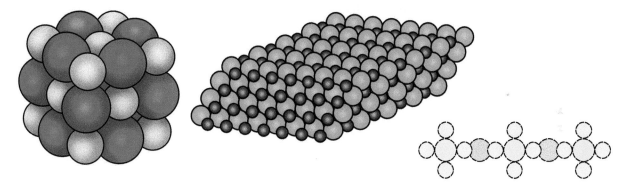

▲ **Figure 7.3** Illustrations of molecules of CaO, $ZnCl_2$, and $Al_2(SO_4)_3$.

Here are some examples:
Step 1: Calcium oxide
Step 2: Ca^{2+} O^{2-}
Step 3: These are balanced.
Step 4: Ca_1O_1
Note: This formula is written as CaO since the ratio of the ions is 1:1. Where the number of an atom or an ion is 1, the 1 is never written down – it is understood.

Step 1: Zinc chloride
Step 2: Zn^{2+} Cl^-
Step 3: Zn^{2+} gives $2+$ Cl^-; Cl^- gives $2-$.
Step 4: Zn_1Cl_2
The formula is $ZnCl_2$.

Step 1: Aluminium sulphate
Step 2: Al^{3+} SO_4^{2-}
Step 3: Al^{3+}, Al^{3+} gives $6+$ SO_4^{2-}; SO_4^{2-}, SO_4^{2-} gives $6-$.
Step 4: $Al_2(SO_4)_3$
The formula is $Al_2(SO_4)_3$.
Note: The subscript 3 is written outside the brackets to indicate three sulphate groups.

Using oxidation numbers

This method can be used for all compounds.

The **oxidation number** is a positive or negative number that indicates the real or hypothetical number of electrons lost or gained by an element in a given compound. Note that the sign ('+' or '−') comes before the number.

Each element in a compound can be given an **oxidation number**. An oxidation number is assigned to an element by a set of rules. A full treatment of oxidation numbers and rules to assign them is given in Section 12.7.

If you are given the oxidation numbers of the elements in any compound, you can work out the formula in much the same way as when you use charges. The sum of the oxidation numbers of elements in a compound is zero.

For some elements, the oxidation number is always or almost always the same. However, some elements, such as iron, have variable oxidation numbers. Roman numerals are used to distinguish between the compounds of such elements. Iron(II) chloride and iron(III) chloride are the names of iron compounds of oxidation number $+2$ and $+3$, respectively.

Writing formulae using oxidation numbers

Step 1: Write the name of the compound.
Step 2: Write the elements in the compound and their oxidation numbers.
Step 3: Balance out the numbers.
Step 4: Write the formula.

▲ **Figure 7.4** Molecules HCl, SO_3, and SO_2

Here are some examples:

Step 1: Hydrogen chloride
Step 2: H^{+1} Cl^{-1}
Step 3: $+1$ and -1 add up to zero.
Step 4: HCl
The formula is HCl.

Step 1: Sulphur(VI) oxide
Step 2: S^{+6} O^{-2}
Step 3: S^{+6} O^{-2}, O^{-2}, O^{-2}; $+6$ and -6 add up to zero.
Step 4: SO_3
The formula is SO_3.

Step 1: Sulphur(IV) oxide
Step 2: $S^{+4} O^{-2}$
Step 3: $S^{+4} O^{-2}$, O^{-2}; $+4$ and -4 add up to zero.
Step 4: SO_2
The formula is SO_2.

Practice

1 Work out the formulae of:

  a nitrogen(I) oxide;

  b nitrogen(IV) oxide;

  c phosphorus(V) chloride.

  The oxidation number of oxygen is -2 and of chlorine is -1.

7.2 Chemical equations

A chemical equation uses chemical symbols to summarise what actually happens in a chemical reaction. When iron metal, for example, is intimately mixed with the non-metallic element sulphur and the mixture is heated, brownish-black iron(II) sulphide is formed.

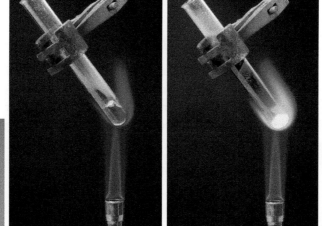

Iron and sulphur   Heating the mixture   Iron(II) sulphide

▲ **Figure 7.5** Formation of iron sulphide from iron and sulphur

We can summarise this reaction in words as:

    iron plus sulphur produces iron(II) sulphide

and by a chemical equation as:

$$Fe(s) + S(s) \xrightarrow{\text{heat}} FeS(s)$$

Note the following points.

- The arrow means 'produces' or 'yields'.
- Information written above or below the arrow indicates the reaction conditions.
- The substances on the left (before the arrow) are the starting materials or the reactants, while the substances on the right (after the arrow) are the products.
- The letters in brackets after the symbols or formulae are state symbols – these indicate the physical state of the reactants or products:
    solid = (s)
    liquid = (l)
    gas = (g)
    aqueous solution = (aq)

What does a chemical equation tell us?

The equation:

$$2HCl(aq) + Na_2CO_3(s) \rightarrow 2NaCl(aq) + H_2O(l) + CO_2(g)$$

tells us the following:

- Hydrochloric acid reacts with solid sodium carbonate to produce sodium chloride solution, water and carbon dioxide (as a gas).
- 2 mol HCl react: with 1 mol Na_2CO_3 to produce 2 mol NaCl, 1 mol H_2O and 1 mol CO_2.
- 2 × 36.5 g (= 73 g) of HCl react with 106 g of Na_2CO_3 to form 2 × 58.5 g (= 117 g) of NaCl, 18 g of H_2O and 44 g of CO_2 (or 22.4 dm^3 of CO_2) measured at STP.

Equations for equilibrium reactions

A pair of reversed arrows is used to indicate that the reaction can proceed simultaneously in both directions and gives rise to an equilibrium mixture of reactants and products (see Section 12.6 for more information about equilibrium reactions).
Note: Left to right is conventionally accepted as the forward direction.

Here is an example:

$$Pt; 600\ °C; 10\ atm$$
$$4NH_3(g) + 5O_2(g) \rightleftharpoons 4NO(g) + 6H_2O(g)$$

The limitations of chemical equations

While chemical equations can give us a lot of information about reactions, there are certain things they cannot tell us about:

- The equation cannot show how long it will take for the reaction to be completed. Some reactions are practically spontaneous, others require minutes, hours, days or even weeks to be completed.

- The equation cannot tell whether the reaction is feasible or not. An equation may be correctly balanced but the reaction it shows may not occur experimentally. For example, one can write a properly balanced equation for the reaction between copper and water:

$$Cu(s) + 2H_2O(l) \rightarrow Cu(OH)_2(s) + H_2(g)$$

but the reaction between copper and water does not occur. Think about copper water pipes!

Writing and balancing equations

A **balanced equation** has the same number of atoms of each element on both sides of the equation.

You must remember that atoms are not changed in a chemical reaction, all that changes is the way in which the atoms are joined together. It follows from this that you cannot 'lose' atoms during a chemical reaction. You must have the same number of the same types of atoms on both sides of an equation. The process of ensuring that this happens is referred to as '**balancing the equation**'.

Some simple equations are already balanced when the formulae for the reactants and products are written, but others have to be balanced. Here are some examples that illustrate the procedure for balancing equations.

Worked example 1

What do you need to do to balance this equation?

$$Fe(s) + S(s) \rightarrow FeS(s)$$

numbers of atoms   Fe 1   Fe 1

S 1   S 1   balanced

$$\boxed{Fe(s) + S(s) \rightarrow FeS(s)}$$

Worked example 2

Do you need to add some numbers to balance this equation?

$$Mg(s) + N_2(g) \rightarrow Mg_3N_2(s)$$

number of atoms   Mg 1   Mg 3

N 2   N 2   not balanced

There are three times as many magnesium atoms on the right as on the left.
Note: Nitrogen is correctly written as $N_2(g)$ since, like most gaseous elements, it exists as diatomic molecules.

To balance: place '3' in front of the magnesium on the left.

$$3Mg(s) + N_2(g) \rightarrow Mg_3N_2(s)$$

number of atoms   Mg 3   Mg 3

N 2   N 2   balanced

$$\boxed{3Mg(s) + N_2(g) \rightarrow Mg_3N_2(s)}$$

Note: Balancing by writing $Mg_3(s)$ is incorrect. This would imply (incorrectly) that magnesium exists as triatomic molecules.

Worked example 3

Is this equation balanced? If not, make the changes necessary to balance it.

$$Al(s) + O_2(g) \rightarrow Al_2O_3(s)$$

number of atoms   Al 1   Al 2

O 2   O 3   not balanced

There are twice as many Al on the right as on the left and $\frac{3}{2}$ as many O on the right as on the left. This equation can be balanced by placing 2 in front of Al, and $\frac{3}{2}$ in front of O_2 on the left .

$$2Al(s) + \tfrac{3}{2}O_2(g) \rightarrow Al_2O_3(s)$$

number of atoms   Al 2   Al 2

O $(\tfrac{3}{2} \times 2) = 3$ O 3   balanced

Note: Fractions are not allowed in equations, so change them to whole numbers by multiplying the whole equation by 2.

$$4Al(s) + 3O_2(g) \rightarrow 2Al_2O_3(s)$$

number of atoms   Al 4   Al 4

O 6   O 6   balanced

Note: A number placed in front of a formula multiplies the whole formula.

$$\boxed{4Al(s) + 3O_2(g) \rightarrow 2Al_2O_3(s)}$$

Worked example 4

Here is another equation to balance. You need to treat the sulphate ion (SO_4^{2-}) and the hydroxide ion (OH^-) as groups rather than as individual atoms.

$$CuSO_4(aq) + NaOH(aq) \rightarrow Cu(OH)_2(s) + Na_2SO_4(aq)$$

number of ions Cu^{2+} 1 Cu^{2+} 1

SO_4^{2-} 1 SO_4^{2-} 1

Na^+ 1 Na^+ 2

OH^- 1 OH^- 2   not balanced

There are twice as many Na^+ ions and OH^- groups on the right as on the left. Place '2' in front of the NaOH on the left.

$$CuSO_4(aq) + 2NaOH(aq) \rightarrow Cu(OH)_2(s) + Na_2SO_4(aq)$$

number of ions Cu^{2+} 1 Cu^{2+} 1

SO_4^{2-} 1 SO_4^{2-} 1

Na^+ 2 Na^+ 2

OH^- 2 OH^- 2   balanced

$$\boxed{CuSO_4(aq) + 2NaOH(aq) \rightarrow Cu(OH)_2(s) + Na_2SO_4(aq)}$$

A summary of the rules for writing equations

- Use chemical symbols and formulae to describe what happens in the reaction. List all reactants on the left of the arrow, and all products on the right.

- Represent elements by their symbols. For solid elements, use the symbol for a single atom, for example Mg, Fe, C, etc. For gases, use the symbol for the molecule, for example, most common elements are diatomic – thus H_2, O_2, N_2, etc.

- Use state symbols (s), (l), (g) and (aq) to indicate the physical state of the substances in the equation.

- Check symbols and formulae to ensure that they are all correct.

- Balance the equation by inserting the whole numbers in front of each term of the equation where necessary. Do not change the numbers within the formula.

Practice

2 Write and balance equations for the following reactions:

a Zinc metal reacts with copper sulphate solution to form zinc sulphate solution and copper metal.

b Lithium metal reacts with fluorine gas to form solid lithium fluoride.

c Ammonium chloride solution and silver nitrate solution react to form ammonium nitrate solution and solid silver chloride.

3 Use the guidance provided above to balance the following equations:

a $Mg(s) + O_2(g) \rightarrow MgO(s)$

b $Al(s) + N_2(g) \rightarrow AlN(s)$

c $NaOH(aq) + H_3PO_4(aq) \rightarrow Na_3PO_4(aq) + H_2O(l)$

d $CH_3CH_2OH(l) + O_2(g) \rightarrow CO_2(g) + H_2O(l)$

e $NH_4Cl(aq) + AgNO_3(aq) \rightarrow AgCl(s) + NH_4NO_3(aq)$

f $Pb(NO_3)_2(aq) \rightarrow PbO(s) + NO_2(g) + O_2(g)$

g $Ca(OH)_2(aq) + HCl(aq) \rightarrow CaCl_2(aq) + H_2O(l)$

h $Al(s) + HCl(aq) \rightarrow AlCl_3(aq) + H_2(g)$

Ionic equations

Ionic equations are equations that show only the species taking part in a reaction between ionic substances.

Reactions involving ions can often be more usefully represented by **ionic equations**. For example, the reaction between silver nitrate and potassium chromate(VI) may be written as:

$$2AgNO_3(aq) + K_2CrO_4(aq) \rightarrow Ag_2CrO_4(s) + 2KNO_3(aq)$$

Silver nitrate and potassium chromate(VI) are both ionic compounds, and so they form ions that are free in aqueous solution. The reaction can, therefore, be represented by the expanded equation:

$$2Ag^+(aq) + 2NO_3^-(aq) + 2K^+(aq) + CrO_4^{2-}(aq) \rightarrow Ag_2CrO_4(s) + 2K^+(aq) + 2NO_3^-(aq)$$

Notice that:

- the $K^+(aq)$ and the $NO_3^-(aq)$ ions occur unchanged on both sides of the equation; these are merely 'spectator ions' and can be omitted from the ionic equation, leaving us with:

$$2Ag^+(aq) + CrO_4^{2-}(aq) \rightarrow Ag_2CrO_4(s)$$

- the $Ag^+(aq)$ ions and the $CrO_4^{2-}(aq)$ ions react to form a solid precipitate, $Ag_2CrO_4(s)$; it is these species that are included in the equation;

- if any other soluble silver salt or any other soluble chromate(VI) were used, the reaction and the ionic equation would be the same.

Points to remember

1 Ionic equations include only those ions involved in the formation of precipitate, gas or weak electrolyte, omitting all 'spectator ions'.

Here are some examples:

Full equation	$Ba(NO_3)_2(aq) + H_2SO_4(aq) \rightarrow BaSO_4(s) + 2HNO_3(aq)$
Ionic equation	$Ba^{2+}(aq) + SO_4^{2-}(aq) \rightarrow BaSO_4(s)$
	precipitate

Full equation	$HNO_3(aq) + NaOH(aq) \rightarrow NaNO_3(aq) + H_2O(l)$
Ionic equation	$H^+(aq) + OH^-(aq) \rightarrow H_2O(l)$
	weak electrolyte

Full equation	$HNO_3(aq) + NaOH(aq) \rightarrow NaNO_3(aq) + H_2O(l)$
Ionic equation	$H^+(aq) + OH^-(aq) \rightarrow H_2O(l)$
	weak electrolyte

This equation represents the neutralisation of any aqueous acid by any hydroxide.

2 Ionic equations must be balanced with respect to charge as well as numbers of each type of atom. For example, the following equation is balanced for the number of atoms, but it is not balanced for charge:

$$\text{not balanced} \quad Al(s) + 6H^+(aq) \rightarrow Al^{3+}(aq) + 3H_2(g)$$

The reactants have a net charge of 6+ while the products have a net charge of 3+. The equation is properly balanced as:

$$2Al(s) + 6H^+(aq) \rightarrow 2Al^{3+}(aq) + 3H_2(g)$$

charges	0	6+	6+	0

net charge	6+		6+	

The net charge on both sides of the equation is the same.

How to balance ionic equations

Step 1: Complete and balance the equation in the form in which it is given. If the equation is given in the molecular form, then complete it in molecular form, balance it and then change it to an ionic equation.

Step 2: Write all strong electrolytes in the ionic form. Write other species in the molecular form.

Step 3: Ensure that there is the same number of each type of atom or polyatomic ion on both sides of the equation.

Step 4: Omit all ions appearing on both sides of the equation unchanged.

Step 5: Write the ionic equation. Include only those ions that have undergone a change.

Step 6: Ensure that the net charges are equal on both sides.

Worked example 5

Follow through these steps to produce a balanced ionic equation.

Step 1: Balance the equation:

$$KI(aq) + Pb(NO_3)_2(aq) \rightarrow PbI_2(s) + KNO_3$$

balanced equations: $\quad 2KI(aq) + Pb(NO_3)_2(aq) \; PbI_2(s) + 2KNO_3(aq)$

Step 2: All the species are ionic, but lead(II) iodide is only slightly soluble in water. Write lead iodide in the unionised form.

$$2K^+(aq) + 2I^-(aq) + Pb^{2+}(aq) + 2(NO_3)^-(aq) \rightarrow PbI_2(s) + 2K^+(aq) + 2NO_3^-(aq)$$

Step 3: Check that the equation balances:

number of ions		
$K^+ 2$	$K^+ 2$	
$I^- 2$	$I^- 2$	
$Pb^+ 1$	$Pb^+ 1$	
$NO_3^- 2$	$NO_3^- 2$	balanced

Step 4: Omit ions that appear unchanged on both sides of the equation:

$K^+(aq)$	to	$K^+(aq)$	(unchanged) ions in solution
$I^-(aq)$	to	$I^-(s)$	(changed) ion no longer in solution
$Pb^{2+}(aq)$	to	$Pb^{2+}(s)$	(changed)
$NO_3^-(aq)$	to	$NO_3^-(aq)$	(unchanged)

Step 5: Write the ionic equation:

$$Pb^{2+}(aq) + 2I^-(aq) \rightarrow PbI_2(s)$$

Step 6: Check that the equation balances for charge:

charges	$2I^-(aq) = 2 \times 1^-$	$2I^-(s) = 2 \times 1^-$
	$Pb^{2+}(aq) = 1 \times 2^+$	$Pb^{2+}(s) = 1 \times 2^+$
	net charge $= 0$	net charge $= 0$ balanced

Worked example 6

Here is another example of writing an ionic equation.

Step 1: Balance the equation:

$$Al(s) + HCl(aq) \rightarrow AlCl_3(aq) + H_2(g)$$

balanced equation $\quad 2Al(s) + 6HCl(aq) \rightarrow 2AlCl_3(aq) + 3H_2(g)$

Step 2: HCl and $AlCl_3$ are written as ions. Al is a free metal and H_2 is a gas, and so are not charged.

$$2Al(s) + 6H^+(aq) + 6Cl^-(aq) \rightarrow 2Al^{3+}(aq) + 6Cl^-(aq) + 3H_2(g)$$

Step 3: Check that the equation balances:

number of atoms	Al 2	Al 2	
	H 6	H 6	
	Cl 6	Cl 6	balanced

Step 4: Omit ions that appear on both sides of the equation:

$Al(s)$	to	$Al^{3+}(aq)$	(changed)
$H^+(aq)$	to	$H_2(g)$	(changed)
$Cl^{-1}(aq)$	to	$Cl^-(aq)$	(unchanged)

Step 5: Write the ionic equation:

$$2Al(s) + 6H^+ \rightarrow 2Al^{3+}(aq) + 3H_2(g)$$

Step 6: Check that the equation balances for charge:

charges	$6H^+(aq) = 6 \times 1^+$	$2Al^{3+}(aq) = 2 \times 3^+$	
	net charge $= 6^+$	net charge $= 6^+$	balanced

Understand it better

Which substances are written in molecular form rather than as separate ions in ionic equations?

Compounds written in the molecular form in ionic equations are those that produce few or no ions in solution. These include:

- non-electrolytes, such as glucose;
- weak electrolytes, such as ethanoic acid, ammonia solution, water and solids;
- solids and precipitates formed by mixing aqueous solutions of ionic compounds, such as $AgCl(s)$, $CaCO_3(s)$ – these most often exist as ions but are only sparingly soluble in water so the ions in them cannot be considered to be available;
- gases, such as $H_2(g)$, $N_2(g)$, $CO_2(g)$;
- metals.

Practice

4 Use the guidance above to help you write balanced ionic equations for the following reactions:

a $Al(s) + HBr(aq) \rightarrow AlBr_3(aq) + H_2(g)$

b $NaOH(aq) + H_3PO_4(aq) \rightarrow Na_3PO_4(aq) + H_2O(l)$

c $Fe(OH)_3(s) + HCl(aq) \rightarrow FeCl_3(aq) + H_2O(l)$

d $MnO_2(s) + HCl(aq) \rightarrow MnCl_2(aq) + Cl_2(g) + H_2O(l)$

e $Ca(OH)_2(aq) + HNO_3(aq) \rightarrow Ca(NO_3)_2(aq) + H_2O(l)$

f $H_2SO_4(aq) + NaHCO_3(aq) \rightarrow Na_2SO_4(aq) + CO_2(g) + H_2O(l)$

g $Ca(CH_3COO)_2(aq) + HCl(aq) \rightarrow CaCl_2(aq) + CH_3COOH(l)$

Summary

- A chemical equation is a summarising statement of what happens in a chemical reaction.

- A balanced equation can tell us:
  - what the reactants and products are;
  - the formulae of these substances;
  - the number of particles or moles of reactants and of products;
  - the masses of reactants and products.

- A chemical equation should include information about the states of the reactants and products.

- A chemical equation can be written so as to provide information about the reaction conditions, for example:

$$CaCO_3(s) \xrightarrow{\text{heat}} CaO(s) + CO_2(g)$$

- Ionic equations include only those species that actually take part in the reaction.

End-of-chapter questions

1 What does the formula C_2H_6 mean?

  A There are two atoms of carbon and hydrogen.

  B There is one atom of carbon and two atoms of hydrogen.

  C There are two atoms of carbon and six atoms of hydrogen.

  D There is one atom of carbon, one 2, one atom of hydrogen and one 6

2 A polyatomic ion:

  A Is made up of diatomic molecules

  B Is a group of atoms with a charge

  C Is a number of atoms in a compound

  D Always has a negative charge

3 What does a balanced equation show?

    A How long it takes for a reaction to occur

    B Only the reactions that can occur

    C The same number of molecules on both sides of the equation

    D The same number of the same types of atoms on both sides of the equation

4 Which equation below is correctly balanced?

    A $2Al + 3O_2 \rightarrow Al_2O_3$

    B $Mg + O_2 \rightarrow 2MgO_2$

    C $CaCO_3 + 2HCl \rightarrow CaCl_2 + H_2O + CO_2$

    D $H_2 + O_2 \rightarrow 2H_2O$

5 Select the statement(s) that are true.

    I   Non-metals have minus charges.

    II Metal ions have positive charges.

    III State symbols indicate the chemical state of substances.

    IV Each element in a compound either gains or loses electrons.

    A I and IV

    B II and III

    C IV only

    D I, III and IV

---

6 Write the correct chemical formula for the following:

    a Magnesium sulphate

    b Calcium hydrogen carbonate

    c Potassium hydrogen sulphate

    d Aluminium nitride

    e Sodium ethanoate

    f Copper(II) hydroxide

    g Copper(I) oxide

    h Silver nitrate

7 Write the correct formula and name for the anion in each of the following compounds:

    a H_3PO_4

    b $(NH4)_2CO_3$

    c CH_3COOH

    d $PbCrO_4$

    e $MgCl_2$

    f HNO_3

    g KF

    h $Ca(NO_3)_2$

8 Indicate the total numbers of the different atoms in one molecule or formula unit of each of the following compounds:

    a $CuSO_4 \cdot 5H_2O$

    b $FeSO_4(NH_4)_2SO_4 \cdot 7H_2O$

    c $Na_2CO_3 \cdot 10H_2O$

    d $Na_2Zn(OH)_4$

    e Na_3AlF_6

9 Give the name of each of the following compounds to show the oxidation number of the <u>underlined</u> element, e.g. P<u>B</u>O lead(II) oxide; NaN<u>O</u>$_3$ sodium nitrate(V):

   a <u>Fe</u>Cl$_2$

   b <u>Mn</u>O$_2$

   c <u>Mn</u>Cl$_2$

   d K$_2$<u>Cr</u>O$_4$

   e Ca<u>S</u>O$_3$

10 Change the following word equations into chemical equations and then balance them:

   a sodium chloride + lead(II) nitrate → lead(II) chloride + sodium nitrate

   b iron(III) oxide + hydrochloric acid → iron(III) chloride + water

   c barium nitrate + sulphuric acid → barium sulphate + nitric acid

   d aluminium sulphate + sodium hydroxide → aluminium hydroxide + sodium sulphate

   e zinc hydroxide + sulphuric acid → zinc sulphate + water

   f hydrogen bromide + calcium hydroxide → calcium bromide + water

   g iron + chlorine → iron(III) chloride

11 Balance each of the following equations, giving the state symbols for each reactant and product. When thinking about the state symbols, consider if the substance is ionic, covalent or a metal.

   a $MnO_2 + Al → Mn + Al_2O_3$

   b $PCl_3 + H_2O → H_3PO_4 + HCl$

   c $CaCO_3 + H_3PO_4 → Ca_3(PO_4)_2 + CO_2 + H_2O$

   d $2AgNO_3 + CuCl_2 → 2AgCl + Cu(NO_3)_2$

   e $Na + H_2O → NaOH + H_2$

   f $Mg + O_2 → MgO$

   g $Mg + Cu(NO_3)_2 → Mg(NO_3)_2 + Cu$

12 Complete and balance the following equations, writing them as (i) total ionic equations and (ii) net ionic equations. Give the state symbols for each substance.

   a $BaCl_2 + (NH_4)_2CO_3 → BaCO_3 + NH_4Cl$

   b $Fe + CuSO_4 → FeSO_4 + Cu$

   c $CaO + HCl → CaCl_2 + H_2O$

   d $Fe(NO_3)_3 + NaOH → NaNO_3 + Fe(OH)_3$

   e $MgCl_2 + Na_2CO_3 → MgCO_3 + NaCl$

   f $Cl_2 + NaBr → NaCl + Br_2$

In this chapter, you will study the following:

- the mole and molar mass;
- how to interconvert mass, moles and number of particles
- the use of the mole in calculations.

This chapter covers
Objectives 6.1 and 6.2 of Section A of the CSEC Chemistry Syllabus.

You already know that substances are made of extremely small particles which may be atoms, ions or molecules. How can scientists determine the quantities of these tiny particles?

The **mole concept** is an important tool used in such calculations.

▲ **Figure 8.1** Avogadro's mole

8.1 What is one mole of a substance?

In everyday life, we use certain terms to represent a fixed number of items. When we refer to one pair of shoes, we always mean two shoes. We say one dozen eggs, rather than 12 eggs, while one pack of playing cards always contains 52 cards. In a similar manner, chemists are referring to a fixed number of particles when they use the term **one mole**. It doesn't matter which particles you are referring to – ions, molecules or atoms.

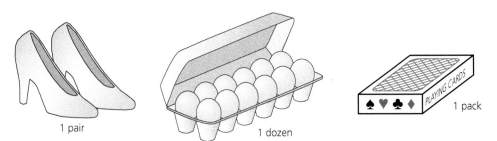

▲ **Figure 8.2** Terms we use to represent a fixed number of items

The mole

The mole is the amount of substance which contains as many particles as there are carbon atoms in 12 g of carbon-12.

We can calculate how many atoms are present in 12 g of carbon-12. This will tell us the number of particles in a mole. (This is the chemical equivalent of knowing that a dozen is 12.)

- The mass of one carbon-12 atom is 1.99×10^{-26} kg $= 1.99 \times 10^{-23}$ g.

- The number of carbon atoms in 12 g of carbon-12 will be:

$$\frac{12 \text{ g}}{1.99 \times 10^{-12} \text{ g}} = 6.023 \times 10^{23}$$

Avogadro's number is the number of particles in one mole of a substance. This number is 6.0×10^{23}.

This value, when rounded off, gives an approximate value of 6.0×10^{23}. This number is referred to as **Avogadro's number** – the number of particles in one mole of a substance. The Avogadro number or constant is 6.023×10^{23}.

Amadeo Avogadro (1776 to 1856) was born in Turin in Italy, into a family of distinguished lawyers. He graduated when he was 20 years old in ecclesiastical law, but then followed his overwhelming interest in natural science. His research has given us Avogadro's number and Avogadro's Law.

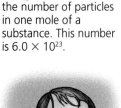

▲ **Figure 8.3**
Amedeo Avogadro

Practice

1 Fill in the gaps in Table 8.1:

▼ **Table 8.1**

Atom	Mass of 1 atom (g)	Mass of 6.0×10^{23} atoms (g)
oxygen	2.66×10^{-23} g	
copper	1.06×10^{-22} g	
hydrogen	1.67×10^{-24} g	
magnesium	4.00×10^{-23} g	

2 Find the A_r values for the atoms in Table 8.1 from a text book. Compare these values to the values you calculated in the last column of the table. What do you notice?

3 a You will also notice that the A_r values of the atoms is also close to the nucleon number (mass number) of the atom. Explain why this is so.
(Hint: What accounts for the mass of the atom?)
   b Account for the fact that while the nucleon (mass) number of ^{35}Cl is 35, the relative atomic mass of chlorine is 35.5.
(Hint: Chlorine has two isotopes.)

Molar mass

The molar mass is the mass in grams of 1 mole of a chemical substance.

From these calculations you should find that the mass of an element that contains the Avogadro's number of atoms is in fact the A_r of the element, expressed in grams. This mass of substance is described as the **molar mass** of the substance.

Therefore, 1 mole of atoms of an element:
- has a mass equal to the molar mass of that element (which is different for each element);
- contains 6.0×10^{23} atoms of that element.

It is important to note that although the molar mass of an element and its relative atomic mass are numerically equal, the terms should not be used interchangeably. Molar mass has units of $g \, mol^{-1}$ but A_r has no units.

As an example, 1 mole of sulphur atoms contains 6.0×10^{23} sulphur atoms and has a mass of 32.0 g. So, A_r of sulphur is 32.0 and the molar mass of sulphur is $32.0 \, g \, mol^{-1}$.

It should now be clear to you that the molar mass of each element is different but corresponds to the same number of atoms. For example:
- 6.0×10^{23} atoms are present in each of 108 g of silver and 12 g of carbon;
- 1.2×10^{24} atoms are present in each of 216 g of silver and 24 g carbon.

A warning!

For molecular elements such as the gases chlorine (Cl_2), oxygen (O_2), hydrogen (H_2) and nitrogen (N_2), and solids such as sulphur (S_8) and phosphorus (P_4), a distinction must be made between one mole of atoms and one mole of molecules:
- 1 mole of atoms contains 6.0×10^{23} atoms.
- 1 mole of molecules contains 6.0×10^{23} molecules.

Practice

4 What mass of sulphur will contain the same number of atoms as 240 g of carbon?
5 What mass of iron will contain the same number of atoms as 780 g of potassium?
6 Which of the following contains more atoms: 50.0 g of aluminium or 50.0 g of iron?
7 What is (a) the number of atoms in 1 mole of oxygen molecules and (b) the mass of 1 mole of oxygen molecules?

Conversions involving the mole

When chemists carry out experiments, they measure the mass and volume of reagents and products. In order to understand reactions, however, and to be able to represent them in equations, chemists need to be able to convert these measurements to moles and vice versa.

Worked example 1

Calculate the number of moles of sodium chloride formula units present in 1.17 g of sodium chloride.

Solution

mass of 1 mole of NaCl = 23 + 35.5 = $58.5 \, g \, mol^{-1}$
58.5 g of NaCl contain 1 mole of NaCl formula units.

1.17 g NaCl will contain $\frac{1.17}{58.5}$ mol NaCl formula units.

$\qquad$ = 0.02 mol formula units

Note that to do this conversion, we divided the mass of the substance mentioned in the question (here this was 1.17 g) by the mass of 1 mole of the substance (here this was 58.5 g mol⁻¹).

To convert mass to moles:

$$\text{number of moles} = \frac{\text{given mass (g) of element or compound}}{\text{mass of 1 mole (g mol}^{-1}\text{) of element or compound (molar mass)}}$$

Worked example 2

What is the mass of 0.25 mole of sulphuric acid?

Solution

This problem involves conversion of a known number of moles to mass:
mass of 1 mole H_2SO_4 = [(2 × 1) + 32 + (4 × 16)] = 98 g mol⁻¹
Therefore 0.25 mole of sulphuric acid will have a mass = 0.25 mol × 98 g mol⁻¹.
= 24.5 g

To convert moles to mass:
mass of a certain number of moles = the mass of 1 mole × number of moles

Worked example 3

What is the number of sodium atoms in 0.5 mole of the element?

Solution

1 mole sodium contains 6.0×10^{23} sodium atoms.
0.5 mole sodium contains $6.0 \times 10^{23} \times 0.5 = 3.0 \times 10^{23}$ sodium atoms.

To convert moles to number of particles:
number of particles = Avogadro number × number of moles

Worked example 4

Calculate the number of moles of magnesium metal that contains 1.2×10^{24} magnesium atoms.

Solution

There are 6.0×10^{23} magnesium atoms in 1 mole of magnesium.

there will be 1.2×10^{24} atoms in $\frac{1.2 \times 10^{24}}{6.0 \times 10^{23}} \times 1 = 2$ moles of magnesium.

To convert number of particles to moles:

$$\text{number of moles} = \frac{\text{number of particles}}{\text{number of particles in one mole of the substance}}$$

Remember: The number of particles in 1 mole is the Avogadro number.

Worked example 5

How many molecules of carbon dioxide are present in 880 g of the compound?

Solution

Step 1: Convert mass to number of moles:

$$\text{number of moles} = \frac{\text{mass (g)}}{\text{mass of 1 mole (g mol}^{-1})}$$

$$\text{mass of 1 mole carbon dioxide} = 12 + (2 \times 16) = 44 \text{ g mol}^{-1}$$

$$\text{number of moles in 880 g of CO}_2 = \frac{880 \text{ g}}{44 \text{ g mol}^{-1}}$$

$$= 20 \text{ mol}$$

Step 2: Convert number of moles to number of particles (see example 3, above):

$$\text{number of particles} = \text{Avogadro number} \times \text{number of moles}$$

$$= 6.0 \times 10^{23} \times 20$$

$$= 1.2 \times 10^{25} \text{ molecules of CO}_2$$

To convert mass to number of particles:

$$\text{Avogadro's number} \times \frac{\text{mass (g)}}{\text{mass of 1 mole (g mol}^{-1})}$$

Worked example 6

What is the mass of 3×10^{21} molecules of nitric acid? (Note that this is the same as example 5, but in reverse.)

Solution

Step 1: Convert number of molecules to moles:

$$6.0 \times 10^{23} \text{ molecules are present in 1 mol HNO}_3$$

$$3 \times 10^{21} \text{ molecules are present in } \frac{3 \times 10^{21}}{6.0 \times 10^{23}} = 5 \times 10^{-3} \text{ mol}$$

Step 2: Convert moles to mass:

$$\text{mass of 1 mol HNO}_3 = 1 + 14 + (3 \times 16) = 63 \text{ g}$$

$$\text{mass of } 5 \times 10^{-3} \text{ mol HNO}_3 = 63 \times 5 \times 10^{-3} \text{ g}$$

$$= 3.15 \times 10^{-1} \text{ g HNO}_3$$

To convert number of particles to mass:

$$\frac{\text{number of particles}}{\text{Avogadro's number}} \times \text{mass of 1 mole}$$

In these calculations, we reinforced the following idea for compounds:

- The mass of 1 mole of the ionic or covalent compound is obtained by expressing the M_r value in gram units.
- 1 mole of the covalent compound, sulphuric acid, will therefore have a mass of 98 g and will contain 6.0×10^{23} sulphuric acid molecules.
- 1 mole of the ionic compound sodium chloride will have a mass of $23 + 35.5 = 58.5$ g and will contain 6.0×10^{23} formula units, each formula unit represented by Na^+Cl^-.

Worked example 7

Consider 13.2 g of ammonium sulphate $(NH_4)_2SO_4$. Calculate the following:

a The number of formula units

b The number of ammonium ions

c The number of sulphate ions

Solution

Step 1: Calculate the mass of 1 mole of ammonium sulphate:

mass of 1 mole of $(NH_4)_2SO_4 = (2 \times 14) + (8 \times 1) + 32 + (4 \times 16) = 132$ g mol^{-1}

Step 2: Convert the mass of 13.2 g to number of moles of $(NH_4)_2SO_4$:

13.2 g of $(NH_4)_2SO_4 = \frac{13.2}{132} = 0.1$ mole $(NH_4)_2SO_4$

Step 3: Convert number of moles to number of particles. This gives the answer to part a.

1 mole $(NH_4)_2SO_4$ has 6.0×10^{23} particles (formula units):

a 0.1 mol $(NH_4)_2SO_4$ will have $6.0 \times 10^{23} \times 0.1 = 6.0 \times 10^{22}$ formula units

Step 4: Examine the formula of $(NH_4)_2SO_4$:

1 formula unit contains two NH_4^+ ions and one SO_4^{2-} ion:

$(NH_4)_2SO_4 \rightarrow 2NH_4^+ + SO_4^{2-}$

Step 5: Use the information from Step 4 to calculate the answers to parts b and c:

b There are $2 \times 6.0 \times 10^{22} = 1.2 \times 10^{23}$ NH_4^+ ions.

c There are 6.0×10^{22} SO_4^{2-} ions.

Summary of conversions

1 To convert mass to moles:

$$\text{number of moles} = \frac{\text{given mass (g) of element or compound}}{\text{mass of 1 mole (g mol}^{-1}\text{) of element or compound (molar mass)}}$$

2 To convert moles to mass:

mass = mass of 1 mole $\times$ the number of moles

3 To convert moles to number of particles:

number of particles = Avogadro number of particles $\times$ number of moles

4 To convert number of particles to moles:

$$\text{number of moles} = \frac{\text{number of particles}}{\text{number of particles in 1 mole (Avogadro number)}}$$

5 To convert mass to number of particles:

$$\text{number of particles} = \frac{\text{mass(g)}}{\text{mass of 1 mole (g mol}^{-1}\text{)}} \times \text{Avogadro's number}$$

6 To convert number of particles to mass:

$$\text{mass} = \frac{\text{number of particles}}{\text{Avogadro's number}} \times \text{mass of 1 mole}$$

8.2 Empirical and molecular formulae

The actual number of atoms or ions in one molecule or one formula unit of a compound is called the **molecular formula** of the compound.

The simplest whole number ratio of atoms or ions in a compound is called the **empirical formula** of the compound.

The mole concept can also be used when working out the formulae of compounds. There are several types of formulae used in describing chemical compounds. You have already come across **molecular formula**. One other type of formula is the **empirical formula**.

You need to note the following.

- The empirical formula and the molecular formula of a substance may be the same or they may be different. For example, the empirical formulae and molecular formulae are the same for water (H_2O), carbon dioxide (CO_2) and methane (CH_4). The molecular formula of butene is C_4H_8, which gives an empirical formula of CH_2. The molecular formula of octadecanoic acid (stearic acid) is $C_{18}H_{36}O_2$, so its empirical formula is $C_9H_{18}O$.

- Compounds with different molecular formulae may have the same empirical formula. Butene is C_4H_8 and pentene is C_5H_{10}, so they both have an empirical formula of CH_2.

The **percentage composition** by mass of a compound is the percentage by mass of each element in 1 mole of the compound.

The empirical formula of a compound can be readily found from its percentage composition by mass, and this can be worked out experimentally. Once the **percentage composition** is known, the empirical formula can be calculated.

ORR
A&I

Experiment 8.1 To work out the percentage composition by mass

Theory

Copper forms two compounds with oxygen, copper(I) oxide (a red solid) and copper(II) oxide (a black solid). The relative numbers of atoms of copper and oxygen in each oxide can be worked out by reducing each oxide separately with hydrogen. Hydrogen combines with oxygen from the oxide to form water, leaving behind metallic copper (Figure 8.4). We can find the mass of the copper.

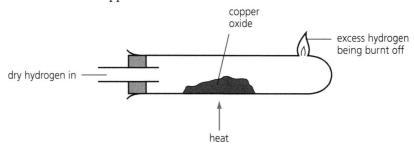

Figure 8.4
Reducing a copper oxide with hydrogen ▶

'Oxidation' and 'reduction' are technical terms in chemistry. They are explained in more detail in Section 12.7.

Procedure

⚠ This experiment should be conducted behind a safety screen.

1 Before beginning this experiment, hydrogen should be passed through the apparatus for some time to drive out all the air.

2 The hydrogen should be dried by passing it through concentrated sulphuric acid.

3 When the oxide has been completely reduced, and copper metal appears, the copper should be allowed to cool. This should happen in a stream of dry hydrogen to prevent reconversion to the oxide. (How can you be sure that all the copper oxide has been converted to copper?)

Results

In an experiment to work out the formula of an oxide of copper, the following data were obtained:

mass of oxide	= 3.975 g
mass of copper	= 3.175 g
mass of oxygen removed	= 0.800 g

Working out an empirical formula

This calculation uses the results from Experiment 8.1.

The formula of the oxide can be worked out by following these steps.

Step 1: Calculate the percentage composition by mass of each element

the percentage by mass of a particular element in a compound

$$= \frac{\text{total mass of the particular element}}{\text{mass of the compound} \times 100}$$

$$\text{percentage composition by mass of Cu} = \frac{3.175 \text{ g}}{3.975 \text{ g}} \times 100 = 79.87\%$$

$$\text{percentage composition by mass of O} = (100 - 79.87) = 20.13\%$$

(alternatively, this can be calculated as $\frac{0.800 \text{ g}}{3.975 \text{ g}} \times 100 = 20.13\%$)

	Cu	**O**
Step 1: Percentage composition by mass	79.87%	20.13%
Step 2: Find the mass of each element in 100 g of compound	79.87 g	20.13 g
Step 3: Calculate the number of moles of each element (divide mass by A_r)	$\frac{79.87}{63.5} = 1.26$	$\frac{20.13}{16} = 1.26$
Step 4: Calculate the relative number of moles of each element (divide by smallest number of moles)	$\frac{1.26}{1.26} = 1$	$\frac{1.26}{1.26} = 1$

The result from this experiment is that 1 mol of copper atoms combines with 1 mol of oxygen atoms. Therefore, the empirical formula of the copper oxide is CuO.

Determination of molecular formula

To work out the molecular formula of a compound, we need to know the empirical formula. From this, we can then work out the mass of 1 mol of empirical formula units. We also need to know the mass of 1 mol of the compound.

number of empirical formula units in 1 mol of a compound

$$= \frac{\text{mass of 1 mol of compound}}{\text{mass of 1 empirical formula unit}}$$

$$= x \text{ empirical units\mol}$$

The next step is to multiply the subscripts of each atom in the empirical formula by x (the number of empirical units per mol). This gives us the molecular formula of the compound.

This continues with the results from Experiment 8.1.

The empirical formula of copper oxide is CuO, which corresponds to a mass of 79.50 g. If the relative formula mass of CuO is given as 79.50 g, then:

$$\text{number of empirical formula units in 1 mol CuO} = \frac{79.50\ g}{79.50\ g} = 1$$

The molecular formula of the copper oxide is therefore Cu_1O_1, which we write as CuO.

Worked example 8

The hydrocarbon benzene contains 92.3% by mass of carbon, the remainder being hydrogen. The mass of 1 mol of benzene is 78 g. What are the empirical and molecular formulae of this compound?

Solution

First work out the empirical formula.

In 100 g of this compound, there are 92.3 g of carbon and 7.7 g of hydrogen. This can be expressed in moles:

$$\text{C:}\ \frac{92.3}{12} = 7.69\ mol \qquad\qquad \text{H:}\ \frac{7.7}{1.0} = 7.7\ mol$$

Divide through by the smallest number of moles:

$$\text{C:}\ \frac{7.69}{7.69} = 1.0 \qquad\qquad \text{H:}\ \frac{7.7}{7.69} = 1.0$$

The empirical formula of benzene is therefore CH. This gives the mass of one empirical formula unit as $(12 + 1) = 13$ g:

The mass of 1 mol of benzene is 78 g, so the number of empirical formula units in 78 g:

$$= \frac{78}{13} = 6$$

Therefore, there are six empirical formula units per mole of benzene and the molecular formula of benzene is C_6H_6.

Practice

8   Find the empirical formula of the following compounds from their composition by mass:

   a   Pb = 92.8%, O = 7.2%

   b   Na = 43.4%, C = 11.3%, O = 45.3%

   c   O = 63.2%, N = 36.8%

   d   K = 41.0%, S = 33.7%, O = 25.3%

   e   K = 42.4%, Fe = 15.2%, C = 19.5%, N = 22.8%

9 Find the empirical and molecular formulae of the following compounds:

a An organic acid containing 26.7% carbon, 2.2% hydrogen and 71.1% oxygen by mass, of relative molecular mass 90

b A gaseous compound, of molar mass 44 g mol^{-1}, containing 27.3% carbon, the other element being oxygen

c An oxide of phosphorus, with a relative molecular mass of 284, containing 43.7% phosphorus and 56.3% oxygen

d A compound of relative molecular mass 545, containing 20.0% carbon, 2.2% hydrogen and 77.8% chlorine by mass

Summary

- Relative atomic mass is the ratio of the average mass of the atoms of an element to one-twelfth of the mass a carbon-12 atom. Relative atomic mass is dimensionless, i.e. has no units.

- Relative molecular mass is the ratio of the average mass of the molecules of a chemical compound to one-twelfth the mass of a carbon-12 atom. Relative molecular mass is equal to the sum of the relative atomic masses of the atoms making up the molecule.

- Relative formula mass is the mass of one mole of a chemical compound, calculated from the relative atomic masses for each of the atoms in the formula.

- The mole is the standard term for the physical quantity of an amount of substance. A mole of any substance contains 6.0×10^{23} particles of the substance.

- The empirical formula is the simplest formula for a compound and gives the atoms in their lowest proportions to each other.

- The molecular formula is the formula of a compound, arranged to show the number of atoms of each element present.

Some conversions involving the mole can be summarised as follows:

- To convert moles to mass:
  multiply the mass of one mole $\times$ the number of moles

- To convert mass to moles:
  divide the given mass by the mass of 1 mole or molar mass

- To convert number of atoms to mass:
  divide given number of atoms (Y) by Avogadro's constant (L), and then multiply by the mass of one mole of the element

$$\text{mass} = \frac{Y \times \text{mass of one mole of element}}{L}$$

End-of-chapter questions

1 What is the empirical formula of a compound?
  A The one found by experiment
  B The simplest ratio of atoms in the compound
  C The same as the molecular formula
  D The elements it consists of

2 How many moles of carbon dioxide are there in 11 g of this gas?
  A 0.25 mole
  B 25 moles
  C 25 dm³
  D 2.5 g

3 What is the mass of 6 moles of Chlorine gas, Cl_2?
  A 213 grams
  B 500 grams
  C 426 grams
  D 426 moles

4 Certain precious stones are measured in 'carats' – 1 carat is equivalent to a mass of 250 mg. If a ring contains 2.5 carats of diamond, how many moles of carbon does it have? How many carbon atoms does this represent?

5 A package of aluminium foil weighs 360 g. How many moles of aluminium will you get when you buy this package of foil?

6 How many electrons are there in the following?
  a A silicon atom
  b 1 mol of silicon atoms
  c 0.25 mol of silicon atoms
  d 0.25 g of silicon atoms

7 Arrange the following in order of increasing number of atoms:
  a 27 g of aluminium
  b 0.50 mol of aluminium
  c 0.27 g of aluminium
  d 1×10^{-2} mol of aluminium
  e 2.1 mol of aluminium

8 What information does the formula of ethanol, CH_3CH_2OH, contain? How many moles of ethanol are present in 920 g of ethanol?

9 Calculate:
  a the number of moles of chlorine atoms in 3.08 g of tetrachloromethane, CCl_4;
  b the number of moles of fluorine atoms in 204 g of boron trifluoride, BF_3;
  c the number of moles of sulphate ions in 144 g of magnesium sulphate, $MgSO_4$.

10 A compound that has one sulphur atom per molecule, contains 17.2% of sulphur by mass. What is the relative molecular mass of this compound?

11 A compound of formula M_3N contains 0.673 g of nitrogen per gram of the metal M. What is the relative atomic mass of M? Identify M.

12 The green pigment found in plants, chlorophyll, contains one magnesium atom per chlorophyll molecule. The percentage by mass of magnesium per chlorophyll molecule is 2.72%. What is the molar mass of chlorophyll?

9 The mole concept applied to gases and solutions

In this chapter, you will study the following:

- Avogadro's Law and its application;
- the Law of Conservation of Matter;
- the standard solution and its preparation;
- volumetric analysis and related calculations.

This chapter covers
Objectives 6.3, 6.4, 6.7 and 7.11 of Section A of the CSEC Chemistry Syllabus.

Avogadro recognised that equal amounts of any gas at the same temperature and pressure will occupy the same volume. Molar volume is the volume of one mole of a gas, that is, the volume of gas containing 6×10^{23} gas particles.

Volumetric analysis is used to determine the amount, volume or concentration of a substance. It makes use of a technique called titration, which involves reacting two solutions, one of which has a known concentration - the standard solution. Titration involves measuring the volume of a solution delivered from a burette that reacts exactly with a precise volume of another solution.

A standard solution can be prepared by dissolving an exact mass of a solid in a known volume of solute, or by diluting an exact volume of solution in a volumetric flask. The standard solution can then be used to standardise other solutions using titration. This chapter focuses on calculations relating to molar volume, and molar and mass concentrations from titration.

The mole concept has many applications in Chemistry. In this chapter, you will see how it can be applied to determine the quantities of substances and particles present when considering gases and solutions. By this time you may have already studied the behaviour of gases and seen examples of solutions.

You may want to look back at Chapter 1 to remind yourself about the three states of matter.

9.1 The mole concept applied to gases
Molar volumes

Gases have very low densities, so it is more practical to express a quantity of gas as a volume rather than as a mass. There are several laws that relate to volumes of gases. **Avogadro's Law** uses the idea of the mole.

Avogadro's Law states that equal volumes of gases under the same conditions of temperature and pressure contain the same number of molecules.

The volume of a gas that contains one mole of molecules of the gas is known as its **molar volume**. This value is the same for all gases measured under the same conditions of temperature and pressure.

> The molar volume of a gas is the volume that contains one mole of molecules of the gas.

As you have already seen, the volume of a gas is influenced by both temperature and pressure. The molar volume of gases is usually quoted under one of two sets of conditions:

- Standard temperature and pressure (STP), which is a temperature of 273 K (0 °C) and a pressure of 101 kPa (1 atm). The approximate volume of 1 mole of any gas at STP is 22.4 dm³ (22 400 cm³).

- Room temperature and pressure (RTP), which is a temperature of 298 K (25 °C) and a pressure of 101 kPa (1 atm.). The approximate volume of 1 mole of any gas at RTP is 24 dm³ (24 000 cm³).

Thus, 1 mol of oxygen gas:
- contains 6.0×10^{23} O_2 molecules;
- has a mass of $(2 \times 16) = 32$ g;
- contains 2 mol of oxygen atoms $(2 \times 6.0 \times 10^{23})$
- occupies a volume of 22.4 dm³ at STP;
- occupies a volume of 24 dm³ at RTP.

▲ **Figure 9.1**   Avagadro's mole

Worked example 1

What volume is occupied by 8 g of hydrogen at STP?

Solution

1 mol of hydrogen, H_2, has a mass of $(2 \times 1) = 2$ g.
2 g of hydrogen occupy 22.4 dm³ at STP

Therefore, 8 g of hydrogen occupy $\frac{8.0}{2.0} \times 22.4$ dm³ at STP

$\quad = 89.6$ dm³

Worked example 2

How many molecules are present in 5.6 dm³ of carbon dioxide at STP?

Solution

1 mol of carbon dioxide occupies 22.4 dm³ at STP and contains 6.0×10^{23} CO_2 molecules.

Hence, 5.6 dm³ of carbon dioxide will contain $\dfrac{5.6}{22.4} \times 6.0 \times 10^{23}$.

$$= 1.5 \times 10^{23} \; CO_2 \text{ molecules}$$

Practice

1   **a**   What is the mass of 44.8 dm³ of hydrogen chloride at STP?
    b   How many molecules are present in 2.24 dm³ of sulphur dioxide at STP?

2   1.5 g of a gaseous compound of carbon and hydrogen occupies 1.12 dm³ at STP. What is the mass of 1 mol of this compound and what is its likely formula?

9.2   Applications of Avogadro's Law

Finding the molar mass of a gas

Avogadro's Law provides a simple method of determining the molar mass of a gas. This is illustrated by the following examples.

Worked example 3

250 cm³ of carbon dioxide ($M_r = 44$) weigh 0.44 g, while the same volume of a gaseous oxide of nitrogen weighs 0.46 g at the same temperature and pressure. Find the mass of 1 mole of this oxide of nitrogen.

Solution

number of moles of CO_2 present in 250 cm³ $= \dfrac{00.44 \text{ g}}{44 \text{ g mol}^{-1}}$

$$= 0.01 \text{ mol}$$

By Avogadro's Law, it follows that 250 cm³ of the oxide of nitrogen contain 0.01 mol. In other words, 0.01 mol of the oxide of nitrogen weigh 0.46 g and 1 mol of the oxide of nitrogen weighs:

$$\frac{1 \text{ mol}}{0.01 \text{ mol}} \times 0.46 \text{ g} = 46 \text{ g}$$

(You could go on to suggest that the formula of the oxide of nitrogen is NO_2, as this would give a molar mass of 46 g.)

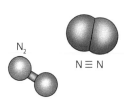

N_2

N≡N

▲ **Figure 9.2**
N_2 molecule

Worked example 4

100 cm³ of nitrogen gas ($M_r = 28$) has a mass of 0.14 g.

a Calculate the molar mass of a gaseous hydrocarbon if 250 cm³ of this gas weighs 0.7 g at the same temperature and pressure as the nitrogen.

b What is the likely formula of the hydrocarbon?

Solution

a number of moles of N_2 present in 100 cm³ $= \dfrac{0.14 \text{ g}}{28 \text{ g mol}^{-1}} = 0.005$ mol

By Avogadro's Law, 250 cm³ of the gaseous hydrocarbon at the same temperature and pressure will contain:

$$\frac{250 \text{ cm}^3}{100 \text{ cm}^3} \times 0.005 \text{ mol} = 0.0125 \text{ mol}$$

So, 0.0125 mol of the hydrocarbon weigh 0.7 g and this means that 1 mol of the hydrocarbon weighs:

$$\frac{1 \text{ mol}}{0.0125 \text{ mol}} \times 0.7 \text{ g} = 56 \text{ g}$$

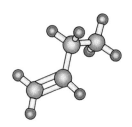

▲ **Figure 9.3**
C_4H_8 molecule

b Since the M_r of the hydrocarbon is 56, it follows that it must contain 4 carbon atoms ($4 \times 12 = 48$) and 8 hydrogen atoms ($8 \times 1 = 8$) per molecule. The molecular formula of the hydrocarbon is C_4H_8.

Predicting the volume of gaseous reactants and products from the balanced equation

We worked out how to write balanced equations in Section 7.2.

The balanced equation gives the number of moles of reactants and products involved in the reaction. Using Avogadro's Law, we can easily switch from 'number of moles' to 'volume', since we know the volume occupied by one mole of gas.

Worked example 5

What volume of oxygen is required for the complete combustion of 100 cm³ of propane? What volume of carbon dioxide is produced?

All gas volumes are measured at the same temperature and pressure.

Solution

Using the equation:

$$C_3H_8(g) + 5O_2(g) \rightarrow 3CO_2(g) + 4H_2O(l)$$

we can deduce that 1 mole of C_3H_8 reacts with 5 moles of O_2 to give 3 moles of CO_2.

By Avogadro's Law, we can say that 1 volume of C_3H_8 reacts with 5 volumes of O_2 to give 3 volumes of CO_2.

Thus, 100 cm³ of C_3H_8 reacts with 500 cm³ of O_2 to give 300 cm³ of CO_2, at the same temperature and pressure.

Note: Avogadro's Law applies *only* to gases!

Using the gas volumes to work out the equation for a reaction

From the volumes of gaseous reactants and products, we can work out the number of moles of reactant and products. From this, we can work out the equation for the reaction, as shown in the following example.

Worked example 6

On strong heating, 40 cm³ of ammonia gives 20 cm³ of nitrogen and 60 cm³ of hydrogen. (All volumes were measured at the same temperature and pressure.) Work out the equation for the decomposition of ammonia by heat.

Solution

40 cm³ of $NH_3 \rightarrow$ 20 cm³ of N_2 + 60 cm³ of H_2
Divide by the smallest volume:

or 2 cm³ of $NH_3 \rightarrow$ 1 cm³ of N_2 + 3 cm³ of H_2
or 2 volumes of $NH_3 \rightarrow$ 1 volume of N_2 + 3 volumes of H_2
or 2 moles of $NH_3 \rightarrow$ 1 mole of N_2 + 3 moles of H_2

$$2NH_3(g) \rightarrow N_2(g) + 3H_2(g)$$

9.3 The Law of Conservation of Matter

> The Law of Conservation of Matter states that matter can neither be created nor destroyed, but simply changed from one form to another.

The Law of Conservation of Matter is also known as the Law of Conservation of Mass. The change of form that can occur may arise as a result of one of the following:

- A physical process, such as heating or cooling, which results in a change of state, but the mass remains constant

- A chemical reaction, in which the identity of the chemical species present changes, but the total mass remains the same

▲ **Figure 9.4**
The same mass of water is shown in the solid, liquid and gaseous forms. How can we verify this?

For example, if the reaction in Worked example 5 ($C_3H_8(g) + 5O_2(g) \rightarrow 3CO_2(g) + 4H_2O(l)$) was performed in a sealed vessel (so that the gas, CO_2, cannot escape), the identity of the compounds present will change, but the total mass of the vessel will remain the same.

The calculations later in this chapter will all verify the Law of Conservation of Matter.

9.4 The mole and solutions

We have already described the concentration of solutions in terms such as dilute, concentrated and saturated (see Section 5.2). Now we will consider more precise ways of stating the concentration of solutions. This will allow us to state or calculate the concentration in terms of the amount of solute dissolved and also in terms of the number of particles (species) of solute in the solution.

The concentration of solutions

There are several ways of expressing concentration:

- Mass concentration: e.g. mass of solute dissolved in 1 000 cm³ of solution
  The unit is g/dm³ or g dm⁻³.

- Molar concentration or molarity: e.g. number of moles of solute in 1 000 cm³ of solution
  The unit is mol/dm³ or mol dm⁻³.

It is important to note that concentration refers to volume of solution, not to the volume of solvent. All the apparatus used for measuring volume (for example, burette, pipette, volumetric flask) measure the volume of solution.

Mass concentration is the mass of the solute (in grams) dissolved in 1 000 cm³ (1 dm³) of solution.

Mass concentration

Worked example 7

Find the mass concentration of the following solutions:

a Sodium carbonate: 25 cm³ of the solution contains 0.5 g Na_2CO_3

b Potassium hydroxide: 5 dm³ of the solution contains 200 g KOH

Solution

a 25.0 cm³ of Na_2CO_3(aq) contains 0.5 g Na_2CO_3.

1 000 cm³ of Na_2CO_3(aq) contains $\frac{1\,000}{25} \times 0.5$ g of Na_2CO_3 = 20 g dm⁻³ Na_2CO_3.

b 5 dm³ of KOH(aq) contains 200 g KOH.

1 dm³ of KOH(aq) contains $\left(\frac{1}{5} \times 200\right)$ g = 40 g dm⁻³ KOH.

From a knowledge of the mass concentration, in g dm⁻³, it is possible to find the mass of solute present in any volume of solution.

Practice

3 Find the mass concentration in g dm⁻³ of the following solutions:

    a 21.25 cm³ of a solution of nitric acid containing 6.25 g HNO_3

    b 25 cm³ of a solution of zinc sulphate containing 20.0 g $ZnSO_4$

    c 10 dm³ of a solution of copper(II) sulphate containing 1 kg $CuSO_4$

    d 1 cm³ of a solution of ammonia containing 2.5 g NH_3

    e 25 cm³ of a solution of sodium hydroxide containing 0.02 mol NaOH

Molar concentration is the number of moles of solute dissolved in 1 000 cm³ (1 dm³) of solution =

$\dfrac{n \text{ (number of moles of solute)}}{V \text{ (dm}^3 \text{ of solution)}}$

Molar concentration

The concentration in moles per dm³ can be used to determine the number of moles (n) present in any other volume of solution, V (cm³).

$$\text{number of moles in volume } V\,(\text{cm}^3) = \left(\frac{V}{1\,000}\right) \text{dm}^3 \times \text{concentration in mol dm}^{-3}$$

Once the concentration (in mol dm⁻³) of a solution is known, the number of solute particles in a measured volume is therefore known.

The following examples illustrate this.

Worked example 8

Find the concentration in mol dm⁻³ of:
a a solution containing 0.015 mol HCl in 25 cm³ of solution;
b a solution containing 4.0 mol HCl in 5.0 dm³ of solution;
c a solution containing 0.05 mol HCl in 500 cm³ of solution.

Solution

a 25 cm³ of solution contains 0.015 mol HCl.

1 cm³ of solution contains $\frac{0.015}{25}$ mol HCl.

1 000 cm³ of solution contains $\frac{0.015 \text{ mol}}{25 \text{ cm}^3} \times 1\,000 \text{ cm}^3 = 0.6$ mol HCl.

Therefore, the concentration of HCl in this solution is 0.6 mol dm⁻³.

b 5.0 dm³ of solution contains 4.0 mol.

1 dm³ of solution contains $\frac{4.0 \text{ mol}}{5 \text{ dm}^3} = 1 \text{ dm}^3 = 0.8$ mol HCl.

Hence, the concentration of the solution is 0.8 mol dm⁻³.

c 500 cm³ of solution contains 0.05 mol.

1 000 cm³ of solution contains $\frac{0.05 \text{ mol}}{500 \text{ cm}^3} \times 1000 \text{ cm}^3 = 0.10$ mol.

Concentration of solution = 0.10 mol dm⁻³.

Practice

4 What is the concentration in mol dm⁻³ of the following solutions?
   a Sulphuric acid, in which 750 cm³ contains 1.5 mol H_2SO_4
   b Sodium hydroxide, in which 7.2 dm³ contains 10.0 mol NaOH
   c Sodium chloride, in which 25 cm³ contains 2.4×10^{-3} mol NaOH

5 Calculate the number of moles of solute in the following volumes of solution:
   a 25.0 cm³ of 0.05 mol dm⁻³ H_2SO_4
   b 200 cm³ of 0.1 mol dm⁻³ NaOH
   c 4.0 dm³ of 0.01 mol dm⁻³ NaCl

Mass concentration and molar concentration

The number of moles of a given substance can be found if the mass of the substance is divided by the mass of 1 mole of it.

$$\text{number of moles} = \frac{\text{mass of substance}}{\text{mass of 1 mol of the substance}}$$

It can be deduced that the number of moles of solute in 1 dm^3 of solution is:

$$\frac{\text{mass of solute in 1 } dm^3 \text{ of solution}}{\text{mass of 1 mol of solute}}$$

i.e. $\text{concentration (in mol dm}^{-3}) = \dfrac{\text{mass concentration of solute (in g dm}^{-3})}{\text{mass of 1 mol of solute}}$

Rearranging the above expression gives two further important relationships:

1 mass concentration of solute (g dm^{-3}) = concentration (mol dm^{-3}) $\times$ mass of 1 mol of solute (g)

2 mass of 1 mol of solute (g) = $\dfrac{\text{mass concentration of solute (g dm}^{-3})}{\text{concentration (mol dm}^{-3})}$

The examples that follow illustrate the use of these relationships.

Worked example 9

What is the concentration in mol dm^{-3} of a solution of nitric acid containing 2.52 g dm^{-3} HNO$_3$?

Solution

The mass concentration is given in the question.
The relative molecular mass of nitric acid is $[1 + 14 + (3 \times 16)] = 63$.

$$\text{concentration in mol dm}^{-3} = \frac{\text{mass concentration of nitric acid)}}{\text{mass of 1 mol of nitric acid}}$$

$$= \frac{2.52}{63} = 0.04 \text{ mol dm}^{-3}$$

Worked example 10

Find the mass of 1 mol of potassium mangante(VII), given that the concentration of its solution is 0.02 mol dm^{-3}, and the mass concentration of this solution is 3.16 g dm^{-3}.

Solution

$$\text{mass of 1 mol KMnO}_4 = \frac{\text{mass concentration (g dm}^{-3})}{\text{concentration (mol dm}^{-3})}$$

$$= \frac{3.16}{0.02} = 158 \text{ g}$$

Worked example 11

250 cm³ of a solution of sodium chloride contain 11.70 g NaCl. Find the concentration in mol dm⁻³ of this solution.

Solution

First determine the mass concentration in g dm⁻³:
250 cm³ of solution contain 11.70 g of solute.

1 000 cm³ of solution contain $\frac{1\,000}{250} \times 11.70 = 46.8$ g of solute (NaCl).

mass of 1 mol of NaCl $= 23 + 35.5 = 58.5$ g

$$\text{concentration in mol dm}^{-3} = \frac{\text{mass concentration (g dm}^{-3})}{\text{mass of 1 mol (g mol}^{-1})}$$

$$= \frac{46.8}{58.5} = 0.80 \text{ mol dm}^{-3}$$

Worked example 12

What is the mass concentration in g dm⁻³ of a solution of 0.2 mol dm⁻³ potassium nitrate (KNO_3)?
(The relative molecular mass of KNO_3 = 101.)

Solution

$$\text{mass concentration (g dm}^{-3}) = \text{concentration in mol dm}^{-3} \times \text{mass of 1 mol (g mol}^{-1})$$
$$= 0.2 \times 101$$
$$= 20.2 \text{ g dm}^{-3}$$

Worked example 13

A solution of $FeSO_4.xH_2O$ has a concentration of 0.22 mol dm⁻³ and a mass concentration of 61.16 g dm⁻³. Find the value of x in the formula, $FeSO_4.xH_2O$.

Solution

$$\text{The mass of 1 mol of } FeSO_4.xH_2O = \frac{61.16 \text{ g dm}^{-3}}{0.22 \text{ mol dm}^{-3}}$$

$$= 278 \text{ g mol}^{-1}$$

$A_r(\text{Fe}) + A_r(\text{S}) + [4 \times A_r(\text{O})] + [x \times M_r(H_2O)] = M_r[FeSO_4.xH_2O]$

$56 + 32 + 64 + 18x = 278$

$18x = 278 - 152 = 126$

$$x = \frac{126}{18}$$

$$= 7$$

Hence, $FeSO_4.xH_2O = FeSO_4.7H_2O$.

Worked example 14

Given a solution of MCl_2 has a concentration of 0.025 mol dm^{-3} and a mass concentration of 2.375 g dm^{-3}, find the relative atomic mass of M.

Solution

$$\text{mass of 1 mol of } MCl_2 = \frac{\text{mass concentration (in g dm}^{-3})}{\text{concentration (in mol dm}^{-3})}$$

$$= \frac{2.375}{0.025}$$

$$= 95.0 \text{ g mol}^{-1}$$

Then work out the relative atomic mass of M as follows:

$$A_r(M) + 2 \times A_r(Cl) = M_r(MCl_2)$$
$$A_r(M) + 2 \times 35.5 = 95$$
$$A_r(M) = 95 - (35.5 \times 2)$$
$$= 95 - 71$$
$$= 24$$

9.5 The standard solution and its preparation

A **standard solution** is a solution of known concentration.

A **standard solution** is a solution of known concentration of an element or a substance made by dissolving a known mass of solute in a solvent to produce a specific volume of solution. It is very important in titrations. A standard solution is measured in mol/dm^3, and we can make a simple standard by diluting a single element or a substance in a solvent.

You can prepare a standard solution in two ways:

1   By using a known mass of a pure solid in a definite volume of solution

2   By diluting a more concentrated solution

How to prepare a standard solution

Preparing a standard solution by diluting a more concentrated solution

For this method, you begin with a more concentrated solution and add definite amounts of water until you have obtained the desired concentration. Many solutions prepared for use in a laboratory are made in this way. The following examples illustrate this method.

Worked example 15

How can you prepare 2.0 dm^3 of a 0.1 mol dm^{-3} solution of NaOH starting with a 5.0 mol dm^{-3} solution of NaOH?

Solution

Follow these steps.

Step 1: Calculate the number of moles of the compound that the diluted solution must contain.

Since the required concentration of the diluted solution is 0.1 mol dm^{-3}, it follows that 1 dm^3 of the diluted solution contains 0.1 mol NaOH and 2 dm^3 of the solution will contain:

$$\frac{2.0}{1.0} \times 0.1 = 0.2 \text{ mol NaOH}$$

Step 2: Calculate the volume of the concentrated solution that contains this number of moles.

The concentration of the original solution is 5.0 mol dm^{-3}, i.e. 5.0 mol of NaOH are present in 1 000 cm^3 of solution and 0.2 mol of NaOH are present in:

$$\frac{0.2}{5} \times 1\,000 = 40 \text{ cm}^3$$

Step 3: Use this volume of concentrated solution to make the required volume of the diluted solution.

Accurately transfer 40 cm^3 of 5.0 mol dm^{-3} NaOH to a 2.0 dm^3 volumetric flask and add water until the volume of solution in the flask is exactly 2.0 dm^3. Stir the resulting solution to ensure that it is homogenous.

Worked example 16

What volume of 11.80 mol dm^{-3} hydrochloric acid (a saturated solution at room temperature and pressure) is needed to make 1 dm^3 of 0.5 mol dm^{-3} of aqueous hydrochloric acid?

Solution

First determine the number of moles of HCl needed to make up the dilute solution.

$$\text{concentration in mol dm}^{-3} = \frac{\text{number of moles}}{\text{volume in dm}^{-3}}$$

$$\text{number of mol} = \text{concentration in mol dm}^{-3} \times \text{volume in dm}^3$$

$$= 0.5 \times 1$$

$$= 0.5 \text{ mol HCl}$$

Preparing a standard solution by using a known mass of a pure solid in a definite volume of solution

M&M
A&I

Experiment 9.1   To prepare a standard solution of sodium carbonate

Procedure

1 Weigh out the required amount of solid as accurately as possible.

2 Transfer all the weighed solid to a volumetric flask.

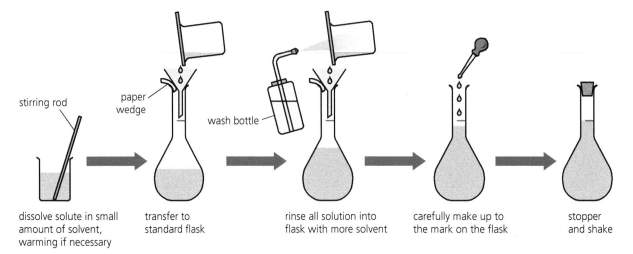

stirring rod

paper wedge

wash bottle

dissolve solute in small amount of solvent, warming if necessary

transfer to standard flask

rinse all solution into flask with more solvent

carefully make up to the mark on the flask

stopper and shake

▲ **Figure 9.5** Preparing a standard solution

3 Add water to dissolve the solid.

4 After the solid has completely dissolved, add more water to bring the volume up to the capacity of the volumetric flask.

5 Stir or shake the solution to ensure that it is completely homogenous.

Worked example 17

What mass of Na_2CO_3 is required to make 500 cm³ of 0.5 mol dm³ aqueous sodium carbonate?

Solution

First calculate the number of moles of the substance (Na_2CO_3) needed to make the required volume of solution.

$$\text{moles required} = \frac{\text{volume required}}{1\ 000} \times \text{required concentration}$$

$$= \frac{500}{1\ 000} \times 0.5$$

$$= 0.25 \text{ mol } Na_2CO_3$$

Next, convert this number of moles to mass.

$$\text{mass required} = \text{moles required} \times \text{mass of 1 mol}$$

$$= 0.25 \times 106$$

$$= 26.5 \text{ g}$$

26.5 g of Na_2CO_3 are required to obtain 500 cm³ of a solution of concentration 0.5 mol dm³.

Then find the volume of the concentrated solution that contains the number of moles of HCl (namely 0.5) required to make up the dilute solution.

1 000 cm^3 of the concentrated HCl solution contain 11.80 mol HCl.

Then, the volume that contains 0.5 mol HCl = $\dfrac{1\ 000 \times 0.5}{11.80}$

$$= 42.37\ \text{cm}^3$$

It follows that 1 dm^3 of 0.5 mol dm^3 can be made up by diluting 42.17 cm^3 of the concentrated HCl to a volume of 1 dm^3.

9.6 Volumetric analysis

Volumetric analysis is one of the experimental methods used to determine the concentration of solutions. We use a standard solution to determine the concentration of another solution of unknown concentration.

Method

The standard solution is usually added to a titration flask via a pipette. The solution whose concentration is to be determined is then added to the standard solution from a burette. The process of adding solution from burette to solution in a titration flask until the reaction is complete is termed a **titration**. The point at which the reaction is complete is known as the **equivalence point** or **end-point**, and the volume of solution added from the burette is the **titre**.

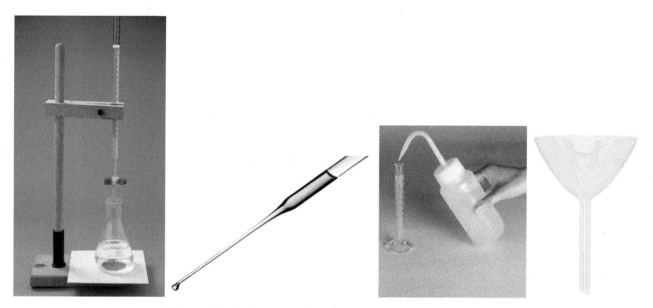

▲ **Figure 9.6** Equipment such as this is used in volumetric analysis.

The apparatus commonly used in volumetric analysis includes a balance, a pipette, a burette and a volumetric flask. Each of these, when properly used, is capable of a certain degree of accuracy, which is fixed when the apparatus is manufactured.

ORR

M&M

A&I

Experiment 9.2   Practising titrations

Get as much practice as possible in using pipettes, burettes and volumetric flasks. The correct use of the pipette is illustrated in Figure 9.7.

Procedure

1 The liquid is drawn to just above the calibration mark using a pipette filler.

2 Use your index finger to adjust the flow of the liquid. Safety pipette fillers have a three-way valve which allows control as the liquid runs out, without removing the pipette filler.

3 Allow the liquid to run out under gravity, finally touching the tip of the pipette onto the surface of the liquid in the flask. This transfers part of the final drop into the flask, and the pipette will be designed to measure volumes accurately assuming you do this.

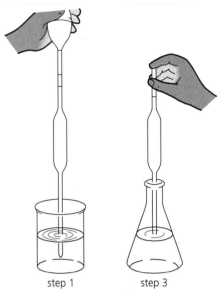

step 1   step 3

▲ **Figure 9.7**   The correct use of a pipette

4 Do NOT blow out the final part of the last drop from the pipette. Look at Figures 9.8 and 9.9 to see how you should use a burette.

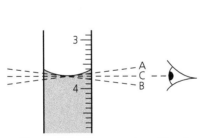

▲ **Figure 9.8**   The correct position for the eye is at C when reading a burette.

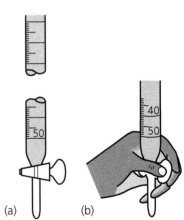

(a)   (b)

▲ **Figure 9.9**   (a) The burette and (b) the correct method of grasping the stopcock

The rough titration

The rough titration is, for the most part, a time-saving strategy. One could proceed with a titration as follows.

Place the solution and indicator (if needed) in the titration flask. Then add the other solution from the burette, shaking continuously until there is a slight excess, as indicated by the first permanent change in the indicator. Note that colour changes are best seen against a white background.

Let us assume that the rough titration in an experiment was 26 cm³. For subsequent titrations, up to 24 cm³ can be added rapidly, with thorough shaking. This is followed by drop-by-drop addition until the first permanent colour change is clearly seen.

Presentation of titration results

Formats like the one in Table 9.1 are commonly used to present titration results.

▼ **Table 9.1** In pipette: 25.0 cm³ portions of base, in burette: acid; indicator: 1 drop of methyl orange

Burette readings (cm³)	Rough titration	Titration 1	Titration 2	Titration 3
final burette reading	26.40	28.20	26.30	
initial burette reading	0.00	1.90	0.00	
volume of acid used	26.40	26.30	26.30	

volume to be used in calculation = 26.30 cm³
summary: 25.0 cm³ of base = 26.30 cm³ of acid

With good pipette and burette techniques, it is possible to achieve titration values that agree to within 0.05 cm³ on a regular basis. If, as in the example given in Table 9.1, readings 1 and 2 agree, there is no need for a third titration. Titrations that agree within experimental error are called concordant.

Acid–base titrations

An acid–base titration can be generally represented by the equation:

$$HX + BOH \rightarrow BX + H_2O$$
$$\text{acid} \quad \text{base} \quad \text{salt} \quad \text{water}$$

Acid–base reactions are neutralisation reactions (see Sections 11.2 and 12.5). They may be represented by an ionic equation:

$$H^+(aq) + OH^-(aq) \rightarrow H_2O$$

The general procedure for the titration of a soluble base (an alkali) with an acid is as follows:

Step 1: The aqueous solution of the alkali is placed in the titration flask.

Step 2: A few drops of a suitable indicator are added.

Step 3: The acid is run into the titration flask from a burette.

In practice, the acid in the burette is slowly added to the alkali until there is a colour change in the contents of the conical flask. The colour change is due to the reaction of the acid with the indicator and is the end-point of the reaction.

▲ **Figure 9.10**
Many laboratories in the Caribbean use titration in their testing processes, such as the Caribbean Industrial Research Institute (CARIRI) based at UWI, St Augustine Campus, Trinidad and Tobago.

A&I

Experiment 9.3   An acid–base titration

Procedure

X is a solution containing 5.6 g dm^{-3} of the substance M(OH)$_n$, the molar mass of which is 106 g mol^{-1}. 25.0 cm^3 portions of X were titrated with 0.10 mol dm^{-3} hydrochloric acid solution, using methyl orange as indicator. The results are shown in Table 9.2.

▼ **Table 9.2**   Titration results.

Burette readings (cm³)	Rough titration	Titration 1	Titration 2
final burette reading	26.50	30.60	33.90
initial burette reading	0.00	4.20	7.50
volume used	26.50	26.40	26.40

volume to be used in calculation = 26.40 cm^3.

Calculations

Q1   How many moles of M(OH)$_n$ are there in 1 dm^3 of solution X?

$$\text{number of moles per dm}^3 = \frac{\text{mass in 1 dm}^3}{\text{mass of one mole}}$$

$$= \frac{5.6 \text{ g dm}^{-3}}{106 \text{ g mol}^{-1}}$$

$$= 0.0528 \text{ mol dm}^{-3}$$

Q2   How many moles of M(OH)$_n$ are there in 25.0 cm^3 solution X?
1 000 cm^3 (1 dm^3) contain 0.0528 mol.

25.0 cm^3 contains $\frac{25}{1000} \times 0.0528 = 1.32 \times 10^{-3}$ mol.

Q3   How many moles of HCl are there in 26.40 cm^3 of a 0.100 mol dm^{-3} solution?
26.4 cm^3 contains $\frac{26.4}{1000} \times 0.1$ mol $= 2.64 \times 10^{-3}$ mol.

Q4   How many moles of the HCl solution will react completely with 1 mol of M(OH)$_n$?
2.64×10^{-3} mol HCl react completely with 1.32×10^{-3} mol M(OH)$_n$.
Dividing by the smaller value (which is 1.32×10^{-3}) gives a ratio of 2 mol HCl to 1 mol M(OH)$_n$.

Q5   What is the most likely value of n?
Since 1 mol H$^+$(aq) reacts with 1 mol OH$^-$(aq), 2 mol H$^+$ will react with 2 mol OH$^-$
The most likely value of $n = 2$.
The formula of M(OH)$_n$ is M(OH)$_2$.

Q6   Write a balanced equation for the reaction:
$$M(OH)_2(aq) + 2HCl(aq) \rightarrow MCl_2(aq) + 2H_2O(l)$$

Q7 What is the most likely value for the relative atomic mass of M?

$M(OH)_2$ has a molar mass of 106 g mol^{-1} so $M_r = 106$.

$$M + [(16 + 1) \times 2] = 106$$
$$M + 34 = 106$$
$$M = 106 - 34 = 72$$

Reduction–oxidation (redox) titrations

Section 12.7 gives you lots of detail about oxidation–reduction reactions. '**Redox**' is an abbreviation of '**reduction–oxidation**'.

The most widely used oxidising agent is potassium manganate(VII), a powerful oxidising agent. Potassium manganate(VII) is often used in determining the concentrations of aqueous solutions of reducing agents – such as iron(II) salts, hydrogen peroxide (H_2O_2), glucose ($C_6H_{12}O_6$), ethanedioic acid ($H_2C_2O_4$) and the ethanedioate ion ($C_2O_4^{2-}$).

Note that aqueous potassium manganate(VII) is nearly always placed in the burette. During these redox reactions, the purple manganate(VII) ion (MnO_4^-) is reduced to the manganese(II) ion (Mn^{2+}). The Mn^{2+} ion is light pink in aqueous solution but at the concentrations used in these titrations, it will appear colourless. In these titrations, when all the reducing agent has been used to convert the manganate ion $MnO_4^-(aq)$ to $Mn^{2+}(aq)$, the addition of the slightest excess of $MnO_4^-(aq)$ gives a pink colour to the solution, signalling that the end-point is reached.

Worked example 18

A solution of an iron(II) salt containing $0.100 \text{ mol dm}^{-3}$ of $Fe^{2+}(aq)$ was freshly prepared. 25.0 cm^3 portions of this solution were titrated with a potassium manganate(VII) solution of unknown concentration to permanent pink end-point. The results obtained are shown in Table 9.3.

▼ **Table 9.3** Titration results

Burette readings (cm³)	Rough titration	Titration 1	Titration 2	Titration 3
final burette reading	24.50	48.90	24.60	
initial burette reading	0.00	24.50	0.20	
volume used	24.50	24.40	24.40	

Volume to be used in calculation = 24.40 cm^3.

Using these results, we can calculate the concentration of the potassium manganate(VII) solution.

Solution

$Fe^{3+}(aq)$ and $Mn^{2+}(aq)$ are formed when $MnO_4^-(aq)$ and $Fe^{2+}(aq)$ ions react.

The half-equations for this reaction are:

$$MnO_4^-(aq) + 8H^+(aq) + 5e^- \rightarrow Mn^{2+}(aq) + 4H_2O(l) \text{ (this is a reduction)}$$
$$5Fe^{2+}(aq) - 5e^- \rightarrow 5Fe^{3+}(aq) \text{ (this is an oxidation)}$$

1 mol of $MnO_4^-(aq)$ reacts completely with 5 mol of $Fe^{2+}(aq)$.

Step 1: First determine the number of moles of $Fe^{2+}(aq)$ in 25.0 cm^3 of 0.100 mol dm^{-3} $Fe^{2+}(aq)$.

1 000 cm^3 of $Fe^{2+}(aq)$ contain 0.1 mol.

25 cm^3 of $Fe^{2+}(aq)$ contain $\dfrac{25}{1\,000} \times 0.1 = 2.5 \times 10^{-3}$ mol.

Step 2: Determine the number of moles of $KMnO_4$ which react with 2.5×10^{-3} mol $Fe^{2+}(aq)$.

From the equation, 1 mol $KMnO_4$ reacts with 5 mol $Fe^{2+}(aq)$.

number of moles of $KMnO_4$ involved in this titration $= \dfrac{2.5 \times 10^{-3}}{5}$

$$= 5 \times 10^{-4} \text{ mol}$$

Step 3: Determine the number of moles of $KMnO_4$ in 1 dm^3 of solution.

This is left for you to complete!

Summary

- The concentration of a solution is the amount of solute dissolved in a known quantity of solution.
- Two common ways of expressing concentration are:
  - mass concentration (mass of solute in 1 dm^3 of solution);
  - molar concentration (number of moles of solute in 1 dm^3 of solution).
- The relationship linking mass concentration to molar concentration can be expressed as follows:

  mass concentration (g dm^{-3}) = molar concentration (mol dm^{-3}) × the mass of 1 mol of solute
- Standard solutions are solutions of known concentration.
- Titrations can be used to determine the concentration of solutions.
- Avogadro's Law states that equal volumes of gases under the same conditions of temperature and pressure contain the same number of molecules.
- The approximate volume of 1 mole of any gas at:
  - standard temperature and pressure (STP) is 22.4 dm^3;
  - room temperature and pressure (RTP) is 24 dm^3.

End-of-chapter questions

1 Calculate the concentration in mol dm^{-3} of the following solutions, given their mass concentration in g dm^{-3}:
   a Sodium sulphate, 45.44 g dm^{-3} Na_2SO_4
   b Sulphuric acid, 294 g dm^{-3} H_2SO_4
   c Potassium hydroxide, 1.4 g dm^{-3} KOH
   d Sodium thiosulphate, 14.88 g dm^{-3} $Na_2S_2O_3.5H_2O$
   e Potassium manganate(VII), 6.32 g dm^{-3} $KMnO_4$

2 What is the mass concentration of the following solutions?
   a 0.75 mol dm^{-3} sulphuric acid, H_2SO_4
   b 0.88 mol dm^{-3} hydrochloric acid, HCl
   c 2.05 mol dm^{-3} nitric acid, HNO_3
   d 0.55 mol dm^{-3} ammonium iron(II) sulphate, $(NH_4)_2.Fe(SO_4)_2.6H_2O$
   e 2.5 mol dm^{-3} hydrogen peroxide, H_2O_2

3 a 10.0 cm^3 of a solution of sulphuric acid contain 6.125 g H_2SO_4.
   (i) How many moles of H_2SO_4 are present in 25.0 cm^3?
   (ii) What is the concentration of this solution, in mol dm^{-3}?
   b 80.0 cm^3 of a solution of hydrochloric acid contain 0.292 g HCl.
   (i) What is the mass concentration of this solution?
   (ii) What is the concentration in mol dm^{-3} of this solution?
   c 100 cm^3 of a solution of sodium chloride contain 0.028 mol NaCl.
   (i) Find the number of individual sodium ions in 200 cm^3 of the solution.
   (ii) Find the mass concentration of the solution.

4 a What volume of 4.50 mol dm^{-3} HNO_3 should be diluted with water to form 2.5 dm^3 of 1.20 mol dm^{-3} nitric acid?
   b If 200 cm^3 of a solution of sodium chloride is diluted to 1 000 cm^3 with water to form a 0.005 mol dm^{-3} solution, what was the concentration in mol dm^{-3} of the original sodium chloride?
   c What volume of water should be added to pure ethanoic acid (CH_3COOH, 1 180 g dm^{-3}), to make 500 cm^3 of 0.2 mol dm^{-3} ethanoic acid?

5 a 40.0 cm^3 of a solution of hydrated barium chloride, $BaCl_2.xH_2O$, contains 1.952 g of the salt.
   (i) What is the mass concentration of the solution?
   (ii) Given that the concentration of the solution is 0.2 mol dm^{-3}, find the value of x in the formula.
   b A 0.004 mol dm^{-3} solution of Epsom salts contains 1.072 g of the salt in 1 dm^3 of solution.
   (i) Find the mass of 1 mol of Epsom salts.
   (ii) If the formula of Epsom salts is represented by $MSO_4.7H_2O$, find the relative atomic mass of M.

6 25.0 cm^3 of iron(II) sulphate solution required 23.0 cm^3 of 0.0192 mol dm^{-3} $KMnO_4$ for oxidation. Calculate the concentration of the iron(II) sulphate solution.

7 S_1 is a solution containing 0.0200 mol dm^{-3} $KMnO_4$. S_2 is a solution of a reducing agent containing 9.00 g dm^{-3}.
   a Given that the relative molecular mass of the reducing agent is 180, determine the concentration of the reducing agent.
   b Further, given that 25.0 cm^3 of S_1 required 25.0 cm^3 of S_2 for complete reduction, determine:
   (i) the number of moles of $KMnO_4$ in 25.0 cm^3 of S_1;
   (ii) the number of moles of reducing agent in 25.0 cm^3 of S_2;
   (iii) the number of moles of reducing agent that react completely with 1 mol $KMnO_4$.

8 Ascorbic acid ($C_6H_8O_6$) has a relative molecular mass of 176. It quantitatively reduces iodine according to the equation:

$$C_6H_8O_6(aq) + I_2(aq) \rightarrow C_6H_6O_6(aq) + 2HI(aq)$$

In a titration between iodine and ascorbic acid, 25.0 cm³ of a solution S_3 containing 8.80 g dm⁻³ of ascorbic acid were titrated with iodine, using starch as an indicator until a faint blue colour developed. The titration results can be summarised as follows:

25.0 cm³ of ascorbic acid = 26.05 cm³ of the iodine solution

  a How many moles of ascorbic acid are present in 25 cm³ of solution S_3?

  b How many moles of iodine are required to react completely with the ascorbic acid?

  c Determine the mass concentration of the iodine solution. (I = 127)

9 If 25.0 cm³ of 0.10 mol dm⁻³ hydrochloric acid is necessary to neutralise 65.0 cm³ of a solution of sodium hydroxide, calculate the mass concentration of the sodium hydroxide solution.

10 25.0 cm³ of a solution of sulphuric acid is required for the neutralisation of 40.0 cm³ of 0.035 mol dm⁻³ KOH. Find the concentration of the H_2SO_4 in mol dm⁻³.

10 Calculations involving equations and moles

This chapter covers
Objective 6.6 of Section A of the CSEC Chemistry Syllabus.

A detailed understanding of a balanced chemical equation for a titration reaction is the beginning of solution chemistry. A chemical equation gives the mole ratio for the acid and the base that have reacted. Stoichiometry refers to the relative amounts of reactants and products in a balanced chemical equation. Determining the stoichiometry of an equation involves finding the number of moles of each reactant, finding the simplest mole ratio and then writing a balanced equation.

The end point of a titration occurs when you see the first permanent colour change in the indicator. You should repeat titrations until you obtain consistent readings (differing by no more than 0.1 ml). This chapter focuses on calculations that involve equations and moles.

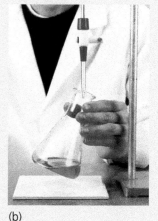

Figure 10.1
Methyl orange in (a) an acidic solution and (b) an alkaline solution ▶   (a)   (b)

A very good knowledge and clear understanding of Chapters 7–9 are essential for this chapter. In all fields of research, processing and manufacturing where chemical reactions are part of the processes, calculations are crucial for ensuring that raw materials, plant conditions, related costs and outputs are optimised.

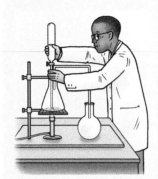
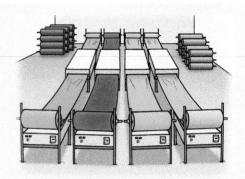

▲ **Figure 10.2**   A chemist in any laboratory, from manufacturing snacks to fabric, must do calculations to ensure that the correct raw materials in optimal quantities under appropriate conditions are used to produce the quality and quantity of products we use every day.

10.1   Quantitative relationships between reactants and products

A completed and balanced equation gives more information than simply which substances are reactants and which are products. It also gives the quantities involved, from which one can convert moles, grams, number of molecules or volumes at STP of one reactant into equivalent numbers of moles, grams, molecules or volumes at STP of products.

Worked example 1

Examine the following balanced equation and list some of the quantitative relationships shown in them:

$$2H_2(g) + O_2(g) \rightarrow 2H_2O(l)$$

Solution

- 2 mol $H_2(g)$ + 1 mol $O_2(g) \rightarrow$ 2 mol $H_2O(l)$
- $2 \times 6.0 \times 10^{23}$ molecules $H_2(g) + 6.0 \times 10^{23}$ molecules $O_2(g)$
$$\rightarrow 2 \times 6.0 \times 10^{23} \text{ molecules } H_2O(l)$$
- 4 g $H_2(g)$ + 32 g $O_2(g) \rightarrow$ 36 g $H_2O(l)$
- 44.8 dm^3 $H_2(g)$ + 22.4 dm^3 $O_2(g) \rightarrow$ 36 g of liquid water (approx. 36 cm^3) at STP

Note that the product is a liquid, and so Avogadro's Law does not apply. Instead, you need to know the density of water (which is 1 g cm^{-3}).

Worked example 2

Give the quantitative relationships in this equation:

$$N_2(g) + 3H_2(g) \rightarrow 2NH_3(g)$$

Solution

- 1 mol $N_2(g)$ + 3 mol $H_2(g) \rightarrow$ 2 mol $NH_3(g)$
- 6.0×10^{23} molecules $N_2(g) + 3 \times 6.0 \times 10^{23}$ molecules $H_2(g)$
$$\rightarrow 2 \times 6.0 \times 10^{23} \text{ molecules } NH_3(g)$$
- 28 g $N_2(g)$ + 6 g $H_2(g) \rightarrow$ 34 g $NH_3(g)$
- 22.4 dm^3 $N_2(g)$ + 67.2 dm^3 $H_2(g) \rightarrow$ 44.8 dm^3 $NH_3(g)$ (at STP)

Determining the mass of reactant needed, or product formed, from a given number of moles of reactant

Since the number of moles of one reactant is given, use the coefficients of the balanced equation to find the number of moles of the unknown.

(Here the term 'coefficient' means the number before a formula in the balanced equation.)

Worked example 3

Calcium carbonate decomposes on heating according to the equation:

$$CaCO_3(s) \rightarrow CaO(s) + CO_2(g)$$

When 0.025 mol of calcium carbonate is heated in this way, find:

a the mass of calcium oxide formed;

b the mass and volume (at STP) of carbon dioxide formed.

Solution

From the equation, 1 mole of $CaCO_3(s)$ produces 1 mole of $CaO(s)$ and 1 mole of $CO_2(g)$.

Thus, 0.025 mol $CaCO_3(s)$ gives 0.025 mol $CaO(s)$ and 0.025 mol $CO_2(g)$.

a mass of $CaO(s)$ formed = number of mol × mass of 1 mol
$$= 0.025 \text{ mol} \times 56 \text{ g mol}^{-1}$$
$$= 1.4 \text{ g}$$

b mass of CO_2 formed = number of mol × mass of 1 mol
$$= 0.025 \times 44 \text{ g}$$
$$= 1.1 \text{ g}$$

volume of $CO_2(g)$ formed (at STP) = number of moles × 22.4 dm³
$$= 0.025 \times 22.4 \text{ dm}^3$$
$$= 0.56 \text{ dm}^3 \text{ or } 560 \text{ cm}^3$$

Determining the mass of reactants required, or mass of products formed, from a given mass (or volume at STP) of reactants

Step 1: Find the number of moles of reactant. Use the relationship:

$$\text{number of moles} = \frac{\text{mass of substance}}{\text{mass of 1 mol of substance}}$$

or

$$\text{number of moles of gas} = \frac{\text{volume of gas at STP}}{22.4 \text{ dm}^3 \text{ mol}^{-1}}$$

Step 2: Use the coefficients in the balanced equation, to find the number of moles of the unknown, and hence the mass or volume at STP.

Worked example 4

On heating, potassium chlorate(V) decomposes into potassium chloride and oxygen:

$$2KClO_3(s) \rightarrow 2KCl(s) + 3O_2(g)$$

9.8 g of $KClO_3(s)$ are heated. Calculate:

a the mass of KCl(s) formed;

b the mass and volume, at STP, of $O_2(g)$ formed.

Solution

$$\text{number of moles of } KClO_3(s) \text{ used} = \frac{\text{mass}}{\text{mass of 1 mol}}$$

$$= \frac{9.8}{122.5}$$

$$= 0.08 \text{ mol}$$

From the equation for the decomposition of $KClO_3(s)$, 2 mol $KClO_3(s)$ yield 2 mol KCl(s) and 3 mol $O_2(g)$. It follows that:

$$0.08 \text{ mol } KClO_3(s) \rightarrow 0.08 \text{ mol KCl(s)} + 0.12 \text{ mol } O_2(g)$$

a mass of KCl(s) formed = number of mol × mass of 1 mol
$$= 0.08 \times 74.5$$
$$= 5.96 \text{ g}$$

b mass of $O_2(g)$ formed = 0.12×32
$$= 3.84 \text{ g}$$

volume at STP of $O_2(g)$ formed = number of moles × 22.4 dm³
$$= 0.12 \times 22.4$$
$$= 2.688 \text{ dm}^3$$

Determining the volumes of gases consumed or produced in a reaction

If the mass or volume at STP of the reactant is given, follow these steps:

Step 1: Convert these quantities to moles.

Step 2: Using the coefficients in the balanced equation, find the number of moles of gas required or produced, and hence the volume at STP.

Worked example 5

Nitrogen(II) oxide combines readily with oxygen gas to produce nitrogen dioxide (nitrogen(IV) oxide):

$$2NO(g) + O_2(g) \rightarrow 2NO_2(g)$$

1.5 g of nitrogen(II) oxide are used. Calculate:

a the volume of oxygen required (at STP);

b the volume of nitrogen(IV) oxide produced.

Solution

Work out the number of moles of NO(g) used (Step 1).

$$\text{number of moles of NO(g)} = \frac{\text{mass of NO(g)}}{\text{mass of 1 mol NO(g)}}$$

$$= \frac{1.5\ \text{g}}{30\ \text{g mol}^{-1}}$$

$$= 0.05\ \text{mol}$$

From the equation, it can be deduced that:

2 mol NO(g) react with 1 mol O_2(g) → 2 mol NO_2(g)

Hence:

0.05 mol NO(g) reacts with 0.025 mol O_2(g) → 0.05 mol NO_2(g).

a The volume of oxygen required is 0.025 mol × 22.4 dm^3
   = 0.56 dm^3 or 560 cm^3

b 0.05 mol of NO_2(g) occupies 0.05 × 22.4 dm^3 at STP.
   = 1.12 dm^3 or 1 120 cm^3

Determining the mass of product formed, given the concentration (in mol dm^{-3}) of the reactants

A calculation such as this is best done in two steps.

Step 1: Work out the number of moles of solute in the given volume of solution of known concentration.

$$\text{number of moles of solute} = \frac{\text{volume in cm}^3 \times \text{concentration (mol dm}^{-3})}{1\,000}$$

$$= \text{volume in dm}^3 \times \text{concentration (mol dm}^{-3})$$

Step 2: Using the coefficients in the balanced equation, work out the number of moles of product formed and hence the mass of product.

Worked example 6

Zinc metal reacts with copper(II) sulphate solution according to the equation:

$$\text{Zn(s)} + \text{CuSO}_4\text{(aq)} \rightarrow \text{ZnSO}_4\text{(aq)} + \text{Cu(s)}$$

a What mass of zinc metal is required to react with 50 cm^3 of 0.5 mol dm^{-3} $CuSO_4$(aq)?

b What mass of copper will be produced in the reaction?

Solution

Work out the number of moles of $CuSO_4$(aq) used (Step 1):

1 000 cm^3 of $CuSO_4$(aq) contain 0.5 mol.

50 cm^3 of $CuSO_4$(aq) contain $\frac{50}{1\,000} \times 0.5 = 0.025$ mol.

From the equation, it can be deduced that:

1 mol $CuSO_4$(aq) reacts with 1 mol Zn(s) → 1 mol Cu(s).

Therefore, 0.025 mol $CuSO_4$(aq) reacts with 0.025 mol Zn(s) → 0.025 mol Cu(s)

a mass of zinc required $= 0.025 \times 65$
$= 1.625$ g

b mass of copper formed $= 0.025 \times 63.5$
$= 1.5875$ g

Worked example 7

Barium chloride reacts with magnesium sulphate to form barium sulphate and magnesium chloride:

$$BaCl_2(aq) + MgSO_4(aq) \rightarrow BaSO_4(s) + MgCl_2(aq)$$

50 cm^3 of 0.2 mol dm^{-3} $BaCl_2$(aq) reacted according to the above equation. Calculate:

a the volume, in cm^3, of 0.4 mol dm^{-3} $MgSO_4$(aq) needed;

b the mass of $BaSO_4$(s) formed.

Solution

Work out the number of moles of $BaCl_2$(aq) used (Step 1).
1 000 cm^3 of 0.2 mol dm^{-3} $BaCl_2$(aq) contain 0.2 mol $BaCl_2$.

50 cm^3 of 0.2 mol dm^{-3} $BaCl_2$(aq) contain $\dfrac{50}{1\,000} \times 0.2 = 0.01$ mol $BaCl_2$.

From the equation we know that:
1 mol $BaCl_2$(aq) reacts with 1 mol $MgSO_4$(aq) $\rightarrow$ 1 mol $BaSO_4$(s).

Hence, 0.01 mol $MgSO_4$(aq) is required to react with 0.01 mol $BaCl_2$(aq) to form 0.01 mol $BaSO_4$(s).

a The volume of 0.4 mol dm^{-3} $MgSO_4$(aq) that contains 0.01 mol of the solute

$= \dfrac{0.01}{0.4} \times 1\,000$

$= 25$ cm^3

b Since 0.01 mol $BaSO_4$ is produced in this reaction, the mass of $BaSO_4$ precipitated

$=$ number of moles $\times$ mass of 1 mol
$= 0.01 \times 233$ g
$= 2.33$ g

10.2   The concept of the limiting reagent

For reactions that go to completion, all reactants are consumed provided that they are mixed in the mass or mole ratio as shown by the balanced equation. For example, in the reaction:

$$2H_2(g) + O_2(g) \rightarrow 2H_2O(l)$$

2 mol hydrogen will react completely with 1 mol oxygen to produce 2 mol water, or, using masses, 4.0 g of hydrogen will react completely with 32.0 g of oxygen to produce 36.0 g of water.

If, in the above reaction, 7.0 g hydrogen were mixed with 32.0 g oxygen, the hydrogen would be in excess. No more water could be produced once all the oxygen had been used up and 3.0 g of hydrogen would be left unreacted.

Similarly, if 4.0 g hydrogen were mixed with 40.0 g oxygen, the amount of water produced is limited by the amount of hydrogen present, since the oxygen is present in excess. 8.0 g of oxygen would be left unreacted.

When quantities of two or more reactants are given, it is necessary to find which is the limiting reagent. It is important to remember that the reactant that produces the smallest yield of products is the limiting reagent. In the first case, oxygen is the limiting reagent while hydrogen is the limiting reagent in the second.

Worked example 8

If 20.0 g calcium carbonate are added to 1 dm³ of a solution containing 10.0 g dm⁻³ of hydrochloric acid, what mass of CO_2 is produced?

Solution

Step 1: Calculate the number of moles of each reactant.

$$\text{mol } CaCO_3 = \frac{\text{mass}}{\text{mass of 1 mol}} = \frac{20.0 \text{ g}}{100 \text{ g mol}^{-1}} = 0.20 \text{ mol}$$

molar concentration $= 10 \text{ g dm}^{-3} = 0.274 \text{ mol dm}^{-3} = 0.274 \text{ mol HCl used}$

Step 2: Calculate the number of moles of product that could be obtained from each reactant.

$$CaCO_3(s) + 2HCl(aq) \rightarrow CaCl_2(aq) + CO_2(g) + H_2O(l)$$

- 0.20 mol $CaCO_3$ will produce 0.20 mol CO_2.
- 0.274 mol HCl will produce 0.137 mol CO_2.

Step 3: Identify the reagent that gives the lower number of moles of product. This is the limiting reagent. In this case it is HCl. So, the mass of CO_2 produced is:
0.137 mol $\times$ 44 g mol⁻¹ = 6.03 g
(M_r for CO_2 is 44 g mol⁻¹.)

Practice

1 How many grams of magnesium chloride can be obtained from 36 g of magnesium via this reaction?

$$Mg(s) + 2HCl(aq) \rightarrow MgCl_2(aq) + H_2(g)$$

2 Calculate the mass of hydrogen produced from 81 g of aluminium in the reaction:

$$2Al(s) + 6NaOH(aq) \rightarrow 2Na_3AlO_3(aq) + 3H_2(g)$$

3 How many grams of barium sulphate could be obtained from 137 g of barium chloride via this precipitation reaction?

$$BaCl_2(aq) + Na_2SO_4(aq) \rightarrow BaSO_4(s) + 2NaCl(aq)$$

10.3 Using chemical equations to find the concentration of solutions

Look at Section 9.6 for more about volumetric analysis.

Worked example 9

25.00 cm³ of a solution of sodium hydroxide required 22.00 cm³ of a solution of hydrochloric acid containing 4.38 g dm⁻³ HCl for a complete reaction. Find (a) the molar concentration and (b) the mass concentration of the sodium hydroxide solution.

Solution

Follow the four steps below for this type of calculation.

Step 1: Find the concentration in mol dm⁻³ of the HCl (the solution of known concentration).

$$\text{concentration of HCl in mol dm}^{-3} = \frac{\text{mass concentration in g dm}^{-3}}{\text{mass of 1 mol of HCl}}$$

$$= \frac{4.38 \text{ g dm}^{-3}}{36.5 \text{ g mol}^{-1}}$$

$$= 0.120 \text{ mol dm}^{-3}$$

Step 2: Find the number of moles present in the volume of HCl used.

concentration of the HCl = 0.120 mol dm⁻³

1 000 cm³ of HCl contain 0.120 mol.

22.00 cm³ of HCl contain $\frac{22}{1000} \times 0.120$ mol HCl.

Step 3: Using the balanced equation for the reaction, find the number of moles of NaOH (solution of unknown concentration) in the volume used.

From the equation:

$HCl(aq) + NaOH(aq) \rightarrow NaCl(aq) + H_2O(l)$

1 mol HCl(aq) reacts with 1 mol NaOH(aq).

$\frac{22}{1000} \times 0.120$ mol HCl react with $\frac{22}{1000} \times 0.120$ mol NaOH.

So, there are $\frac{22}{1000} \times 0.120$ moles of NaOH in 25 cm³ of solution.

Step 4: Find the number of moles of NaOH in 1 dm³ and then the mass concentration.

25 cm³ of NaOH contains $\frac{22}{1000} \times 0.120$ mol.

1 000 cm³ of NaOH contains $\frac{1000}{25} \times \frac{22}{1000} \times 0.120$

$= 0.106 \text{ mol dm}^{-3}$

a Concentration of the NaOH solution is 0.106 mol dm⁻³.
mass concentration of NaOH
= concentration in mol dm⁻³ × mass of 1 mol NaOH in g mol⁻¹
= 0.106 × 40
= 4.24 g dm⁻³

b Mass concentration of the NaOH is 4.24 g dm⁻³.

Worked example 10

25.00 cm³ of a solution of sodium carbonate, of mass concentration 4.24 g dm⁻³, are exactly neutralised by 27.80 cm³ of nitric acid. Find the concentration of the nitric acid (a) in mol dm⁻³ and (b) in g dm⁻³.

Solution

Step 1: First we find the concentration of the sodium carbonate solution (the solution of known concentration).

$$\text{concentration (in mol dm}^{-3}) = \frac{\text{mass concentration (in g dm}^{-3})}{\text{mass of 1 mol of Na}_2\text{CO}_3}$$

$$= \frac{4.24 \text{ g dm}^{-3}}{106 \text{ g mol}^{-1}}$$

$$= 0.04 \text{ mol dm}^{-3}$$

Step 2: Find the number of moles of Na_2CO_3 in the volume used in the titration.

1000 cm³ of the solution contain 0.04 mol Na_2CO_3.

25.00 cm³ of solution contain $\frac{25}{1000} \times 0.04$ mol Na_2CO_3.

Step 3: Find the number of moles of HNO_3 used, from the balanced equation.

$$Na_2CO_3(aq) + 2HNO_3(aq) \rightarrow 2NaNO_3(aq) + H_2O(l) + CO_2(g)$$

From the equation:
1 mol of Na_2CO_3 reacts with 2 moles of HNO_3.

So, $\frac{25}{1000} \times 0.04$ mol Na_2CO_3 react with $2 \times \frac{25}{1000} \times 0.04$ mol HNO_3.

= the number of moles of HNO_3 in 27.80 cm³ of solution

Step 4: Find the concentration and mass concentration of the HNO_3.

27.80 cm³ of solution contain $2 \times \frac{25}{1000} \times 0.04$ mol HNO_3.

1 000 cm³ of solution contain $\frac{1000}{27.80} \times 2 \times \frac{25}{1000} \times 0.04$ mol HNO_3.

= 0.072 mol dm⁻³

a Concentration of HNO_3 is 0.072 mol dm⁻³.
mass concentration of HNO_3 = concentration (in mol dm⁻³) × mass of 1 mol
= 0.072 × 63
= 4.536 g dm⁻³

b Mass concentration of HNO_3 is 4.536 g dm⁻³.

Summary

- The chemical equation can be used to:
  - determine the mass of reactants needed or product formed from a given number of moles of reactants;
  - determine the mass of reactants required or mass of products formed from a given mass or volume at STP (or RTP) of reactants;
  - determine the volume of gases (at STP) consumed or produced in a reaction;
  - determine the mass of product formed, given the concentration (in mol dm^{-3}) of the reactants;
  - determine the concentrations of solutions.
- The limiting agent is the reactant that limits or determines the amount of product formed. The limiting reagent is not available in excess.
- The amount of limiting reagent can be used to determine the theoretical yield of materials formed in a given reaction.

End-of-chapter questions

1 In a redox titration, 25 cm^3 of a 0.02 mol dm^{-3} solution of potassium manganate(VII) reacted with an equal volume of 0.1 mol dm^{-3} iron(II) sulphate. The molar ratio of iron(II) sulphate to potassium manganate(VII) in this titration is:

A 1 mol $FeSO_4$: 1 mol $KMnO_4$
B 1 mol $FeSO_4$: 5 mol $KMnO_4$
C 5 mol $FeSO_4$: 1 mol $KMnO_4$
D 1 mol $FeSO_4$: 2 mol $KMnO_4$

2 The sulphate of an element X is represented by the formula X_2SO_4. The sulphate has a molar mass of 370. The relative atomic mass of X is:

A 145
B 274
C 306
D 137

3 Which of the following statements is/are correct about 12 dm^3 of ammonia gas at room temperature?

I   It contains 0.5 mol ammonia molecules.
II It contains 9 g of ammonia.
III It contains 3×10^{23} ammonia molecules.
IV It contains 8.5 g ammonia.

A I only
B I and II only
C I, II and III only
D I, III and IV only

4 The equation for the reaction of magnesium with sulphuric acid may be written as:

$$Mg(s) + H_2SO_4(aq) \rightarrow MgSO_4(aq) + H_2(g)$$

12 g of magnesium was added to 50 cm^3 of 2.0 mol dm^{-3} sulphuric acid. Which of the following statements is **not true** of this reaction?

A All the magnesium will react.
B 2.4 dm^3 of H_2 gas (measured at RTP) will be evolved.
C The acid is the limiting reagent.
D All of the acid will be used up.

5 1 mole of calcium chloride contains:

A 75.5 g calcium chloride
B 35.5 g of chloride ions
C 6×10^{23} formula units of calcium chloride
D 6×10^{23} chloride ions

6 Which one of the following **does not** represent one mole of substance?

A 23 g sodium metal
B 35.5g of chlorine gas
C 132 g of ammonium sulphate
D 160 g of iron(III) oxide

7 Zinc reacts with hydrochloric acid as shown in the following equation:

$$Zn(s) + 2HCl(aq) \rightarrow ZnCl_2(aq) + H_2(g)$$

An excess of hydrochloric acid was added to 6.5 g of pure zinc metal at room temperature and pressure. Which one of the following correctly describes this reaction?

A The solution left contained no H^+ ions.

B The mass of zinc chloride dissolved in the final solution is 1.36 g.

C 24 dm^3 of hydrogen gas was evolved.

D 0.2 moles of acid was used up in the reaction.

8 Hydrogen gas and oxygen gas combine when ignited to form water. The equation for the reaction is:

$$2H_2(g) + O_2(g) \rightarrow 2H_2O(l)$$

If equal volumes of hydrogen and oxygen gas are used, which of the following is **true**?

A All the hydrogen gas will be used up.

B All the oxygen gas will be used up.

C The number of moles water formed will be equivalent to the number of moles oxygen used up.

D The number of moles water formed will be half of the number of moles hydrogen used up.

9 188 g of copper(II) nitrate contains:

A 1 mol nitrate ions

B 1 mol copper(II) ions

C 1 copper(II) ion

D 2 nitrate ions

10 Hydrogen and chlorine combine in the presence of sunlight to form hydrogen chloride gas. The equation for the reaction is:

$$H_2(g) + Cl_2(g) \rightarrow 2HCl(g)$$

Indicate, by means of calculations, which of the following statements or deductions can be made from the equation.

a 1 mol H_2 will react with 0.5 mol Cl_2 to form 1 mol HCl.

b At RTP, 24 dm^3 of H_2 will react with 24 dm^3 Cl_2 to form 24 dm^3 HCl.

c At RTP, 12 dm^3 of H_2 will react with 24 dm^3 Cl_2 to form 24 dm^3 of HCl.

d 3×10^{23} molecules of hydrogen will react with 12 dm^3 of Cl_2 to form 1 mole of hydrogen chloride.

e When equal volumes of hydrogen and chlorine react, the volume of hydrogen chloride formed is equivalent to the sum of the volumes of hydrogen and chlorine which reacted.

f 1 mol hydrogen chloride has the same number of molecules as there are hydrogen molecules in 2 g of hydrogen gas.

11 The ionic equation for the reaction of copper(II) sulphate, $CuSO_4$, and sodium hydroxide is:

$$Cu^{2+}(aq) + 2OH^-(aq) \rightarrow Cu(OH)_2(s)$$

In an experiment, an excess of sodium hydroxide was added to 25 cm^3 of a solution of copper(II) sulphate with a concentration of 6.35 g dm^{-3}. The precipitate of copper(II) hydroxide was collected by filtration, washed, dried and weighed.

Calculate the following:

a The molar concentration of the copper(II) sulphate solution

b The number of moles of copper(II) sulphate in the 25 cm^3 solution used in the experiment.

c The number of moles of copper(II) ions used in the experiment? Justify your answer.

d The number of moles of hydroxide ions needed to react with the number of moles of Cu^{2+} ions used in the experiment?

e **(i)** The number of moles of copper(II) hydroxide formed

   (ii) The mass of copper(II) hydroxide that should be formed

12 Sulphur burns in oxygen to form sulphur dioxide gas, as shown by the following equation:

$$S(s) + O_2(g) \rightarrow SO_2(g)$$

3.2 g of <u>impure</u> sulphur was burnt in an excess of oxygen and 2 dm^3 of sulphur dioxide gas were formed, measured at room temperature and pressure.

Calculate the following:
a The number of moles of sulphur dioxide in 2 dm³ of the gas
b The number of moles of sulphur that would be needed to produce this quantity of gas
c The mass of sulphur in grams that was needed to produce the quantity of gas formed in the experiment
d The percentage of sulphur in the 3.2 g of impure sulphur used in the experiment

13 Use the following information to work out the equation for the reaction between magnesium chloride and sodium carbonate:

25 cm³ of a solution containing 9.5 g magnesium chloride per dm³ reacted exactly with 10 cm³ of a solution of sodium carbonate containing 26.5 g per dm³. The products of the reaction were magnesium carbonate and sodium chloride.

Calculate the following.
a The molar concentration of:
   (i) the magnesium chloride solution
   (ii) the sodium carbonate solution
b The number of moles of:
   (i) magnesium chloride
   (ii) sodium carbonate that reacted
c The equation for the reaction by using the following procedure:
   (i) From the experiment, the molar ratio of magnesium chloride to sodium carbonate is _____ to _____.
   (ii) Each mole of magnesium chloride produces _____ moles magnesium ions and _____ moles chloride ions.
   (iii) Each mole of sodium carbonate produces _____ moles sodium ions and _____ moles carbonate ions.
   (iv) Each mole of magnesium carbonate contains _____ moles magnesium ions and _____ moles carbonate ions.
   (v) Each mole of sodium chloride contains _____ moles sodium ions and _____ moles chloride ions.
   (vi) Therefore, when _____ moles magnesium chloride react with _____ moles sodium carbonate, _____ moles magnesium carbonate and _____ moles sodium chloride will be formed.
   (vii) Write the balanced equation for the reaction from your deductions in part (vi).

14 Use the relevant balanced equations to calculate the following:
a **(i)** The mass of iron(II) chloride formed
   (ii) The volume of hydrogen gas produced at RTP when 5.6 g of iron reacts with an excess of hydrochloric acid
$$Fe(s) + 2HCl(aq) \rightarrow FeCl_2(aq) + H_2(g)$$
b **(i)** The volumes of oxygen gas formed
   (ii) The volume of nitrogen dioxide gas formed
   (iii) The mass of zinc oxide left, when 3.78 g of zinc nitrate is completely decomposed by heating
$$2Zn(NO_3)_2(s) \rightarrow 2ZnO(s) + 4NO_2(g) + O_2(g)$$
c **(i)** The maximum volume of chlorine gas that can be formed from 1 dm³ of 1.0 mol dm⁻³ concentrated hydrochloric acid
   (ii) The mass of manganese(IV) oxide needed for the reaction
$$MnO_2(s) + 4HCl(aq) \rightarrow MnCl_2(aq) + 2H_2O(l) + Cl_2(g)$$

Multiple-choice questions for Chapters 7–10

You will need to use this data in these questions:

A_r values: H = 1, C = 12, O = 16, Mg = 24, Al = 27, S = 32, Cl = 35.5, K = 39, Ca = 40, Cu = 64, Zn = 65, Na = 23, N = 14, Fe = 56

Gas volumes: 1 mol at STP is 22.4 dm³, 1 mol at RTP is 24.0 dm³

Avogadro number = 6.0×10^{23}

1 The reaction between barium chloride solution and sulphuric acid can be represented by the equation:

$$BaCl_2(aq) + H_2SO_4(aq) \rightarrow BaSO_4(s) + 2HCl(aq)$$

The correct ionic equation for this reaction is:

A $Ba^{2+}(aq) + SO_4^{2-}(aq) \rightarrow BaSO_4(s)$

B $2Cl^-(aq) + 2H^+(aq) \rightarrow 2HCl(aq)$

C $Ba_2^{2+}(aq) + 2SO_4^-(aq) \rightarrow 2BaSO_4(s)$

D $Ba^{2+}(aq) + H_2SO_4(aq) \rightarrow BaSO_4(s) + 2H^+$

2 The mass of 1 mole of the compound $Al_2(SO_4)_3$ is:

A 186 g
B 366 g
C 250 g
D 342 g

3 The correct formula for the compound iron(III) nitride is:

A $Fe(NO_3)$
B FeN
C Fe_3N_3
D $Fe(NO_2)_3$

4 The mass of magnesium which contains the same number of atoms as 36 g of carbon is:

A 36 g
B 24 g
C 72 g
D 6.0×10^{23} g

5 The percentage by mass of magnesium in magnesium sulphate ($MgSO_4$) is:

A 17%
B 20%
C 24%
D 80%

6 Which of the following does NOT contain the Avogadro number of atoms?

A 40 g calcium
B 2 g hydrogen
C 12 g carbon
D 24 g magnesium

7 When limestone is heated it decomposes according to the equation:

$$CaCO_3(s) \rightarrow CaO(s) + CO_2(g)$$

The mass and volume of carbon dioxide gas at STP produced by heating 200 g of limestone is:

	Mass (g)	Volume (dm³)
A	88	22.4
B	44	44.8
C	88	44.8
D	44	22.4

8 The number of moles of sodium hydroxide in 25 cm³ of a solution containing 0.5 mol dm⁻³ is:

A 0.0125
B 0.25
C 5
D 12.5

9 What is the correct way to balance the following equation?

$$Al(s) + O_2(g) \rightarrow Al_2O_3(s)$$

A $2Al(s) + O_3(g) \rightarrow Al_2O_3(s)$
B $Al_2(s) + 3O_2(g) \rightarrow Al_2(O_3)_2(s)$
C $2Al(s) + 3O(g) \rightarrow Al_2O_3(s)$
D $4Al(s) + 3O_2(g) \rightarrow 2Al_2O_3(s)$

10 100 cm³ of a solution of potassium hydroxide contains 0.56 grams of the dissolved solute. The molar concentration of this solution is:

A 0.01 mol dm⁻³
B 0.1 mol dm⁻³
C 1.0 mol dm⁻³
D 10 mol dm⁻³

11 1 mol of an oxide of copper produces 128 g of metallic copper and 18 g water when reduced by hydrogen gas. What is the formula for the oxide of copper?

A CuO
B Cu_2O
C CuO_2
D Cu_2O_2

11 Acids, bases and salts

In this chapter, you will study the following:

- acidity and alkalinity in relation to the pH scale;
- acids, acid anhydrides, and characteristic reactions of non-oxidising acids;
- acids in living systems;
- bases, alkalis and characteristic reactions;
- basic, amphoteric and neutral oxides;
- salts and preparations of salts;
- acidity and alkalinity of salts;
- uses and dangers of salts.

This chapter covers
Objectives 7.1–7.9 of Section A of the CSEC Chemistry Syllabus.

Your early studies of acids and bases might have focused on their colourful reactions with indicators, or their corrosive properties, or applications such as 'acid rain' or even 'acid indigestion'.

(a)

(b)

(c)

▲ **Figure 11.1**
An example of the effects of acid rain

▲ **Figure 11.2** Acids, bases and salts in everyday life. (a) Acid rain has damaged these trees in Scandinavia. (b) This fertiliser contains ammonium nitrate. (c) Suffering from indigestion? You may need to take an 'antacid', but be sure to read the instructions on the packet!

11.1 Acids and bases – what is the difference?

Indicators are dyes that are one colour in acidic solution and another colour in a solution of a base (that is, an alkali).

It is easy to detect if a substance is an acid or a base by using an **indicator** or doing a pH test.

Indicators

Table 11.1 shows the colour changes of some common indicators. The photographs show in Figure 11.3 the colour changes with methyl orange. Generally most indicators can only be used to determine whether a solution is acidic or alkaline.

▼ **Table 11.1** The colour changes of some common indicators.

	Litmus	Methyl orange	Screened methyl orange	Phenolphthalein
Colour with acid	red	red	light red	colourless
Colour with water	purple	orange	grey	colourless
Colour with base	blue	yellow	green	pink

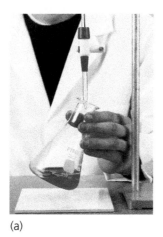

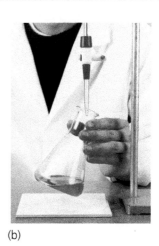

(a)   (b)

Figure 11.3
(a) Methyl orange in an acidic solution.
(b) Methyl orange in an alkaline solution. ▶

Testing pH

The pH of a solution can tell you how acidic or how alkaline it is.

- The stronger the acid the lower the pH.
- The stronger the base the higher the pH.

pH is a substance's potential or power for hydrogen ion dissociation.

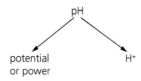

The **pH scale** is a number scale that indicates whether a solution is alkaline, acidic or neutral:

- Acidic solutions have a pH of less than 7.
- Neutral solutions have a pH of 7.
- Alkaline (basic) solutions have a pH of more than 7.

You can get a quick estimate of the pH of a solution by using full range or universal indicators. These are mixtures of indicators that give a range of colours that can be matched to pH. A pH meter is used to measure pH precisely.

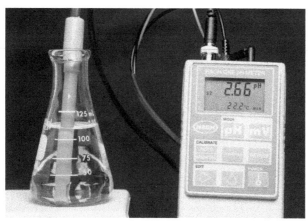

▲ **Figure 11.4** A pH meter

pH	1	2	3	4	5	6	7	8	9	10	11	12	13	14
colour	RED		ORANGE		YELLOW		GREEN			BLUE		PURPLE-VIOLET		
strength	Strong		ACIDS		Weak		Neutral	Weak		ALKALIS		Strong		

▲ **Figure 11.5** A pH scale

M&M
A&I

Experiment 11.1   Testing toothpaste

Obtain five different brands of toothpaste.

▲ **Figure 11.6** Different brands of toothpaste

Procedure

1 Using universal indicator solution or pH paper, determine the pH of each brand.

2 What is the pH range of the brands of toothpaste examined?

3 Why do you think manufacturers make toothpaste in this pH range?

Everyday acid and base chemistry

Table 11.2 describes some acidic and basic substances used in everyday life.

▼ **Table 11.2** Some everyday acids and bases.

Common name	Description/comments	pH
lye	0.1 mol dm^{-3} NaOH, removes grease and stubborn dirt	13
household	effectively a dilute solution of ammonia, used in cleaning glass panes, counter tops, etc.	11–12
lime water	a saturated solution of calcium hydroxide, used to detect the presence of carbon dioxide	10.6
borax	a compound of boron, used to remove wine and tea stains	9.3
baking soda	sodium hydrogencarbonate, used as a raising agent in baking, etc., it reduces the acidity of citrus fruits, also used in stain removal and in cleaning jewellery	8.5
vinegar	a dilute solution of ethanoic acid, used in preserving (pickling) foods	3
lemon juice	contains citric acid, a widely distributed plant (organic) acid, used in foods to provide a tart taste	2
gastric fluid	found in your stomach, contains hydrochloric acid, aids digestion of food	1.3–1.5

Q1 Which of the substances listed is:

    a the strongest base;   **c** the weakest base;

    b the strongest acid;   **d** the weakest acid?

The pigments from some plant materials can be used as indicators. Extract the pigment from red sorrel, red hibiscus flower or purple cabbage by boiling some of the plant material in a small volume of water. Use the extracts to test some of the substances above.

Q2 What colour changes do these 'indicators' show in acids and bases?

Q3 Use your home-made indicators to discover two more fluids used in everyday life that are acids and two more that are bases.

11.2 Acids

What are acids?

An **acid** is a substance which in solution produces hydrogen ions (protons) (H$^+$) as the only positive ions.

▲ **Figure 11.6** Types of acids

The word acid usually brings to mind a fuming, corrosive liquid, but not all acids are corrosive, nor do all acids fume. Acids, though, should be handled with care.

A simple explanation of acid behaviour is that all acids liberate hydrogen ions (H^+) when they are dissolved in water. The 'free' hydrogen ions (protons) are responsible for the observed behaviour of acids. However, the hydrogen ions produced do not remain separate in solution. They become attached to the oxygen atoms in the polar water molecule forming 'hydronium' (or 'hydroxonium ions'), as follows:

$$H_2O + \underset{\text{proton}}{H^+} \longrightarrow \underset{\text{hydronium ion}}{H_3O^+}$$

For simplicity, this ion is often written as H^+(aq).

When acids come into contact with water then, they do not merely dissolve – chemical changes take place. Considerable heat is evolved when concentrated sulphuric acid is dissolved in water, for example. This is an indication that new bonds are being formed.

When not dissolved in water (that is, in their anhydrous state), acids may be:

- solids, e.g. citric acid and ascorbic acid (vitamin C);
- liquids, e.g. nitric acid and phosphoric acid;
- gases, e.g. HCl (which dissolves in water to form hydrochloric acid), SO_3 (which dissolves in water to form sulphuric acid) and NO_2 (which dissolves in water to form a mixture of nitric acid and nitrous acid).

Acids only exhibit their acidic properties when dissolved in water.
Note: All acids have hydrogen in their formulae, but not all hydrogen-containing compounds are acids.

It takes water to make acids!

- Anhydrous ethanoic acid does not give a colour reaction with litmus or other indicators. Ethanoic acid only behaves like an acid when dissolved in water.

- Hydrogen chloride does not show acidic properties when it is dissolved in a non-aqueous solvent such as methylbenzene, but it does when dissolved in water.

Both water and hydrogen chloride molecules are polar covalent molecules.

- The oxygen atom in a water molecule has a slight negative charge.
- The hydrogen atom in a hydrogen chloride molecule has a slight positive charge.

The negatively charged oxygen atoms are attracted to the positively charged hydrogen atoms. The hydrogen proton is thus pulled away from the hydrogen chloride molecule, forming hydronium ions and liberating chloride ions.

hydroxonium ion

$$H_2O(l) + HCl(g) \rightarrow H_3O^+(aq) + Cl^-(aq)$$

In methylbenzene, no hydronium ions are formed.

Q1 Draw diagrams to illustrate what happens when hydrogen chloride dissolves in water (as described in the paragraph above).

Q2 What is a 'non-aqueous solvent'?

Q3 What conclusions can you draw about the structure of methylbenzene?

Preparation of acids

Acids are generally produced by dissolving oxides of non-metals in water. These acidic oxides are often referred to as **acid anhydrides**.

An acid anhydride is a non-metallic oxide that dissolves in water to form an acid.

$$SO_2(g) + H_2O(l) \rightarrow H_2SO_3(aq)$$
sulphurous acid (sulphuric(IV) acid)

$$SO_3(g) + H_2O(l) \rightarrow H_2SO_4(aq)$$
sulphuric acid (sulphuric(VI) acid)

$$CO_2(g) + H_2O(l) \rightarrow H_2CO_3(aq)$$
carbonic acid

$$N_2O_5(g) + H_2O(l) \rightarrow 2HNO_3(aq)$$
nitric(V) acid

$$2NO_2(g) + H_2O(l) \rightarrow HNO_2(aq) \quad + \quad HNO_3(aq)$$
nitrous acid   nitric(V) acid
(nitric(III) acid

Classifying acids

Acids may be grouped in different ways. They may be referred to as:

- mineral or organic;
- strong or weak;
- monobasic, dibasic or tribasic.

▲ **Figure 11.7** Foods that contain organic acids

Acids	

Mineral acids

These were originally obtained from minerals.

Examples:

- hydrochloric acid (HCl)
- nitric acid (HNO_3)
- sulphuric acid (H_2SO_4)
- phosphoric acid (H_3PO_4)

Organic acids

These were originally obtained from plant and animal materials.

Examples:

- ethanoic acid (CH_3COOH) – vinegar
- tartaric acid ($C_4H_6O_6$) – in grapes
- citric acid ($C_6H_8O_7$) – in citrus fruits
- lactic acid ($C_3H_6O_3$) – in sour milk
- ethanedioic acid ($C_2H_2O_4$) – in beet and rhubarb leaves

```
                            ┌─────────┐
                            │  Acids  │
                            └────┬────┘
                  ┌──────────────┴──────────────┐
```

Strong acids

These acids are completely ionised in aqueous solutions.

Their aqueous solutions conduct electricity well.

Examples:
- hydrochloric acid
- sulphuric acid
- nitric acid
- phosphoric acid

Weak acids

These acids are only partially ionised in aqueous solutions. Their aqueous solutions conduct electricity to a small extent.

Examples:
- ethanoic acid
- tartaric acid
- citric acid
- lactic acid
- ethanedioic acid

```
                            ┌─────────┐
                            │  Acids  │
                            └────┬────┘
          ┌──────────────────────┼──────────────────────┐
```

Monobasic acids

These acids yield one free H^+ ion for each molecule of acid in aqueous solution.

Examples:
- hydrochloric acid
- nitric acid
- ethanoic acid
- lactic acid

Dibasic acids

These acids yield two free H^+ ions per molecule of acid in aqueous solution.

Examples:
- sulphuric acid
- ethanedioic acid
- tartaric acid

Tribasic acids

These acids yield three free H^+ ions per molecule of acid in aqueous solution.

Examples:
- phosphoric acid
- citric acid

Note: Some acids contain more than one hydrogen atom in their formula, but not all the hydrogen atoms necessarily form hydrogen ions. Ethanoic acid, CH_3COOH, has four hydrogen atoms in each formula unit, but produces only one hydrogen ion and is monobasic:

$$CH_3COOH(aq) \rightleftharpoons CH_3COO^-(aq) + H^+(aq)$$

Acids in living systems

Vitamin C

Vitamin C (chemical name: ascorbic acid) has the molecular formula $C_6H_8O_6$. It is obtained from a variety of sources, e.g. West Indian cherries, blackcurrants, citrus fruits, cabbage and spinach.

Here are some of the roles of vitamin C:
- It controls the formation of dentine, cartilage and bone.
- It helps in the formation of red blood cells.
- It helps in the healing of wounds.

▲ **Figure 11.8** These fruits all contain vitamin C, which is an organic acid.

Some people believe that large doses of vitamin C reduce the risk of viral infections, including the common cold. There is, as yet, no conclusive evidence to support these claims.

Vitamin C is water-soluble and easily oxidised. The extent of oxidation is increased by:

- cutting or crushing – this releases enzymes called oxidases;
- increasing the temperature;
- the action of alkalis;
- traces of copper.

Losses of vitamin C occur when fruits and vegetables are stored. Further losses occur during preparation and cooking, because vitamin C breaks down on cooking.

Lactic acid

Have you noticed that after doing an unusual amount of exercise, your muscles begin to feel tired and cramp up? This is due to the accumulation of lactic acid ($C_3H_6O_3$) in the cells of your muscles. When doing a usual amount of exercise, aerobic respiration occurs because enough oxygen is being supplied to the muscle cells by circulation. The chemical equation for aerobic respiration is:

$$C_6H_{12}O_6 + 6O_2 \longrightarrow 6CO_2 + H_2O + energy$$

$$glucose + oxygen \longrightarrow carbon\ dioxide + water + energy$$

However, during prolonged periods of exercise, the blood does not supply enough oxygen to the cells and aerobic respiration takes place. The chemical equation for anaerobic respiration is:

$$C_6H_{12}O_6 \longrightarrow C_3H_6O_3 + energy$$

$$glucose \longrightarrow lactic\ acid + energy$$

Methanoic acid

Methanoic acid is also called formic acid, and it is the irritant that ants inject when they sting. Some ants bite and some sting, and some do both. Ants use this as a defence mechanism against predators. Since methanoic acid is a weak acid, it can be neutralised using a weak base. Therefore, you can treat ant stings using washing soda (sodium carbonate). Mix the sodium carbonate into a paste and apply it to the affected area.

However, note that you should not treat burns caused by strong acids by neutralising then with a strong base. Neutralisation is an exothermic reaction, and the heat produced will make the burn worse. You should just run cool water over the burn, apply an antiseptic (burn cream) and seek proper medical attention.

Strength of acids

When thinking about acids you must be careful not to confuse the terms 'strong' and 'weak' with the terms 'concentrated' and 'dilute'.

- The terms 'concentrated' and 'dilute' indicate how much acid is dissolved in water (i.e. the concentration).
- The terms 'strong' and 'weak', when used to describe acids, have a different meaning.

A 'strong' acid dissociates completely in aqueous solution. Nitric acid, for example, dissociates as follows:

$$HNO_3(l) \xrightarrow{H_2O} H^+(aq) + NO_3^-(aq)$$

The arrow '→' means that the reaction goes to completion. It goes in one direction only.

A 'weak' acid, for example, only partially dissociates in aqueous solution. Ethanoic acid dissociates as follows:

$$CH_3COOH(l) \underset{}{\overset{H_2O(l)}{\rightleftharpoons}} H^+(aq) + CH_3COO^-(aq)$$
ethanoic acid

The arrow '⇌' means that the reaction is reversible (see Section 12.6). This means that hydrogen ions produced by the forward reaction (from left to right) are continually being removed by the backward reaction (from right to left). At any one time, the solution will contain undissociated ethanoic acid molecules as well as ethanoate ions and hydrogen ions from the dissociated acid. Overall, therefore, there are not as many free hydrogen ions (protons) in a solution of a weak acid as is the case with a strong acid.

Practice

1 Give the name and formula of the acid from which each of the following anions is formed:

a NO_2^-
b ClO_3^-
c SO_3^{2-}
d S^{2-}
e CO_3^{2-}
f PO_4^{3-}

Reactions of non-oxidising acids

A **non-oxidising acid** is one in which the anion of the acid is a weaker oxidising agent than H^+. Examples of non-oxidising acids are most organic acids (like CH_3COOH), phosphoric acid (H_3PO_4), sulphuric acid (H_2SO_4) when cold and dilute, and hydrochloric acid (HCl).

An **oxidising acid** is one in which the anion of the acid is a stronger oxidising agent than H^+. Examples of non-oxidising acids are nitric acid (HNO_3) and concentrated sulphuric acid (H_2SO_4). You will learn more about this in Chapter 21.

Acids have the following characteristic reactions:

1 They give particular colours with indicators. With universal indicator, the exact colour depends on the pH of the acid.

2 With active metals (e.g. magnesium and zinc), acids react to give hydrogen gas and a salt.

$$metal \; + \; acid \; \rightarrow \; salt \; + \; hydrogen$$
$$Zn(s) \; + \; H_2SO_4(aq) \; \rightarrow \; ZnSO_4(aq) \; + \; H_2(g)$$

3 Acids react with metal oxides (and hydroxides) to form a salt and water only.

$$CuO(s) + H_2SO_4(aq) \rightarrow CuSO_4(aq) + H_2O(l)$$
$$NaOH(aq) + HNO_3(aq) \rightarrow NaNO_3(aq) + H_2O(l)$$

This is a neutralisation reaction. (See Sections 12.5 and 9.4.)

4 Acids react with carbonates and hydrogencarbonates to yield a salt, water and carbon dioxide.

$$Na_2CO_3(s \; or \; aq) + 2HCl(aq) \rightarrow 2NaCl(aq) + H_2O(l) + CO_2(g)$$
$$NaHCO_3(aq) + HNO_3(aq) \rightarrow NaNO_3(aq) + H_2O(l) + CO_2(g)$$

The reaction between calcium carbonate and sulphuric acid soon stops. Can you think why? It is because insoluble calcium sulphate is formed, which coats the calcium carbonate and stops the reaction.

Acids have the following properties too:

* Aqueous solutions of acids conduct electric currents.
* Acids have sour tastes (beware).
* All acids have hydrogen in their formulae; however, not all hydrogen-containing compounds are acids.

Experiment 11.2 Testing acids

⚠ Be careful when using all acids. Use protective eyewear and avoid contact of acids with the skin and with clothes.

Procedure
The results of some tests on three common acids are given in Table 11.3. Complete the table by carrying out those tests for which results are not given.

▼ **Table 11.3**

	Hydrochloric acid (HCl)	Sulphuric acid (H_2SO_4)	Ethanoic acid (CH_3COOH)
indicators			
litmus	turns red		turns red
universal or full range indicator			
methyl orange			
metals			
powdered copper	no reaction	no reaction	

powdered iron			
granulated zinc			
magnesium ribbon	rapid reaction, hydrogen evolved; reaction faster than for zinc or iron		hydrogen evolved at moderate rate
metal oxide copper oxide (a black powder)	reaction; forming a green solution of copper(II) chloride		no reaction in the cold; a blue-green solution forms on warming
carbonates sodium carbonate (solid or aqueous)	effervescence; CO_2 evolved		
calcium carbonate (solid)	as above		slight effervescence; CO_2 evolved
sodium hydrogencarbonate (baking soda) (solid or aqueous)			

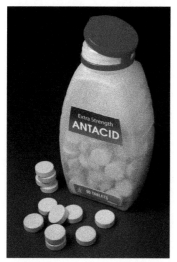

▲ **Figure 11.9** Think of ways in which the above are useful in everyday life.

Questions

Q1 Write a summary of the typical reactions of acids based on the results in Table 11.3.

Q2 Write balanced equations for the following reactions:
- **a** Copper oxide and ethanoic acid
- **b** Magnesium and hydrochloric acid
- **c** Calcium carbonate and nitric acid

Find out more

How are the reactions of acids used in everyday life?

a Crush some antacid tablets and add some dilute hydrochloric acid or vinegar to the powder in a test-tube.

b Add a few ml of dilute acid to a spatula of baking powder in a test-tube.

Q1 Use tests to identify any gases evolved in (a) and (b).

Q2 Find out the ingredients in antacid tablets and baking soda, and explain the reactions observed.

Q3 Find out how a typical reaction of acids is used in fire extinguishers.

11.3 Bases

What are bases?

A base is a proton acceptor.

Many **bases** are oxides and hydroxides of metals. Bases are also defined as substances that react with acids to form salts and water only, as shown in the following equations:

$$\text{base} \quad + \quad \text{acid} \quad \rightarrow \quad \text{salt} \quad + \quad \text{water}$$
$$MgO(s) \quad + \quad 2HCl(aq) \rightarrow MgCl_2(aq) \quad + \quad H_2O(l)$$
$$NaOH(aq) \quad + \quad HNO_3(aq) \rightarrow NaNO_3(aq) \quad + \quad H_2O(l)$$
$$2KOH(aq) \quad + \quad H_2SO_4(aq) \rightarrow K_2SO_4(aq) \quad + \quad 2H_2O(l)$$
$$NH_3.H_2O(aq) \quad + \quad HCl(aq) \quad \rightarrow NH_4Cl(aq) \quad + \quad H_2O(l)$$

The ionic equations for these reactions are:

$$O^{2-}(s) + 2H^+(aq) \rightarrow H_2O(l) \text{ for the first equation}$$
$$OH^-(aq) + H^+(aq) \rightarrow H_2O(l) \text{ for the other equations}$$

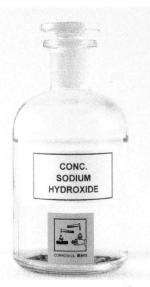
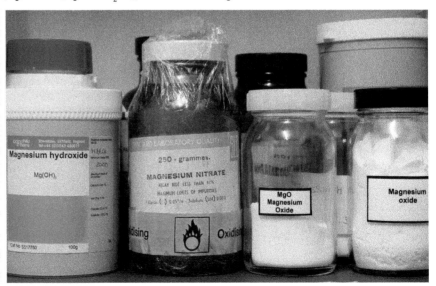

▲ **Figure 11.10** Bases and salts in the chemistry laboratory

These equations clearly show that bases accept protons.

In fact, any substance that accepts protons is a base. Though it is not an oxide or hydroxide, sodium carbonate is a base often used in the laboratory and in industry:

$$CO_3^{2-}(aq) + 2H^+(aq) \rightarrow CO_2(g) + H_2O(l)$$

Some bases are soluble in water (e.g. KOH, NaOH and Na_2O), while others are not (e.g. Cu_2O, Fe_2O_3 and $Cu(OH)_2$). Soluble bases are commonly known as **alkalis**.

> An alkali is a base that is soluble in water. Commonly used alkalis are NaOH, $Ca(OH)_2$, KOH, and a solution of ammonia in water ($NH_3.H_2O$, often written as NH_4OH).

Alkalis may be strong or weak (like acids).

- Strong alkalis ionise completely in aqueous solutions:

$$NaOH(s) \xrightarrow{H_2O} Na^+(aq) + OH^-(aq)$$

- Weak alkalis are only partially ionised in aqueous solutions. Ammonia, for example, only provides few OH^- ions (proton acceptors).

Practice

2 Give the formulae and names of bases that contain each of the following cations:
   a Rb^+
   b Ba^{2+}
   c Al^{3+}
   d Fe^{2+}
   e Cu^{2+}
   f Ca^{2+}

Characteristics of bases

1 Bases give characteristic colours with indicators. For example, they turn litmus blue and phenolphthalein from colourless to pink.

2 They conduct electricity when in aqueous solution.

3 They neutralise acids to form a salt and water.

4 They react with ammonium salts to produce ammonia gas. For example:
$$(NH_4)_2SO_4(s) + Ca(OH)_2(s) \rightarrow 2NH_3(g) + 2H_2O(l) + CaSO_4(s)$$

5 Alkalis are slippery to the touch. This is due to the conversion of oils on the skin to soap by alkalis.

(Caution: Strong alkalis can damage the skin.)

Some bases, such as sodium carbonate and sodium hydroxide, are widely used in industry.

Basic amphoteric and neutral oxides

An oxide is a compound made up of oxygen bonded to a metal or non-metal. We classify oxides based on their reactions with acids and bases. There are acidic, basic, amphoteric and neutral oxides.

Basic oxides

Basic oxides react with acids to produce a salt and water. They are usually metallic oxides. Basic oxides that are soluble in water form solutions called alkalis, for example, Na_2O, CaO, K_2O, MgO and CuO. For example, the basic insoluble black copper oxide reacts with dilute sulphuric acid to form blue copper sulphate solution:

$$CuO(s) + H_2SO_4(aq) \rightarrow CuSO_4(aq) + H_2O(l)$$

Soluble sodium oxide reacts with water to form the alkali sodium hydroxide:

$$Na2O(s) + H2O(l) \rightarrow NaOH(aq)$$

Neutral oxides

A neutral oxide will not react with either bases or acids. Some examples are carbon monoxide (CO), nitrogen oxide (NO), dioxide (NO_2) and water:

$$2C(s) + O_2(g) \rightarrow CO(g)$$

Amphoteric oxides

Amphoteric oxides can react with both acids and bases. Examples are aluminium oxide (Al_2O_3), lead oxide (PbO) and zinc oxide (ZnO).

Lead oxide (PbO) reacts with hydrochloric acid to form lead chloride and water:

$$PbO (s) + HCl(aq) \rightarrow PbCl_2(aq) + H_2O(l)$$

Lead oxide (PbO) also reacts with sodium hydroxide to form sodium plumbate:

$$PbO(s) + NaOH(aq) \rightarrow Na_2Pb(OH)_4(aq)$$

This reaction can also be written as:

$$PbO(s) + NaOH(aq) \rightarrow Na_2PbO_2(s) + H_2O(aq)$$

11.4 Salts

What are salts?

Salts are formed when metal ions or the ammonium ion take the place of the hydrogen ion (or ions) of an acid:

$$NaOH(aq) + HCl(aq) \rightarrow NaCl(aq) + H_2O(l)$$
$$\qquad\qquad\quad acid \qquad\quad salt$$

Here the sodium ion takes the place of the hydrogen ion.

In dibasic and tribasic acids, the hydrogen ions of the acid may be either partially or totally replaced:

- A normal salt is formed if all the hydrogen ions are replaced.
- An acid salt is formed if the hydrogen ions are partially replaced.

For example, the dibasic acid H_2SO_4 forms two different salts:

$$NaOH(aq) + H_2SO_4(aq) \rightarrow NaHSO_4(aq) + H_2O(l)$$
acid salt

$$2NaOH(aq) + H_2SO_4(aq) \rightarrow Na_2SO_4(aq) + 2H_2O(l)$$
normal salt

With the tribasic acid H_3PO_4 three different salts are possible:

$$KOH(aq) + H_3PO_4(aq) \rightarrow KH_2PO_4(aq) + H_2O$$
acid salt

$$2KOH(aq) + H_3PO_4(aq) \rightarrow K_2HPO_4(aq) + 2H_2O$$
acid salt

$$3KOH(aq) + H_3PO_4(aq) \rightarrow K_3PO_4(aq) + 3H_2O$$
normal salt

How many different salts will ethanoic acid form with potassium hydroxide?

Practice

3 Identify each of the following as an acid, a base, a normal salt or an acid salt:

a H_2S

b $NaHSO_3$

c $LiOH$

d $Ca(HCO_3)_2$

e $Fe(OH)_2$

f H_3PO_3

g $CaCl_2$

h $CuSO_4$

We use salts in a variety of ways in everyday life.

Find out more

Salts used in everyday life

▼ **Table 11.4** The uses of some everyday salts

Salt	Colour and other characteristics	Uses
ammonium chloride	white crystals	dry cells (batteries), fertilisers
ammonium sulphate (sulphate of ammonia)	white crystals	fertilisers
calcium carbonate (marble, limestone)	white but can be coloured	decorative stones, manufacture of cement and lime
calcium sulphate (plaster of Paris, gypsum)	white crystals	plastering walls, making casts, etc.
magnesium sulphate (Epsom salts)	white crystals	purgative
copper(II) sulphate	blue crystals	fungicides
sodium carbonate (washing soda)	white crystals or powder	in cleaning, in laundry as a water softener, in the manufacture of glass

Q1 Check the labels of substances found in your home and identify as many other salts as you can.

Q2 Some salts are used for their hygroscopic properties.
  a What does the term 'hygroscopic' mean?
  b Find out the names of two hygroscopic salts and how they are used.
  c Silica gel (in small parcels) is often included in shoe boxes because it is hygroscopic. What is the chemical formula of silica gel? Is it a salt?

▲ **Figure 11.11** Somes uses of salts.

Practice

4 Which of the salts listed in Table 11.4 are soluble in water?

5 Write the formulae for all of the salts listed in Table 11.4.

6 What type of bonding would you expect salts such as NaCl and KNO_3 to have?

7 Name two physical characteristics you expect salts such as NaCl and KNO_3 to have.

Preparation of salts

The method used to prepare a salt depends on whether the salt is soluble in water or not. Table 11.5 contains a guide to the solubility of salts.

▼ **Table 11.5** The solubility of common salts in water

Salts	Solubility characteristics
nitrates	all nitrates are soluble
chlorides, bromides and iodides (halides)	all halides are soluble, except the halides of silver and lead; lead chloride and lead bromide, however, are soluble in hot water
sulphates	all sulphates are soluble, except barium sulphate and lead sulphate; calcium sulphate and silver sulphate are slightly soluble
carbonates	all carbonates are insoluble, except sodium carbonate, potassium carbonate and ammonium carbonate
hydrogencarbonates	most soluble

Note: All common salts containing the ions Na^+, K^+ and NH_4^+ are soluble.

Preparing soluble salts

Soluble salts can be prepared in the following ways.

1 The action of acids on alkalis.

$$NaOH(aq) + HNO_3(aq) \rightarrow NaNO_3(aq) + H_2O(l)$$
$$Ba(OH)_2(aq) + 2HCl(aq) \rightarrow BaCl_2(aq) + 2H_2O(l)$$

2 The action of acids on excess metal, insoluble metal oxide or metal hydroxide, metal carbonate or hydrogencarbonate.

$$Zn(s) + H_2SO_4(aq) \rightarrow ZnSO_4(aq) + H_2(g)$$
$$PbO(s) + 2HNO_3(aq) \rightarrow Pb(NO_3)_2(aq) + H_2O(l)$$
$$Mg(OH)_2(s) + 2HNO_3(aq) \rightarrow Mg(NO_3)_2(aq) + 2H_2O(l)$$
$$CaCO_3(s) + 2HCl(aq) \rightarrow CaCl_2(aq) + H_2O(l) + CO_2(g)$$
$$2NaHCO_3(s) + H_2SO_4(aq) \rightarrow Na_2SO_4(aq) + 2H_2O(l) + 2CO_2(g)$$

It is common practice whenever a solid and an acid are used in such preparations to use excess of the solid. This ensures that all the acid is used up in the reaction. Under these conditions, the acid is the limiting reagent. On completion of the reaction, excess (unreacted) solid is removed by filtration.

A sample of the salt crystals can then be obtained from the filtrate by carrying out the following steps:

1 Concentrate the filtrate by gentle evaporation (use a water bath).
2 Cool the concentrate, testing for crystal formation – slow cooling leads to the formation of big crystals.
3 Filter to collect the crystals.
4 Carefully wash and dry the crystals.

Water of crystallisation

Water molecules may become part of the crystal structure of salts when salts are formed by crystallisation from aqueous solution. Such water is known as 'water of **crystallisation**'.

The 'water of crystallisation' should be included as part of the formula of the salt. The dot in the middle of the formula, just before the H_2O, indicates water of crystallisation.

Salts with water of crystallisation are also known as hydrated salts.

Examples of salts with water of crystallisation include:
- $CuSO_4.5H_2O$ – hydrated copper sulphate; more correctly named copper(II) sulphate pentahydrate;
- $CaCl_2.6H_2O$ – calcium chloride-6-water;
- $FeSO_4.7H_2O$ – iron(II) sulphate-7-water.

If solutions of these salts are evaporated to dryness, a powdery (non-crystalline) form of the salt is produced.

Salts without water of crystallisation are described as anhydrous.

Experiment 11.3   To prepare copper(II) sulphate from copper(II) oxide

⚠ Avoid contact of acids and alkalis with the skin, eyes and clothing.

Procedure

1 Warm 50 cm³ of 2.0 mol dm⁻³ sulphuric acid in a suitable container.

2 Add copper(II) oxide, stirring until the reaction is complete. Warm if necessary to help the reaction. Unreacted copper(II) oxide will clearly be present in the container.

3 Filter the mixture. Divide the filtrate into two portions. The filtrate is the liquid that goes through the filter paper.

4 Concentrate one portion of the filtrate to saturation using a water bath.

5 Cool the resulting solution to obtain crystals.

6 Filter and then wash crystals with a minimum of water before drying the crystals.

7 Evaporate the other portion of the filtrate to dryness.

8 Compare the product in the two cases.

Questions

Q1 Account for the difference in the appearance of the products in steps 5 and 7.

Q2 What type of reaction is the reaction between copper(II) oxide and sulphuric acid?

Q3 Write a balanced equation for the reaction.

Experiment 11.4   To prepare a sample of sodium chloride – a neutralisation reaction

⚠ Avoid contact of acids and alkalis with the skin, eyes and clothing. This reaction can be carried out more accurately by titration (see Section 9.6).

Procedure

1 Measure 25 cm³ of 2.0 mol dm⁻³ sodium hydroxide solution into a clean conical flask.

2 Add two drops of (screened) methyl orange indicator.

3 Add about 20 cm³ of 2.0 mol dm⁻³ hydrochloric acid, followed by the addition of drops of hydrochloric acid until the solution is neutral.

4 Measure the total volume of acid added and repeat the experiment but without using the indicator.

5 Concentrate the mixture in an evaporating basin.

6 Allow the concentrate to cool slowly.

7 Examine the crystals formed using a hand lens or low power microscope.

Questions

Q1 Why is step 4 necessary?

Q2 Write a balanced equation for the reaction.

Preparing insoluble salts

This method is also used to prepare soluble salts like $AlCl_3$ and $FeCl_3$, which are hydrolysed in water.

Insoluble salts can be prepared in the following ways:

• Direct combination, for example:

$$Fe(s) + S(s) \rightarrow FeS(s)$$

• Ionic precipitation. A precipitate sometimes forms when two solutions of soluble salts are mixed. The soluble salts, between them, provide the ions needed to make the insoluble salt. Here is an example:

$$Ba(NO_3)_2(aq) + K_2SO_4(aq) \rightarrow BaSO_4(s) + 2KNO_3(aq)$$

The procedure is as follows:
• The aqueous solutions are mixed.
• The mixture is then warmed, if necessary, and filtered.
• The residue is washed and dried.

Sometimes an acid may be used in place of one of the soluble salts. For example, sulphuric acid can replace potassium sulphate in the preparation of barium sulphate. The equation for the reaction is:

$$Ba(NO_3)_2(aq) + H_2SO_4(aq) \rightarrow BaSO_4(s) + 2HNO_3(aq)$$

The formation of a white precipitate of barium sulphate is used to identify the sulphate ion in qualitative analysis.

Summary of preparing salts

The method to use for preparing a particular salt can be seen in Figure 11.12.

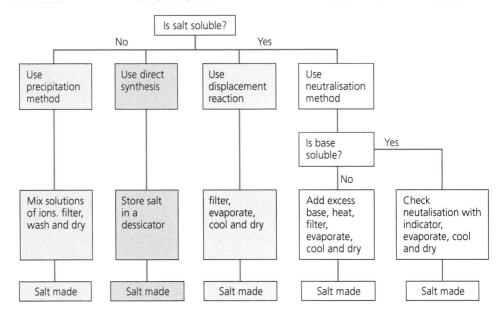

Figure 11.12
Choosing the method
to prepare a salt ▶

Acidity and alkalinity of salts

You might expect that all salts are neutral and have a pH of 7. For example, when sodium chloride is dissolved in water, a solution of pH 7 is produced. However, not all salts form neutral aqueous solutions. Some salts in solution have a pH greater than 7 (sodium ethanoate and sodium carbonate are examples) and some have a pH less than 7 (ammonium sulphate and ammonium chloride are examples).

Salts formed from a strong acid and a strong base are neutral (pH = 7).

Salts formed from a weak base (e.g. ammonia) and a strong acid have a pH less than 7 and are acidic.

Salts formed from a strong base (e.g. NaOH and KOH) and a weak acid have a pH greater than 7 and are alkaline.

It should not be surprising that soils become acidic after they have been repeatedly treated with ammonium fertilisers such as ammonium nitrate and ammonium sulphate. Farmers call such soils 'sour soils'. Sour soils can be made neutral by treating them with lime (calcium oxide – a base).

Acid salts are usually acidic in solution. They still have replaceable hydrogen from their parent acids.

Practice

8 Predict the change in pH when the following substances are dissolved in water. Select your answers from 'increases', 'decreases', 'stays the same'.
   a Sodium chloride
   b Carbon dioxide
   c Ammonium chloride
   d Magnesium sulphate
   e Sodium carbonate

The uses and dangers of salts

Baking powder

Baking powder is a 'raising agent', which is used in making cakes, some breads, roti and bake (fried flat bread). Baking powder is sodium hydrogencarbonate and also typically contains tartaric acid. When baking powder is added to dough, the tartaric acid reacts with the sodium hydrogencarbonate to form carbon dioxide gas, which causes the dough to rise.

The action of heat on sodium hydrogencarbonate also produces CO_2 gas. The reaction in the use of baking powder is:

$$2NaHCO_3(s) \rightarrow Na_2CO_3(s) + H_2O(g) + CO_2(g)$$

Cement

The raw materials used to make cement include limestone (calcium carbonate), clay and sand. Cement is a material that is used in building and construction, and is used to produce concrete and mortar. When concrete is freshly mixed, it is plastic and malleable so that it can be cast in almost any shape. In construction, it can be used both as a structural element, since it is strong and durable, and as a decorative element as it can be coloured, painted and decorated.

The most important type of cement produced today is Portland cement. Its approximate chemical composition is:

- 60–70% calcium oxide;
- 17–25% silica;
- 3–8 % aluminium oxide;
- small amounts of other materials.

Plaster of Paris (calcium sulphate)

Plaster of Paris is a white powdery mixture of gypsum (calcium sulphate, $CaSO_4$). It is called plaster of Paris because the first deposit of gypsum was found in Paris. When mixed with water, this powder solidifies without losing its volume. Plaster of Paris is widely used to create soft bandages to treat bone fractures. The fractured bone is wrapped with a bandage together with plaster of Paris. The casting holds the broken bone together until it heals.

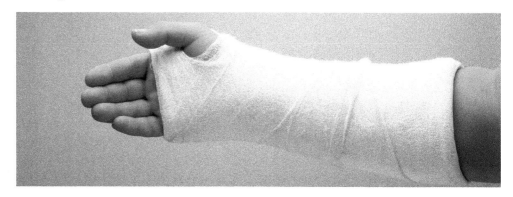

Figure 11.13
One medical use of salt is setting a broken bone using plaster of Paris. ▶

▼ **Table 11.6** Uses and dangers of salts

Salt	Use	Mechanism of action	Comments including dangers
Sodium chloride (common salt) NaCl	Used to flavour foods, and to cure meats and fish	Anti-microbial, restricts bacterial growth, reduces the amount of available water, changes osmotic pressure	Linked to high blood pressure, hardens the flesh of meat and fish and decreases its ability to absorb water
Sodium nitrite ($NaNO_2$) Sodium nitrate ($NaNO_3$)	Preserving heat-processed meat, poultry, fish and cheese	Controls the growth and toxin production of *Clostridium botulinum*	Nitrites produce nitric oxide, which gives a bright red colour to meats, they also combine with other products to produce nitrosamines, which can be the cause of some cancer in humans. Sodium nitrate may cause brain damage in infants and it is also suspected to be carcinogenic.
Sodium benzoate	Used in acidic foods such as salad dressings, carbonated drinks, jams and fruit juices, pickles and condiments, and medicines and cosmetics	Antimicrobial, mainly inhibits yeast and bacterial growth, effective only in acidic conditions (pH < 4)	Sodium benzoate can, in the presence of ascorbic acid, form benzene (a carcinogen).

Summary

- pH is a measure of how acidic or basic (alkaline) an aqueous solution of a substance is:
  - Substances with pH less than 7 are acidic.
  - Substances with pH greater than 7 are alkaline.
  - A pH of 7 indicates that a solution under test is neither acidic nor alkaline.

Acids

- Apart from their pH, acids can be detected by the following properties:
  - They liberate carbon dioxide from carbonates and hydrogencarbonates.
  - They liberate hydrogen when they react with active metals.
  - They turn moist blue litmus paper red and develop definite colours with other indicators.
- Acids are substances that produce 'free' hydrogen ions in aqueous solution.
- The basicity of an acid refers to the number of 'free' hydrogen ions that are produced for every molecule of the acid which dissolves in aqueous solution.
- Acids are conductors of electricity when in aqueous solution.
- Acids react with bases (alkalis) to form a salt and water as the only products. This reaction is called a neutralisation reaction.
- Acids are described as strong if they dissociate completely in aqueous solutions, and weak if they dissociate incompletely.

Bases

- Some bases are water-soluble. Water-soluble bases are called alkalis.
- Alkalis may be strong or weak, depending on the extent of their dissociation.
- Apart from their pH, alkalis have the following properties:
  - They are slippery to the touch.
  - They conduct electricity in aqueous solutions.
  - They turn moist litmus paper blue and give definite colours with other indicators.
- Whereas the oxides of metals are mainly bases, those of non-metals are mainly acid anhydrides.

Salts

- Insoluble salts can be prepared in the following ways:
  - By direct combination
  - By precipitation
- Soluble salts can be prepared by reaction between:
  - an acid and an alkali;
  - an acid and (an excess of) an active metal;
  - an acid and (an excess of) a carbonate or hydrogencarbonate;
  - an acid and (an excess of) an insoluble metal oxide or metal hydroxide.
- Salts are widely used in the home and in industry.

End-of-chapter questions

1 The reaction of bases with ammonium salts produces:

  A NH_3

  B CO_2

  C O_2

  D H_2

2 What is an acid anhydride?

  A An acidic oxide

  B An acid

  C A base

  D A salt

3 The compound formed when a metal cation replaces the H^+ ion of an acid is:

  A An alkali

  B A salt

  C An acid anhydride

  D A base

4 What is an oxide called that can react with both bases and acids?

  A An acidic oxide

  B A basic oxide

  C An amphoteric oxide

  D A neutral oxide

5 Which one of the following is a weak acid?

  A HCl

  B H_2SO_4

  C H_3PO_4

  D CH_3COOH

6 Which one of these salts is soluble?

  A Sodium carbonate

  B Calcium carbonate

  C Barium sulphate

  D Silver chloride

7 How is copper sulphate commonly prepared in the lab?

  A Ionic precipitation

  B Direct synthesis

  C A displacement reaction

  D Neutralisation

8 What are phosphoric acid and citric acid examples of?

  A Monobasic acids

  B Dibasic acids

  C Tribasic acids

  D Mineral acids

9 Which of these salts is used as a food preservative, but is also carcinogenic?

  A Sodium sulphite

  B Sodium nitrate

  C Ascorbic acid

  D Sodium chloride

10 What does anaerobic respiration in muscle cells produce?

  A Lactic acid

  B Citric acid

  C Acetic acid

  D Ethanol

---

11 An unknown solution was tested with universal indicator paper and was found to have a pH of 3. Is it an acid or alkali?

12 a Using suitable examples, distinguish between:

  (i) strong acids and weak acids;

  (ii) acid oxides, basic oxides and amphoteric oxides.

  b What is meant by the term 'acid anhydride'? Give suitable examples.

13 a Write balanced equations to show the neutralisation of:

  (i) calcium hydroxide by hydrochloric acid;

  (ii) sulphuric acid by potassium hydroxide;

  (iii) sodium hydroxide by ethanoic acid;

  (iv) hydrogen bromide by magnesium oxide.

  b Write three equations illustrating the stepwise neutralisation of H_3PO_4 with NaOH. Classify the salt formed in each stage as a normal or acid salt.

14 Outline the preparation of pure samples of:
   a zinc sulphate starting from metallic zinc;
   b magnesium chloride starting from magnesium hydroxide;
   c copper(II) nitrate starting from copper(II) oxide;
   d sodium sulphate starting from sodium carbonate;
   e lead(II) chloride starting from lead(II) nitrate.

15 A pack of baking powder (a raising agent) contains sodium hydrogencarbonate, tartaric acid and rice starch.

   a Write an equation for the action of heat on sodium hydrogencarbonate.

   b Suggest a reason for the presence of tartaric acid in the mixture.

   c Describe a test you would carry out to indicate the presence of rice starch in the mixture.

16 A popular 'health salt' is sold as a solid mixture. The label on the bottle reads:

sodium hydrogencarbonate	56%
tartaric acid	26.7%
citric acid	16.3%
flavouring, etc.	1.0%

   a Why is there a fizzing action when water is added to the health salt?

   b Write a word equation for the reaction in water of the ingredients of the health salt.

   c Describe how you would identify the gaseous product in the reaction described in part (b).

   d If the net mass of the ingredients in a bottle of the health salt is 330 g, determine the number of moles of sodium hydrogencarbonate in the bottle.

12 Types of chemical reactions

- different types of chemical reactions;
- the neutralisation reaction: effect of temperature change;
- common oxidising and reducing substances in everyday life;
- oxidation and reduction (redox) reactions;
- oxidation number deduction from formulae;
- oxidising and reducing agents;
- tests for oxidising and reducing agents.

This chapter covers
Objectives 7.10 and 8.1–8.6 of Section A of the CSEC Chemistry Syllabus.

We have seen that changes involving a change of state, such as melting, freezing and boiling, are called physical changes. Chemical changes, however, involve the rearrangement of atoms so that different substances are formed. Signs of these changes include the production of gases, or the formation of precipitates or substances that have different colours to the starting reactants. Some chemical reactions give off heat, while some absorb heat. Some reactions are explosive and some are very gradual. However, the vast number of chemical reactions of elements and compounds can be grouped into seven basic reaction types.

12.1 Direct combination reactions

In **direct combination reactions**, two or more elements react to form a single product.

Direct combination reactions can be represented by the general equation:

$$A + B \rightarrow AB$$

where A and B are elements.

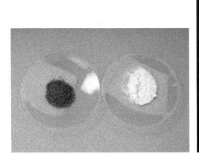

▲ Iron and sulphur

▲ Heating the mixture

▲ Iron(II) sulphide

▲ **Figure 12.1** The direct combination reaction between iron and sulphur to form iron sulphide.

The following are examples of direct combination reactions.

- The reaction of a metal with oxygen to produce the oxide of the metal:

  metal + oxygen $\longrightarrow$ metal oxide

  $2Mg(s) + O_2(g) \longrightarrow 2MgO(s)$

  Some metals combine with the oxygen of the air at room temperature, others have to be heated to form the oxides.

- Heating a non-metal in air or oxygen to produce the non-metal oxide:

  non-metal + oxygen $\longrightarrow$ non-metal oxide

  $$S(s) + O_2(g) \xrightarrow{\text{heat}} SO_2(g)$$

  $$C(s) + O_2(g) \xrightarrow{\text{heat}} CO_2(g)$$

- The reaction of a metal with a non-metal to produce a salt:

  metal + non-metal $\longrightarrow$ salt

  $2Na(s) + Cl_2(g) \longrightarrow 2NaCl(s)$

  $Fe(s) + S(s) \longrightarrow FeS(s)$

The direct combination reaction is also called the synthesis reaction.

Practice

1 Complete and balance the equations for the following combination reactions:

   a $Ca(s) + O_2(g) \rightarrow$

   b $Mg(s) + S(s) \rightarrow$

   c $Al(s) + Cl_2(g) \rightarrow$

12.2 Decomposition reactions

In **decomposition reactions**, one substance undergoes a reaction to form two or more substances.

In **decomposition reactions**, the substance broken down is always a compound, and the products may be elements or compounds. Heat is often necessary for the process, in which case the reaction can be described as **thermal decomposition**. The thermal decomposition of some metallic hydroxides, carbonates and nitrates are examples. Other examples include:

- Loss of some or all of their water of crystallisation by some hydrated salts on heating:

  $CuSO_4.5H_2O(s) \rightarrow CuSO_4(s) + 5H_2O(l)$

     hydrated salt   anhydrous

                  salt

  $MgSO_4.7H_2O(s) \rightarrow MgSO_4(s) + 7H_2O(l)$

The decomposition of water by electrolysis is an example of decomposition that does not require heat.

Practice

2 Complete and balance the equations for the following decomposition reactions that occur on heating:

a $H_2O_2(l) \rightarrow$

b $KHCO_3(s) \rightarrow$

c $MgCO_3(s) \rightarrow$

d $Cu(OH)_2(s) \rightarrow$

e $CaSO_4.2H_2O(s) \rightarrow$

f $BaCl_2.2H_2O(s) \rightarrow$

12.3 Substitution (displacement) reactions

In **substitution** or **displacement reactions**, one element displaces another element from a compound.

Displacement reactions can be represented by the general equation:

$$A + BX \rightarrow AX + B$$

A and B may be metals or non-metals.

Displacement reactions can generally be divided into one of three types:

1 A more reactive metal displaces a less reactive metal from an aqueous solution of its salt.

2 A reactive metal displaces hydrogen from an acid or water.

3 A more reactive non-metal displaces a less reactive non-metal.

1 More reactive metal/less reactive metal in aqueous solution

$$Zn^0(s) + CuSO_4(aq) \rightarrow ZnSO_4(aq) + Cu^0(s)$$
$$Cu^0(s) + 2AgNO_3(aq) \rightarrow Cu(NO_3)_2(aq) + 2Ag^0(s)$$

Zinc metal will displace copper from a solution of a copper salt because zinc is more reactive. Similarly, copper displaces silver from a solution of a silver salt.

2 Reactive metal/hydrogen in acid or water

For a metal to displace hydrogen ions from an acid, the metal must be above hydrogen in the reactivity series. Thus, both magnesium and zinc will react with hydrochloric acid to produce hydrogen gas, but copper will not react with dilute hydrochloric acid:

$$Mg^0(s) + 2HCl(aq) \rightarrow MgCl_2(aq) + H_2(g)$$
$$Zn^0(s) + 2HCl(aq) \rightarrow ZnCl_2(aq) + H_2(g)$$
$$2Na(s) + 2H_2O(l) \rightarrow 2NaOH(aq) + H_2(g)$$

3 More reactive non-metal/less reactive non-metal

$$F_2(g) + 2NaCl(aq) \rightarrow 2NaF(aq) + Cl_2(g)$$
$$Cl_2(g) + 2KI(aq) \rightarrow 2KCl(aq) + I_2(aq)$$

Practice

3 Complete and balance the equations for the following substitution (displacement) reactions:

a $Zn(s) + AgNO_3(aq) \rightarrow$

b $Mg(s) + HCl(aq) \rightarrow$

c $Br_2(l) + KI(aq) \rightarrow$

d $Al(s) + H_2SO_4(aq) \rightarrow$

e $Ni(s) + CuSO_4(aq) \rightarrow$

f $Fe(s) + Cu(NO_3)_2(aq) \rightarrow$

12.4 Ionic precipitation reactions

Ionic precipitation reactions are sometimes referred to as double decomposition reactions.

> In **ionic precipitation reactions**, two soluble ionic compounds react to form one soluble and one insoluble compound.

Ionic precipitation reactions can be represented generally by the equation:

$$AB + CD \rightarrow AD + CB$$

In this type of reaction, two compounds 'exchange' ions/radicals.

(a)   (b)

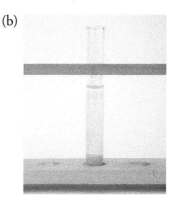

Figure 12.2
Precipitation, using colourless aqueous lead(II) nitrate and potassium iodide (a) can be used to produce (b) lead(II) iodide, a bright yellow solid. ▶

You can predict if a precipitate will form or not by studying the solubility characteristics of compounds (see Section 11.4). Examples of ionic precipitation reactions and their ionic equations are shown below.

$$AgNO_3(aq) + NaCl(aq) \rightarrow AgCl(s) + NaNO_3(aq)$$
$$Ag^+(aq) + Cl^-(aq) \rightarrow AgCl(s)$$

| colourless solution | colourless solution | white precipitate in colourless solution |

$$BaCl_2(aq) + H_2SO_4(aq) \rightarrow BaSO_4(s) + 2HCl(aq)$$
$$Ba^{2+}(aq) + SO_4^{2-}(aq) \rightarrow BaSO_4(s)$$

colourless solution   colourless solution   white precipitate in colourless solution

$$FeSO_4(aq) + 2NaOH(aq) \rightarrow Fe(OH)_2(s) + Na_2SO_4(aq)$$
$$Fe^{2+}(aq) + 2OH^-(aq) \rightarrow Fe(OH)_2(s)$$

green solution   colourless solution   green precipitate

12.5 Neutralisation reactions

All **neutralisation reactions** involve reaction between an acid (or acidic oxide) and a base. Water is produced in neutralisation reactions; heat is also given out, i.e. neutralisation reactions are exothermic.

A reaction between an acid and a base producing water as one of the products is called a **neutralisation reaction**.

You can investigate neutralisation reactions in the laboratory by following the change in pH or in temperature as the acid is added to the alkali.

ORR
A&I
M&M

Experiment 12.1 Thermometric titration: Investigating neutralisation using temperature changes

The solutions used are usually concentrated.

⚠ Use a pipette filler to fill the pipette with alkali.
Place a funnel in the neck of the burette when filling it with the acid.

Procedure

1 Fill the burette with the acid solution.

2 Pipette 25 cm^3 of the alkali solution into a plastic or polystyrene cup.

3 Use a thermometer to measure the temperature of the alkali in the cup.

4 Record this temperature.

5 Do not remove the thermometer from the liquid in the cup throughout the experiment.

6 Add 5 cm^3 of acid from the burette to the solution in the cup. Stir the mixture and then record the highest temperature reached.

7 Without delay, add a further 5 cm^3 of acid to the solution in the cup. Stir and again record the highest temperature reached.

8 Continue this procedure, until you record three successive drops in temperature.

Results – table

Fill your data into a two-column table, as shown in Table 12.1.

▼ **Table 12.1** Experimental results

Volume of acid added (cm³)	Temperature (°C)
0	
5	
10	
15	
20	
25	
30	
35	

Note: The first temperature reading represents the temperature of the alkali before the reaction of the alkali and acid.

Results – graph

1 Plot a graph of temperature on the vertical axis versus volume of acid added on the horizontal axis.

2 Draw two best-fit lines, one through the section where the temperature is increasing, the other through the section where the temperature is falling.

3 Use the point of intersection of these two lines to work out the volume of acid needed for the neutralisation reaction.

Limitations

This method of following neutralisation reactions is less accurate than using indicators as in volumetric analysis, which we explained in detail in Chaper 9. Can you suggest two reasons why this is so?

You may want to consider the following:
* The volume of acid added in each step of the titration
* How accurate your readings of the temperature were
* Any drop in temperature between successive additions of acid

Calculations

You are able to work out the volume of acid needed to neutralise the volume of alkali used in this titration. You need to be told the concentration of one of the solutions. Now you can carry out any of the calculations associated with volumetric analyses (see Sections 9.6 and 10.1).

12.6 Reversible reactions

All the reactions described so far proceed in one direction only, that is, from reactants (on the left) to products (on the right).

> **Reversible reactions** can proceed in both directions, that is, from reactants to products and from products to reactants.

See Section 7.2 for more information about reversible reactions.

Figure 12.3
Yara Trindad Ltd operates a three-plant ammonia production facility in central Trindad. ▶

An important reversible reaction that you will study in more detail later is the reaction between nitrogen and hydrogen, in a closed container, to produce ammonia. The equation for the reaction is:

$$N_2(g) + 3H_2(g) \rightleftharpoons 2NH_3(g)$$

At first the reaction proceeds in the forward direction only (that is, from left to right). As the quantity of product increases, some of the ammonia dissociates and a backward reaction begins as well. The reaction proceeds in both directions until the rate of the forward reaction equals the rate of the backward reaction. At this point, the reaction is said to be in dynamic equilibrium. The reaction vessel will contain both products and reactants. As you will see when you study the industrial production of ammonia, to get a good yield of a particular product in a reversible reaction, the conditions must be carefully controlled.

12.7   Redox reactions or oxidation–reduction reactions

Redox reactions are chemical reactions in which oxidation and reduction take place simultaneously. **Oxidation** and **reduction** are reactions that involve the loss and gain of electrons. Sometimes there is total transfer of electrons (from metal to non-metal) but in some cases (between non-metals) there is only partial transfer. Chapter 4 provides more information.

Redox reactions involve oxidation and reduction:
- Oxidation involves loss of electrons (OIL = oxidation is loss).
- Reduction involves gain of electrons (RIG = reduction is gain).

Total transfer of electrons

We saw earlier in this textbook that metal atoms lose their valence electrons in forming compounds. Looking at the definitions above, we can see that electron loss by an atom or ion is oxidation.

- When a sodium atom loses its valence electron, we say the sodium atom is oxidised to a sodium ion:
    $$Na \rightarrow Na^+ + e^-$$
- Similarly, when an iron(II) ion (Fe^{2+}) loses an electron, it is oxidised to an iron(III) ion (Fe^{3+}):
    $$Fe^{2+} \rightarrow Fe^{3+} + e^-$$
- Non-metal ions can also be oxidised to atoms:
    $$2Cl^- \rightarrow Cl_2 + 2e^-$$
    Each chloride ion loses an electron forming a chlorine atom. Then the chlorine atoms pair up to form chlorine molecules.

Non-metals gain electrons when combining with metals. Again, using the definitions above, electron gain by an atom or ion is reduction.

- When a chlorine atom gains an electron, we say the chlorine atom is reduced to the chloride ion:
    $$Cl_2 + 2e^- \rightarrow 2Cl^-$$
    Each chlorine atom in the molecule gains an electron, forming a chloride ion.

- Metal ions can also be reduced. Copper(II) ions can be reduced to copper atoms by gaining electrons:

$$Cu^{2+} + 2e^- \rightarrow Cu$$

- Similarly, Fe^{3+} ions can be reduced to Fe^{2+} ions:

$$Fe^{3+} + e^- \rightarrow Fe^{2+}$$

You will realise that for electron loss to occur, there must be an atom or ion to receive the electrons. Similarly, for electron gain to occur, there must be an electron donor. Therefore, reduction and oxidation occur simultaneously; this is where the name 'redox' comes from.

In the reaction between sodium metal and chlorine gas to form the ionic compound sodium chloride, sodium atoms lose electrons and chlorine atoms gain electrons:

$$2Na(s) + Cl_2(g) \rightarrow 2Na^+Cl^-(s)$$

The sodium atoms are oxidised to sodium ions, and, at the same time, the chlorine atoms are reduced to chloride ions.

Partial transfer of electrons

You will remember that when two different non-metal atoms share electrons, the molecules may be polar (see Section 4.5). The electrons are pulled closer to the more electronegative atom, which is thus reduced. The less electronegative atom in the compound partially loses the bonding electrons to the more electronegative atom and is oxidised.

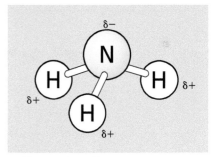

▲ **Figure 12.4** The more electronegative nitrogen atom is reduced and the less electronegative hydrogen atom is oxidised.

Oxidation state or **oxidation number** is a number assigned to an ion in an ionic compound or an atom in a molecule.

- For ions, oxidation state is the number of electrons lost or gained when the ion is formed from its element.
- For atoms in covalent compounds, oxidation state is the number of electrons that an atom partially gains or loses in a molecule.

The oxidation number is positive if electrons are totally or partially lost and is negative if electrons are totally or partially gained. The oxidation number of uncombined elements is zero.

Oxidation states or oxidation numbers

Another way by which you can recognise oxidation or reduction is by using **oxidation numbers** (also known as 'oxidation states'). You can read more about oxidation numbers in Section 7.1.

Determining the oxidation states of elements

When considering oxidation state, you need to note the following:

- Oxidation state does not represent an actual charge on the atom and should not be confused, or used interchangeably, with formal ionic charges.
- In describing oxidation states, the sign always precedes the number, e.g. the oxidation state of oxygen in the oxide ion is -2, while the formal charge is written as $2-$ and the oxidation state of Fe in Fe^{3+} is $+3$ while its charge is $3+$.

The chemical notation indicating the charge and oxidation number is as follows:

$$X^c \rightarrow$$

^{c} is the charge or oxidation number
e.g. Fe^{3+} (charge) or Fe^{+3} (oxidation number)

See Section 2.4 for more information.

Practice

4 Write the following in the appropriate chemical notation:
  a The element sodium of mass number 23 and atomic number 11
  b The sodium ion with a charge of 1+
  c The sodium ion with an oxidation number of +1
  d A molecule of chlorine
  e Two molecules of sulphur made up of eight atoms each
  f An Fe(II) ion

Rules for determining oxidation numbers

▼ **Table 12.2** Rules for determining the oxidation numbers of elements.

Atom or ion	Oxidation number	Example
elements in their elemental or free state	0	Al in Al (the uncombined metal) N in N_2 (nitrogen gas) O in O_2 (oxygen gas)
monatomic ions	equal to the charge on the ion	Mg^{2+} is +2 Al^{3+} is +3 Br^- is −1
atoms in polyatomic ions	the sum of the oxidation numbers of all the atoms equals the charge on the ion	NO_3^- is −1 (N = +5, O = −2) SO_4^{2-} is −2 (S = +6, O = −2) NH_4^+ is +1 (N = −3, H = +1)
Group I metals in their compounds	+1	Na in NaCl is +1 K in K_2SO_4 is +1
Group II metals in their compounds	+2	Mg in $MgCl_2$ is +2 Ca in CaO is +2
Group VII elements in their compounds	−1 in halides, sometimes positive for Cl, Br and I	F in NaF is −1 Cl in $CaCl_2$ is −1 Cl in ClO is +1
hydrogen in its compounds	+1 (usually) −1 in ionic hydrides	H in H_2O is +1 H in NH_3 is +1 H in NaH is −1
oxygen in its compounds	−2 (usually) except in peroxides and superoxides	O in Na_2O is −2 O in CO_2 is −2 O in H_2O_2 is −1
compounds between non-metals	+ for the less electronegative part − for the more electronegative part	H in HCl is +1 Cl in HCl is −1
sum of all the oxidation numbers in a neutral compound	0	for NaCl Na is +1, Cl is −1, sum = 0

Using the rules outlined in Table 12.2, it is relatively simple to determine the oxidation state of an element in a compound.

Worked example 1

Find the oxidation state of (a) sulphur in H_2SO_4 and (b) nitrogen in $Mg(NO_3)_2$.

Solution

a The oxidation states of H and O are $+1$ and -2, respectively. The sum of the oxidation states of the elements in $H_2SO_4 = 0$.

$2 \times$ oxidation state of H $+$ oxidation state of S $+$ $4 \times$ oxidation state of O $= 0$
$(2 \times +1)$ $+$ (oxidation state of S) $+$ (4×-2) $= 0$

The oxidation state of S in $H_2SO_4 = +6$.

b The oxidation states of Mg and O are $+2$ and -2, respectively. The sum of the oxidation states of the elements in $Mg(NO_3)_2 = 0$.

oxidation state of Mg $+$ $2 \times$ oxidation state of N $+$ $6 \times$ oxidation state of O $= 0$
$(+2)$ $+$ $(2 \times$ oxidation state of N$)$ $+$ (6×-2) $= 0$

The oxidation state of N in $Mg(NO_3)_2 = +5$.

Practice

5 Find the oxidation state of the particular elements in the following:
  a Cr in CrO_4^{2-}
  b Mn in Mn_2O_7
  c Cl in ClO_3^{4-}
  d I in $Ca(IO_3)_2$
  e S in Na_2SO_3
  f V in VO_2^+
  g N in NH_4^+
  h N in N_2H_4
  i C in $C_2O_4^{2-}$
  j P in H_3PO_3

Oxidation numbers change during oxidation and reduction reactions.

Some examples of oxidation

1 The formation of Fe^{2+} ions from uncombined iron:
- The oxidation number of uncombined iron is 0.

$$Fe(s) \rightarrow Fe^{2+} + 2e$$

- In the formation of the Fe^{2+} ion from the Fe atom, two electrons are lost (oxidation).
- The oxidation state of iron in the Fe^{2+} ion is $+2$.
- The oxidation number increases in oxidation.

2 The conversion of sodium to sodium ions:

When sodium is oxidised by losing electrons (see above):

$$Na(g) \rightarrow Na^+(g) + e^-$$

its oxidation number increases from 0 in the uncombined atom to $+1$ in the sodium ion.

Some examples of reduction

1 The formation of S^{2-} ions from uncombined sulphur:

- The oxidation number of uncombined sulphur is 0.

$$S + 2e^- \rightarrow S^{2-}$$

- In the formation of the S^{2-} sulphide ion from the sulphur atom, two electrons are gained (reduction).
- The oxidation state of sulphur in the S^{2-} ion is -2.
- The oxidation number decreases in reduction.

2 The conversion of chlorine to chloride ions:

When chlorine is reduced by gaining electrons (see above):

$$Cl_2(g) + e^- \rightarrow 2Cl^-(s)$$

its oxidation number decreases from 0 in the element to -1 in the chloride ion.

In summary

Reduction:
- Gain of electrons
- decrease in oxidation number

Oxidation:
- Loss of electrons
- Increase in oxidation number

Naming compounds using oxidation numbers

As described in Section 7.1, oxidation numbers are used to distinguish between compounds containing the same elements but in different whole number ratios.

For example, sodium, chlorine and oxygen form four compounds that are all called sodium chlorate. In each of these compounds, the chlorine has a different oxidation number. The oxidation number is written in brackets after the chemical formula to distinguish them from each other. See Table 12.3.

▼ **Table 12.3** Naming compounds using oxidation numbers

	$NaClO$	$NaClO_2$	$NaClO_3$	$NaClO_4$
Oxidation state of chlorine	$+1$	$+3$	$+5$	$+7$
Name	sodium chlorate(I)	sodium chlorate(III)	sodium chlorate(V)	sodium chlorate(VII)

Transition metals characteristically exhibit variable oxidation numbers (Table 12.4).

▼ **Table 12.4** Examples of transition metal compounds showing variable oxidation states. Again, the number in brackets gives the oxidation number.

Compound	Oxidation state of the transition metals	Name	Colour
MnO_2	+4	manganese(IV) oxide	brown
$KMnO_4$	+7	potassium manganate(VII)	purple
K_2CrO_4	+6	potassium chromate(VI)	orange
$K_2Cr_2O_7$	+6	potassium dichromate(VI)	orange
CuO	+2	copper(II) oxide	blue
Cu_2O	+1	copper(I) oxide	green
$K_4[Fe(CN)_6]$	+2	potassium hexacyanoferrate(II)	pale green
$K_3[Fe(CN)_6]$	+3	potassium hexacyanoferrate(III)	yellow-brown

Practice

6 Write balanced ionic half-equations for the following changes. Then determine whether oxidation or reduction occurs by using changes in oxidation states:

  a Chlorine to chloride ions

  b Oxygen to oxide ions
  c Aluminium ions to aluminium
  d Magnesium to magnesium ions
  e Tin(IV) ions to tin(II) ions
  f Chromium(II) ions to chromium(III) ions
  g Calcium to calcium ions
  h Iodide ions to iodine

Oxidising and reducing agents

Oxidising agents bring about oxidation. They are easily reduced.

Reducing agents bring about reduction. They are easily oxidised.

- An **oxidising agent** brings about the oxidation of another substance. In the process it is reduced (it accepts electrons).

- A **reducing agent** brings about the reduction of another substance. In the process it is oxidised (it donates electrons).

Consider the redox reaction between zinc metal and aqueous copper(II) sulphate. The equation for the reaction is:

$$Zn(s) + CuSO_4(aq) \rightarrow ZnSO_4(aq) + Cu(s)$$

The ionic equation is:

$$Zn(s) + Cu^{2+}(aq) \rightarrow Zn^{2+}(aq) + Cu(s)$$

Let's now look at the ionic half-equations:

$$Zn(s) - 2e^- \rightarrow Zn^{2+}(aq)$$

The zinc atom loses electrons; it is oxidised to a zinc ion:

$$Cu^{2+}(aq) + 2e^- \rightarrow Cu(s)$$

The copper ion gains electrons; it is reduced to a copper atom:

- The zinc atom brings about the reduction of the copper ion. Zinc is the reducing agent.
- The copper ion brings about the oxidation of the zinc atom. Copper sulphate is the oxidising agent.

Now consider the redox reaction between chlorine and iron(II), which can be summarised as:

$$Cl_2(g) + 2Fe^{2+}(aq) \rightarrow 2Cl^-(aq) + 2Fe^{3+}(aq)$$

The Fe^{2+} ion loses an electron and is oxidised to an Fe^{3+} ion:

$$2Fe^{2+}(aq) - 2e^- \rightarrow 2Fe^{3+}(aq)$$

The Fe^{2+} ion reduces the chlorine atoms in molecular chlorine to chloride ions:

$$Cl_2(g) + 2e^- \rightarrow 2Cl^-(aq)$$

Each Cl atom in $Cl_2(g)$ gains an electron and is reduced to $Cl^-(aq)$. The chlorine oxidises the Fe^{2+} ions to Fe^{3+} ions.

In summary

Oxidising agents oxidise other substances, and are themselves reduced. They gain electrons and their oxidation number decreases.

Reducing agents reduce other substances, and are themselves oxidised. They lose electrons and their oxidation number increases.

Some common oxidising and reducing agents

▼ **Table 12.5** Some common oxidising agents

Oxidising agent	Colour change or other observable sign	Products
nitric acid (conc.)	brown gas evolved	$NO_2(g)$, water, nitrate(aq)
hot concentrated sulphuric acid	gas produced; has characteristic smell	$SO_2(g)$, sulphate(aq), formed with metals or their compounds
potassium manganate(VII)/ dilute sulphuric acid	colour changes from purple to colourless	$Mn^{2+}(aq)$
potassium dichromate(VI)/ acid	colour changes from orange to green	$Cr^{3+}(aq)$
iron(III) salts: $Fe^{3+}(aq)$	yellow to pale green	$Fe^{2+}(aq)$
hydrogen peroxide	effervescence, colourless oxygen gas evolved	O_2
$I_2(aq)$	from brown to colourless	$I^-(aq)$

Note: Sodium thiosulphate is used in reactions involving iodine. The iodine changes from brown/ yellow to colourless iodide ions.

▼ **Table 12.6**   Some common reducing agents

Reducing agent	Colour change or other observable sign	Products
hydrogen sulphide	yellow colloidal suspension formed	sulphur and water
sulphur dioxide	no significant observable change	H_2SO_4 or SO_4^{2-}(aq)
sulphite	no observable change	H_2SO_4 or SO_4^{2-}(aq)
conc. HCl	yellow-green gas	chlorine
KI/H$^+$(aq)	brown solution or black precipitate of I_2	iodine, water
Fe^{2+}(aq)	turns yellow or brown	Fe^{3+}(aq)
hydrogen peroxide	effervescence, colourless hydrogen gas evolved	H_2(g)

A few substances can function both as oxidising and reducing agents. Examples include the following:

- Hydrogen peroxide (H_2O_2) oxidises iron(II), i.e. Fe^{2+}(aq), to iron(III), i.e. Fe^{3+}(aq), but reduces acidified potassium manganate(VII) solution.
- Sodium nitrate ($NaNO_3$) oxidises the iodide ion to iodine, but reduces purple potassium manganate(VII) to the colourless manganese(II) ions.
- Sulphur dioxide (SO_2) oxidises hydrogen sulphide to sulphur, but reduces many other substances.

Tests for oxidising and reducing agents

In some experiments, it may help to identify a substance by testing to see whether it is an oxidising or a reducing agent. The following tests are chosen since they are easy to carry out and give an observable change.

Tests for oxidising agents

▼ **Table 12.7**   Tests for oxidising agents

Reactant	Observable change	Equation
KI(aq)	solution changes from colourless to brown (iodine)	$2I^-(aq) - 2e^- \rightarrow I_2(aq)$
Fe^{2+}(aq)	colour change from pale green to yellow-brown	$Fe^{2+}(aq) - e^- \rightarrow Fe^{3+}(aq)$
Na$_2$S or H$_2$S	colloidal suspension of sulphur	$S^{2-}(aq) - 2e^- \rightarrow S$

Tests for reducing agents

A reducing agent should decolorise acidified potassium manganage(VII) ion:

$$MnO_4^-(aq) \quad + \quad 5e^- \quad \rightarrow \quad Mn^{2+}(aq)$$

purple   from   colourless

reducing

agent

A reducing agent should also change acidified potassium dichromate(VI) (orange) to the green chromium(III) ion:

$$Cr_2O_7^{2-}(aq) \quad + \quad 3e^- \quad \rightarrow \quad 2Cr^{3+}(aq)$$

orange   from   green

reducing

agent

Solutions should be acidified, usually with dilute sulphuric acid, if oxidising and reducing agents are to function efficiently.

Reactions between oxidising and reducing agents

The reactions between some common oxidising and reducing agents are shown in Table 12.8. Note the behaviour of H_2O_2.

▼ **Table 12.8**   Reactions between some common oxidising and reducing agents

	$SO_2(g)/$ $SO_3^{2-}(aq)$	conc. HCl	$KI/H^+(aq)$	Fe^{2+}	H_2O_2
conc. H_2SO_4		complex reaction; chlorine gas evolved; SO_4^{2-} ions converted to $SO_2(g)$, $S(s),H_2S(g)$	iodine liberated; $SO_2(g)$ evolved	$Fe^{3+}(aq)$ formed	effervesence; $SO_2(g)$ evolved
$KMnO_4/H^+$	H_2SO_4 and Mn^{2+} formed	chlorine evolved; purple $MnO_4^-(aq)$ converted to colourless Mn^{2+}	iodine liberated; $Mn^{2+}(aq)$ formed*	$Fe^{3+}(aq)$ formed	$MnO_4^-(aq)$ decolorised
$K_2Cr_2O_7/H^+$	H_2SO_4 and $Cr^{3+}(aq)$ formed; colour change from orange to green	chlorine evolved; orange $Cr_2O_7^{2-}$ converted to green Cr^{3+}	iodine liberated; $Cr^{3+}(aq)$ formed*	$Fe^{3+}(aq)$ formed	$Cr_2O_7^{2-}$ converted to Cr^{3+}

continued

▼ **Table 12.8** *continued*

H_2O_2(aq)	H_2SO_4 and H_2O formed	iodine liberated		Fe^{3+}(aq) formed
Fe^{3+}	green Fe^{2+}(aq) formed; also H_2SO_4	iodine liberated; Fe^{2+}(aq) formed		

* I_2 formed may mask the expected colour change.

 Reducing agents;

Oxidising agents.

Summary

- Combination reactions are those reactions in which two or more substances react to produce one substance. This can be represented generally as:
    $$A + B \rightarrow AB$$
- In decomposition reactions, a compound undergoes a change to form two or more substances. Heat is often necessary for the process. This can be represented generally as:
    $$AB \rightarrow A + B$$
- In substitution or displacement reactions, one element or group in a compound is replaced by another element or group. This reaction can be represented generally:
    $$A + BX \rightarrow AX + B$$
- Ionic precipitation reactions can be represented generally as:
    $$AB + CD \rightarrow AD + BC$$
    In this type of reaction two compounds exchange radicals.
- Neutralisation involves a reaction between an acid (or acidic oxide) and a base (or basic oxide):
    - Water is produced in these reactions.
    - Neutralisation reactions are exothermic.
- Oxidation is a process involving a loss of electrons.
- Reduction is a process involving a gain of electrons.
- Oxidation is any process in which there is an increase in oxidation state.
- Reduction is any process in which there is a decrease in oxidation state.
- The oxidation state represents the number of electrons lost or gained, by total or partial electron transfer, when a compound is formed from its elements.
- The oxidation state of an element in its free state is taken as zero.

End-of-chapter questions

1 What type of chemical reaction occurs when solid lead nitrate is heated and gives off nitrogen(IV) oxide and oxygen?
  A Decomposition
  B Neutralisation
  C Redox
  D Ionic precipitation

2 A common oxidising agent is:
  A KI
  B $FeSO_4$
  C $KMnO_4/H^+$
  D cHCl

3 A common reducing agent is:
  A $K_2Cr_2O_7/H^+$
  B $FeSO_4$
  C cH_2SO_4
  D $cHNO_3$

4 A substitution (displacement) reaction is one in which:
  A A more reactive metal displaces a less reactive metal from an aqueous solution of its salt
  B Two or more elements react to form a single product
  C One substance undergoes a reaction to form two or more substances
  D Two soluble ionic compounds react to form one soluble and one insoluble compound

5 When a metal atom loses electrons it is:
  A Oxidised to a cation
  B Oxidised to an anion
  C Reduced to a cation
  D Reduced to an anion

6 What is the oxidation number of hydrogen in NaH?
  A +2
  B +1
  C −2
  D −1

7 Which oxidising agent changes from orange to green?
  A $K_2Cr_2O_7/H^+$
  B $FeSO_4$
  C cH_2SO_4
  D HNO_3

8 Which oxidising agent changes from purple to colourless?
  A $K_2Cr_2O_7/H^+$
  B cH_2SO_4
  C HNO_3
  D $KMnO_4/H^+$

9 Which of the following occurs as a result of oxidation?
  I   Stain removal
  II Browning of cut fruits
  III Rusting
  IV Food preservation

  A I
  B I +II
  C I, II +III
  D I, II, III + IV

10 Which one of the following is an antioxidant?
  A $NaNO_3$
  B SO_2
  C NaCl
  D CH_3COOH

11 Which of the following equations represent redox reactions?

  a $Ca(s) + Cl_2(g) \rightarrow CaCl_2(s)$

  b $Cl_2(g) + H_2S(g) \rightarrow 2HCl(g) + S(s)$

  c $Ba(NO_3)_2(aq) + Na_2SO_4(aq) \rightarrow BaSO_4(s) + 2NaNO_3(aq)$

  d $CH_4(g) + 2O_2(g) \rightarrow CO_2(g) + 2H_2O(l)$

  e $Ca(OH)_2(aq) + CO_2(g) \rightarrow CaCO_3(s) + H_2O(l)$

  f $Ni(s) + CuSO_4(aq) \rightarrow NiSO_4(aq) + Cu(s)$

  g $Cl_2O_7(g) + H_2O(l) \rightarrow 2HClO_4(aq)$

  h $2CuCl(aq) \rightarrow CuCl_2(aq) + Cu(s)$

12 For the redox reactions in question 1, identify the oxidising agent and the reducing agent and then write ionic half-equations for each, where appropriate.

13 Classify the following reactions into one of the following types: (i) neutralisation, (ii) precipitation, (iii) thermal decomposition, (iv) redox, (v) direct combination and (vi) hydrolysis. Each reaction type may be used once, more than once or not at all.

  a $NaBr(aq) + AgNO_3(aq) \rightarrow AgBr(s) + NaNO_3(aq)$

  b $CaO(s) + 2HCl(aq) \rightarrow CaCl_2(aq) + H_2O(l)$

  c $2Fe^{3+}(aq) + 2I^-(aq) \rightarrow 2Fe^{2+}(aq) + I_2(aq)$

  d $N_2(g) + 3H_2(g) \rightleftharpoons 2NH_3(g)$

  e $2Pb(NO_3)_2(s) \rightarrow 2PbO(s) + 4NO_2(g) + O_2(g)$

  f $2H_2O_2(l) + 2I^-(aq) \rightarrow 2H_2O(l) + I_2(aq) + O_2(g)$

  g $HCl(g) + NH_3(g) \rightleftharpoons NH_4Cl(s)$

In this chapter, you will study the following:

- the difference between conductors and non-conductors using experiments;
- the difference between metallic and electrolytic conduction;
- classification of electrolytes;
- the electrolysis process including reactions at the electrodes;
- electrolysis of select substances;
- Faraday constant and related calculations;
- electrolysis in industry.

This chapter covers
Objectives 9.1–9.11 and part of 8.4 of Section A of the CSEC Chemistry Syllabus.

Electrical conductivity can be used to divide matter into three distinct groups:
- Electrical conductors – substances that conduct an electric current
- Non-conductors – substances that do not conduct an electric current (non-conductors are usually non-metals and non-polar covalent compounds)
- Semi-conductors – substances whose conducting properties are intermediate between those of conductors and non-conductors; they are fairly good conductors only under certain conditions

▲ **Figure 13.1** Conductors and non-conductors are both important. Can you explain why?

13.1 Electrical conductors

Many materials conduct electricity, as you can find out in Experiment 13.1.

Electrical conductors allow a current to flow through them. **Insulators** do not allow the passage of an electric current.

M&M
P&D
A&I

Experiment 13.1 Distinguishing conductors from non-conductors

Apparatus and materials

- Strips of different metals, such as magnesium, copper, zinc
- Solid non-metals, such as sulphur, iodine, a graphite rod
- Solid ionic compounds, such as sodium chloride and lead(ii) bromide
- Aqueous solutions of ionic compounds, such as sodium chloride solution
- Covalent compounds, such as ethanol
- Equipment so you can make the experimental set-ups shown in these photographs

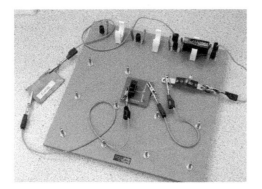
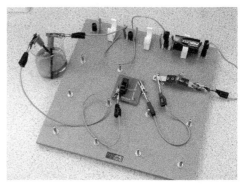

▲ **Figure 13.2** The apparatus needed to test (a) solids and (b) liquids

Procedure

Test each material in turn, to see if the bulb lights up.

⚠ Safety: In a fume cupboard, melt the lead bromide using a Bunsen burner and test the molten bromide.

Questions

Q1 Which of the materials caused the bulb to light up when included in the circuit?

Q2 How can you subdivide the materials that are conductors into smaller groups?

Q3 Compare the conductivity of (a) solid sodium chloride and sodium chloride solution and (b) solid lead(II) bromide and molten lead(II) bromide.

Electrical conductors can be divided into two groups:
- Metals and graphite
- **Electrolytes**

Electrolytes are compounds that conduct an electric current and are decomposed by it.

The differences between these two groups of electrical conductors are summed up in Table 13.1.

▼ **Table 13.1** Differences between electrical conductors

Metallic conductors (and graphite)	Electrolytes
conduct in the solid and liquid states	do not conduct in the solid state; conduct in the molten state or in aqueous solutions
no chemical changes occur when they conduct electricity	chemical changes occur when they conduct an electric current

We now need to consider what accounts for these differences.

▲ **Figure 13.3** In which everyday objects do we find graphite?

Conduction in metals and graphite

Conductivity in metals (and graphite) is due to the presence of mobile electrons (see Sections 4.6 and 4.13). Mobile electrons are present in solid metals and graphite. The metals remain chemically unchanged by the passage of the current.

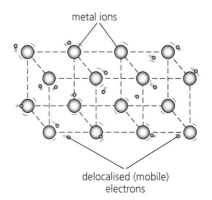

Figure 13.4
Delocalised electrons move through the metal and carry the negative charge. ▶

metal ions

delocalised (mobile) electrons

Conduction in electrolytes

Conductivity in electrolytes is due to the presence of mobile ions (see Section 4.8). The positive and negative ions are separated, and it is this separation that results in the decomposition of the electrolyte.

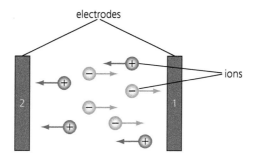

Figure 13.5
Ions can move in the molten state or in solution. ▶

Ionic compounds do not conduct electricity in the solid state, because the ions are rigidly and tightly held in a crystal lattice. In the solid state, the ions are not free to move. In the molten state or in solution, the lattice breaks down and the ions are free to move.

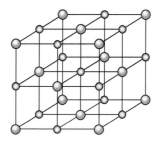

Figure 13.6
Ions are not free to move in a solid. ▶

Electrolytes are molten or aqueous ionic compounds and also some polar covalent compounds that react with water to produce ions.

Classification of electrolytes

Electrolytes may be classified as strong or weak, based on how completely they ionise (see also Chapter 11 about the classification of acids).

Strong electrolytes are completely ionised in aqueous solution. Examples are ionic compounds and strong acids, such as sulphuric acid and hydrochloric acid.

Weak electrolytes are partially ionised in aqueous solution. Examples are weak acids, such as ethanoic acid and methanoic acid, and weak alkalis, such as aqueous ammonia.

$$CH_3COOH(aq) \rightleftharpoons CH_3COO^-(aq) + H^+(aq)$$
$$NH_3(g) + H_2O(l) \rightleftharpoons NH_4^+(aq) + OH^-(aq)$$

Practice

1 Which of the following do you expect to be electrolytes?
  a Graphite
  b Solid copper(II) sulphate
  c Diamond
  d Brass
  e An aqueous solution of calcium chloride
  f Molten candle wax
  g Vinegar

2 a Draw diagrams to show the structure of graphite and copper metal.
  b Compare the electrical conduction of graphite with that of copper.

13.2 Electrolysis

When electrolytes conduct electricity, they are decomposed by the current. This is
known as **electrolysis**. Electrolysis is carried out in an apparatus known as an electrolytic
cell (Figure 13.7).

> **Electrolysis** is the decomposition of an electrolyte by the passage of an electric current
> through it.

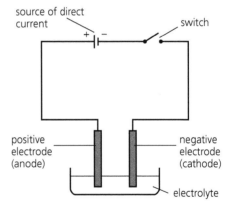

Figure 13.7
An electrolytic cell.
Conduction in the
external circuit is by
electrons. Conduction
in the electrolyte is
by ions. ▶

Electrodes are the
points where current
enters and leaves an
electrolyte:
- The **anode** is the
  positive electrode.
- The **cathode** is the
  negative electrode.

Electrodes are conducting rods, often made of graphite or platinum. Generally,
electrodes are inert and usually do not take part in the chemical changes occurring during
electrolysis. Some electrodes are active and these do take part in these chemical changes.
For example, platinum reacts with chlorine gas and graphite reacts with oxygen gas.

- The positive electrode, or **anode**, is connected to the positive terminal of the power
  supply or battery.

- The negative electrode, or **cathode**, is connected to the negative terminal of the
  power supply or battery.

Ions are positive or negative:
- Anions are negative ions (–).
- Cations are positive ions (+).

During electrolysis:
- anions (negative ions) move towards the anode;
- cations (positive ions) move towards the cathode.

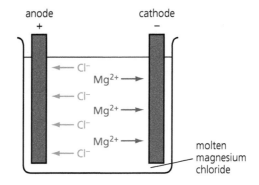

Figure 13.8
The movement of ions in the electrolysis of molten magnesium chloride. ▶

Reactions at the electrodes

Discharge is the process by which ions gain or lose electrons and become atoms or molecules. Ions lose their charge during discharge.

During electrolysis there is **discharge** of ions. Table 13.2 compares discharge at the cathode and at the anode.

▼ **Table 13.2**   Discharge at the cathode and at the anode.

	Cathode	Anode
Movement of ions	positive ions (cations) move towards the cathode	negative ions (anions) move towards the anode
Electrons	cations gain electrons from the cathode	anions lose (give up) electrons to the anode
Result	cations become neutral atoms or molecules	anions become neutral atoms or molecules
Half-equation	$M^{n+} + ne^- \rightarrow M$	$X^{n-} - ne^- \rightarrow X$
Reduction or oxidation?	this is reduction – gain of electrons	this is oxidation – loss of electrons
Example: Electrolysis of fused potassium iodide		
	$2K^+(l) + 2e^- \rightarrow 2K(s)$	$2I^-(l) - 2e^- \rightarrow I_2(g)$
	potassium ion → neutral atoms	iodide ion → neutral molecules

Preferential discharge of ions

When a molten salt is electrolysed, no competition for discharge occurs at the electrodes as a single type of cation and a single type of anion are present in the melt. However, when an aqueous solution is electrolysed, hydrogen ions and hydroxide ions from the partial ionisation of water are also present, in addition to the ions from the electrolyte.

$$H_2O(l) \rightleftharpoons H^+(aq) + OH^-(aq)$$

When the current is passed through this aqueous solution, more than one type of ion moves towards each electrode. When this happens, one type of ion is discharged in preference to the other. For example, when a current is passed through aqueous sodium sulphate, hydrogen ions and sodium ions move to the cathode, while hydroxide ions and sulphate ions move to the anode. The ion that is preferentially discharged is determined mainly by its position in the electrochemical or reactivity series.

The electrochemical series

The electrochemical series is a list of anions and cations arranged in order of decreasing activity. The most reactive ion is at the top and the least reactive ion is at the bottom. When there is more than one ion at an electrode, the one that is lower in the electrochemical series is preferentially discharged. That is, the cation that is less reactive will gain an electron or electrons at the cathode and become an atom. Similarly at the anode, the less reactive anion will give up its electron(s).

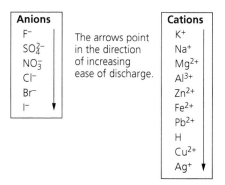

▲ **Figure 13.9**   The order of preferential discharge of ions from an electrolyte

There are, however, exceptions to these general rules, particularly when the electrolyte solution is concentrated. The following examples will illustrate these principles.

Electrolysis of pure water using inert electrodes

Pure water is not a good conductor of electricity because it ionises only to a very small extent.

$$H_2O \rightleftharpoons H^+(aq) + OH^-(aq)$$

With pure water in the apparatus shown in Figure 13.10, no current will flow.

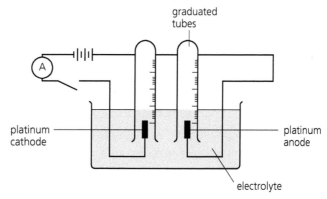

▲ **Figure 13.10**   Apparatus for the electrolysis of aqueous solutions

Electrolysis of dilute sulphuric acid using inert (platinum) electrodes

In this example, the cell is filled with dilute sulphuric acid. Gases are liberated at each electrode, the volume of the gas at the cathode (the negative electrode) being twice that liberated at the anode (the positive electrode), provided the apparatus (Figure 13.11) has been run before (and the gases let out) to saturate the acid with dissolved gases.

The gas at the cathode gives a pop with a lighted splint, indicating that it is hydrogen.

The gas at the anode relights a glowing splint; it is therefore oxygen.

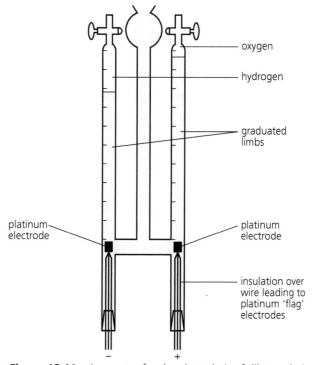

▲ **Figure 13.11**   Apparatus for the electrolysis of dilute solutions

The electrolyte becomes more concentrated as electrolysis proceeds.
Can you suggest why?

The ions in solution are:

$H^+(aq)$	$OH^-(aq)$	from water
$H^+(aq)$	$SO_4^{2-}(aq)$	from sulphuric acid
cations	anions	

$H^+(aq)$ ions move towards the cathode during electrolysis, where they are discharged according to the reactions:

$$H^+(aq) + e^- \rightarrow H^0(g)$$
$$2H^0(g) \rightarrow H_2(g)$$

$OH^-(aq)$ and $SO_4^{2-}(aq)$ both move towards the anode, but the $OH^-(aq)$ ions are preferentially discharged according to the equation:

$$4OH^-(aq) - 4e^- \rightarrow 2H_2O(l) + O_2(g)$$

1 mol of hydrogen molecules is produced for every 2 mol of electrons transferred, whereas 4 mol of electrons are transferred for every mol of oxygen produced. Therefore, 2 volumes of hydrogen are produced for each volume of oxygen produced during the electrolysis of dilute sulphuric acid (see Section 9.1).

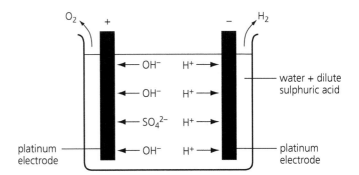

Figure 13.12
Electrolysis of dilute
sulphuric acid using
inert (platinum)
electrodes ▶

Electrolysis of dilute sodium hydroxide using inert electrodes

In this example, the ions in solution are:

$H^+(aq)$	$OH^-(aq)$	from water
$Na^+(aq)$	$OH^-(aq)$	from sodium hydroxide
cations	anions	

$OH^-(aq)$ ions move towards the anode, where they are discharged according to the equation:

$$4OH^-(aq) - 4e^- \rightarrow 2H_2O(l) + O_2(g)$$

$H^+(aq)$ and $Na^+(aq)$ move towards the cathode, where H^+ is preferentially discharged.

The electrolysis of dilute sulphuric acid and of dilute sodium hydroxide gives the same products at the electrodes. In both cases, the electrolyte becomes more concentrated as the reaction proceeds.

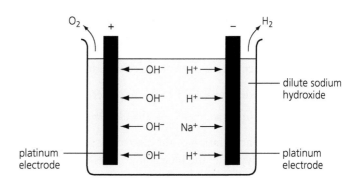

Figure 13.13
Electrolysis of dilute
sodium hydroxide using
inert electrodes. ▶

Electrolysis of aqueous sodium chloride using inert electrodes

The ions in solution are:

$H^+(aq)$	$OH^-(aq)$	from water
$Na^+(aq)$	$Cl^-(aq)$	from sodium chloride
cations	anions	

$H^+(aq)$ ions and $Na^+(aq)$ ions migrate towards the cathode, where hydrogen is preferentially discharged according to the equation:

$$2H^+(aq) + 2e^- \rightarrow H_2(g)$$

What happens at the anode is determined by the concentration of the sodium chloride solution. If the solution is dilute, hydroxide ions are discharged, resulting in the evolution of oxygen at the anode:

$$4OH^-(aq) - 4e^- \rightarrow 2H_2O(l) + O_2(g)$$

Very little chlorine is evolved under these conditions.

If the solution is concentrated, chloride ions are preferentially discharged at the anode, while very little oxygen is obtained:

$$2Cl^-(aq) - 2e^- \rightarrow Cl_2(g)$$

The remaining electrolyte is sodium hydroxide.

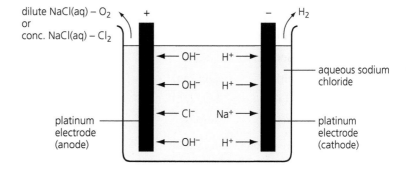

Figure 13.14
Electrolysis of sodium
chloride (dilute or
concentrated) using
inert electrodes ▶

Electrolysis of aqueous copper(II) sulphate using inert electrodes

The ions in solution are:

$H^+(aq)$	$OH^-(aq)$	from water
$Cu^{2+}(aq)$	$SO_4^{2-}(aq)$	from copper(II) sulphate
cations	anions	

Copper is discharged in preference to hydrogen at the cathode:

$$Cu^{2+}(aq) + 2e^- \rightarrow Cu^0(s)$$

As the copper is deposited at the cathode, the blue colour of the solution gradually fades.

Hydroxide ions, $OH^-(aq)$, are preferentially discharged at the anode:

$$4OH^-(aq) - 4e^- \rightarrow 2H_2O(l) + O_2(g)$$

The concentration of the electrolyte, with respect to copper, decreases during the electrolysis (i.e. the blue colour fades); the electrolyte becomes more acidic.

The reactions are the same if a copper cathode is used.

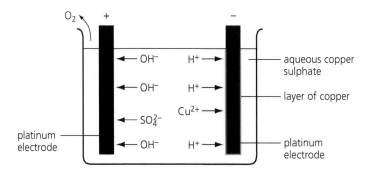

Figure 13.15
Electrolysis of dilute copper(II) sulphate using inert electrodes ▶

Practice

3 Which ions are present in a concentrated solution of hydrochloric acid? Predict the products at the electrodes when a concentrated solution of hydrogen chloride is electrolysed.

4 Arrange $Cu^{2+}(aq)$, $H^+(aq)$ and $Na^+(aq)$ in an order that reflects the ease with which they are discharged.

5 Arrange $OH^-(aq)$, $Cl^-(aq)$ and $SO_4^{2-}(aq)$ in an order that reflects the ease with which they are discharged.

Electrolysis of aqueous copper(II) sulphate using a copper anode and copper cathode

The ions in solution are:

$H^+(aq)$	$OH^-(aq)$	from water
$Cu^{2+}(aq)$	$SO_4^{2-}(aq)$	from copper(II) sulphate
cations	anions	

The reaction at the cathode is:

$$Cu^{2+}(aq) + 2e^- \rightarrow Cu^0(s)$$

i.e. copper atoms are deposited at the cathode.

The reaction at the anode is:

$$Cu^0(s) - 2e^- \rightarrow Cu^{2+}(aq)$$

i.e. copper atoms leave the anode and enter the electrolyte as copper(II) ions, $Cu^{2+}(aq)$. This is the least energetic process that can happen at the anode. The mass of the anode decreases, and the mass of the cathode increases by an equivalent amount. The concentration of the electrolyte is unchanged, i.e. the blue colour stays the same. This is the principle of the method used in the refining of copper and in the electroplating of objects (see Section 13.4).

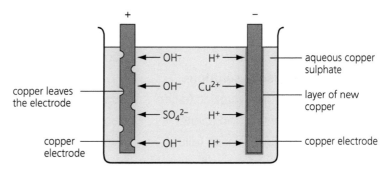

Figure 13.16
Electrolysis of dilute copper(II) sulphate using copper electrodes ▶

Electrolysis of molten lead(II) bromide using inert electrodes

The reaction at the cathode is:

$$Pb^{2+}(l) + 2e^- \rightarrow Pb^0(s)$$

The reaction at the anode is:

$$2Br^-(l) - 2e^- \rightarrow Br_2(g)$$

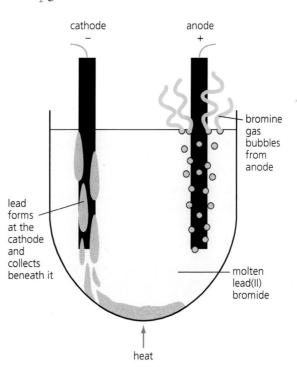

Figure 13.17
Molten lead bromide being electrolysed. Lead forms on the cathode and bromine is released at the anode ▶

Table 13.3 summarises the products of some common examples of electrolysis.

▼ Table 13.3

Electrolyte	Electrodes		Products	
	cathode	anode	cathode	anode
molten				
sodium chloride	carbon	carbon	sodium	chlorine
lead bromide	carbon	carbon	lead	bromine
aqueous solutions				
sodium hydroxide (dilute)	carbon	platinum	hydrogen	oxygen
sulphuric acid (dilute)	carbon	platinum	hydrogen	oxygen
sodium chloride (dilute)	carbon	platinum	hydrogen	oxygen
hydrochloric acid (concentrated)	carbon	carbon	hydrogen	chlorine
sodium chloride (concentrated)	carbon	carbon	hydrogen	chlorine
copper sulphate	carbon	platinum	copper	oxygen
copper sulphate dissolved	copper	copper	copper deposited	copper

It should be obvious by now that the ease with which ions are discharged from solutions or melts is usually opposite to the order of reactivity of the elements themselves (see Figure 13.6).

Practice

6 Write electrode reactions for the electrolysis of molten sodium chloride.

7 What is meant by each of the following?
 a Electrode
 b Electrolyte
 c Electrolysis
 d Preferential discharge

8 Predict what happens at the anode and cathode in the electrolysis of each of the following:
 a Molten potassium iodide
 b Aqueous dilute copper(II) chloride, using carbon electrodes
 c Aqueous silver nitrate, using silver electrodes

9 Part of the reactivity series is shown below:

 Mg Zn Pb Sn H Cu Ag

 decreasing activity

 a Which metals would replace zinc metal from a solution of a zinc salt?
 b Would you expect silver metal to displace copper metal from a solution of a copper salt? Give reasons for your answer.

13.3 The Faraday constant and related calculations

So far, we have considered qualitative aspects of electrolysis. We can use the mole concept (see Chapter 9) to calculate the mass of substances discharged at an electrode during electrolysis.

Much of the early work on electrolysis was done by a British scientist named Michael Faraday (1791–1867). In 1834, Faraday discovered that the mass of substance discharged at an electrode during electrolysis is directly proportional to the quantity of electricity passing through the electrolytic cell.

$$m \propto Q$$

where Q = quantity of electricity in coulombs and m = mass of substance

A coulomb is the amount of electricity that passes a given point in a wire when an electric current of one ampere flows for one second. The unit of electric current is the **ampere**.

1 **coulomb** = 1 **ampere** $\times$ 1 second

$$Q = I \times t$$

where Q = quantity of electricity in coulombs
I = current strength in amperes
t = time in seconds

One mole of electrons has a charge of **96 500 C** and contains the Avogadro number of electrons.

The Faraday constant, F, is the total electric charge carried by 1 mole of electrons, where F = 96 500 C mol^{-1}.

Experimentally, it has been found that one mole of electrons has a charge of 96 500 C. Therefore, when 1 mole of a singly charged ion is discharged at an electrode, the following happens:

$$M^+ + e^- \rightarrow M$$
or $\quad X^- \rightarrow X + e^-$

This process requires the passage of 96 500 C (or the addition or removal of 1 mole of electrons).

Similarly, in the discharge of 1 mole of a doubly charged ion at an electrode, this happens:

$$M^{2+} + 2e^- \rightarrow M$$
or $\quad X^{2-} \rightarrow X + 2e^-$

This process requires the passage of 2 $\times$ 96 500 C (or the addition or removal of 2 moles of electrons).

In general then, to discharge 1 mole of an ion, M^{n+} or X^{n-}, at an electrode, $n \times 96\,500$ C must be passed through the electrolyte. For example, the passage of 96 500 C of electricity through molten sodium chloride liberates 1 mole of sodium at the cathode:

$$Na^+ + e^- \rightarrow Na$$

and $\frac{1}{2}$ mole of chlorine molecules at the anode:

$$Cl^- \rightarrow \tfrac{1}{2}Cl_2 + e^-$$

The passage of $2 \times 96\,500$ C of electricity through molten lead(II) bromide simultaneously liberates 1 mole of lead and 1 mole of bromine molecules. The reactions occurring at the respective electrodes are:

$$Pb^{2+} + 2e^- \rightarrow Pb \text{ at the cathode}$$
$$2Br^- \rightarrow Br_2 + 2e^- \text{ at the anode}$$

Worked example 1

What mass of magnesium is deposited at the cathode by the passage of 2.00 amperes through molten magnesium chloride for 30 minutes?

Solution

quantity of electricity passed (in coulombs) $=$ current (in amperes) $\times$ time (in seconds)
$$= 2 \times 30 \times 60$$
$$Q = 3\,600 \text{ C}$$

From the equation for the discharge of magnesium:

$$Mg^{2+} + 2e^- \rightarrow Mg$$

2 moles of electrons are required for the formation of 1 mole of magnesium, so:
$2 \times 96\,500$ C $\rightarrow$ 24 g of magnesium.

Thus $3\,600$ C $\rightarrow \dfrac{3\,600}{2 \times 96\,500} = 0.45$ g of magnesium.

Worked example 2

How many moles of chlorine molecules are liberated by the passage of 4.32×10^4 C? What mass of chlorine is liberated by this charge? (A_r Cl = 35.5)

Solution

Chlorine is discharged according to the reaction $2Cl^- - 2e^- \rightarrow Cl_2(g)$.
$2 \times 96\,500$ C liberates 1 mol chlorine molecules.

4.32×10^4 C liberates $\dfrac{4.32 \times 10^4}{2 \times 96\,500} = 0.224$ mol chlorine molecules.

The mass of 1 mol of chlorine is 71 g.
0.224 mol chlorine has mass $0.224 \times 71 = 15.9$ g.

Worked example 3

Find (a) the number of moles of sodium atoms and (b) the mass of sodium liberated when a current of 4 A flows through a cell containing molten sodium chloride for 2 h. (A_r Na = 23)

Solution

Sodium is discharged according to the reaction $Na^+ + e^- \rightarrow Na(s)$.

a total charge = current (A) × time (s)

$$= 4 \times 2 \times 60 \times 60 \ C$$

$$= 2.88 \times 10^4 \ C$$

1 mol of electrons (96 500 C) liberates 1 mol of sodium atoms.

2.88×10^4 C liberates $\dfrac{2.88 \times 10^4}{96\,500} = 0.3$ mol of sodium atoms.

b mass of sodium atoms = 0.3×23 g = 6.9 g

As an additional question, what quantity (in mol) of chlorine is simultaneously liberated in the above experiment?

Worked example 4

What cell current will liberate 0.5 kg of copper in 1 h? (A_r Cu = 63.5)

Solution

$$\text{mol of copper} = \frac{500 \ g}{63.5 \ g \ mol^{-1}}$$

copper is discharged according to the reaction $Cu^{2+} + 2e^- \rightarrow Cu^0(s)$

$\dfrac{500}{63.5}$ mol of copper ions need:

$$\frac{2 \times 500}{63.5} \text{ mol of electrons} = \frac{500 \times 2 \times 96\,500}{63.5} \ C$$

total charge = current (A) × time (s)

$$= \text{current} \times 3600$$

$$\text{current} = \frac{500 \times 2 \times 96\,500}{63.5 \times 3\,600} = 422 \ A$$

Note that you need to be confident of your mathematical skills to be able to manipulate equations like this. Do you need some help?

13.4 Electrolysis in industry

Corrosion occurs when a metal reacts with substances in the environment, forming oxides and sometimes sulphides, carbonates, hydroxides and sulphates.

Electrolysis is used commercially in a number of ways. Not only does it play a part in extraction of some elements, but it is very useful in enhancing the appearance of metals and protecting them from **corrosion**. (See also Chapter 21.)

Examples of the use of electrolysis include:

- the extraction of reactive metals such as sodium and aluminium from their ores or compounds (Chapter 22);
- the extraction of active non-metals such as the halogens (Chapter 25);
- electroplating, e.g. chrome plating, nickel plating and galvanising;
- anodising aluminium;
- electrorefining, e.g. in obtaining pure copper from impure copper.

Anodising aluminium

When exposed to the air, aluminium combines with oxygen to form aluminium oxide (Al_2O_3). The oxide forms an even coat and seals the surface, thus protecting the metal from further corrosion. Electrolysis is used to make this protective layer thicker and tougher. This process is called **anodising**.

Anodising is the process by which aluminium is given a thick protective coat of aluminium oxide by electrolysis.

Aluminium is made at the anode (positive electrode) of an electrolytic cell that contains dilute sulphuric acid or dilute chromic(VI) acid as the electrolyte. (Any electrolyte that releases oxygen gas at the anode can be used.)

The reaction at the anode is:

$$4OH^- - 4e^- \rightarrow 2H_2O(l) + O_2(g)$$

The liberated oxygen combines with the aluminium anode, coating it with oxide.

Another advantage of anodising is that this protective layer can be made to absorb dyes, which then are permanently fixed by treatment with boiling water.

Figure 13.18
Some objects we use every day are made from anodised aluminium. ▶

Electroplating

Electroplating is the process of covering one metal with a layer of another by electrolysis.

Electroplating involves covering one metal with a protective coat of another metal. In practice, the outer layer is made of a less reactive metal. There are several reasons for electroplating metal objects, including:

- to enhance the appeal of the plated article – the electroplated article is more decorative than the metal underneath;
- to protect the covered metal from corrosion;
- to avoid using expensive metals for the object.

In electroplating, the object to be covered is made at the negative electrode (cathode) and the pure plating metal is made at the anode. The solution needs to contain ions of the plating metal.

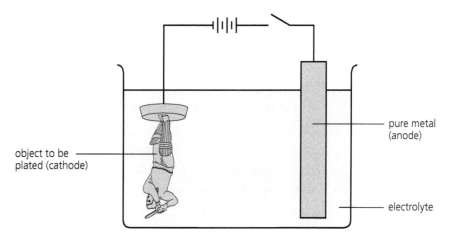

Figure 13.19
Electroplating a
sporting award ▶

During plating, ions pass into solution from the pure anode and are discharged as a thin layer on the cathode.

Chromium plating

Objects that are plated with chromium are often referred to as chrome-plated. Chromium is often used to plate steel objects. Chromium gives an attractive shiny, silver-looking finish, protects the underlying metal (often steel) from corrosion, and increases the hardness of tools and machinery.

- Plating solution: Usually chromic(VI) acid, to which chromium(III) sulphate and chromium(III) fluorosilicate are added

- Anode: Pure chromium

- Cathode: The metal/object to be plated. In decorative chrome plating, this is first given an undercoat of copper and nickel.

Nickel plating

In nickel plating, the mixture varies with the purposes for which the plated articles are to be used. A common set of conditions is the following:

- Plating solution: Nickel(II) sulphate or nickel(II) chloride solution, to which boric acid and a chemical wetting agent are added

- Anode: Nickel

- Cathode: The metal to be coated

▲ **Figure 13.20**
Steel pans are sometimes chrome-plated to prevent them from rusting and to give them an attractive finish.

Silver plating

Nickel objects are often themselves electroplated with a coat of silver. This makes the objects appear more expensive than they really are.

Electrorefining

Electrorefining is a process in which the purity of metals, such as copper, is improved by electrolysis.

Refining of copper by electrolysis is an example of **electrorefining**.

Copper of high quality is needed for some purposes, such as wiring. This is because electrical resistance increases with the quantity of impurities in the copper. Copper obtained by methods such as chemical reduction is not sufficiently pure.

- Electrolyte: Mixture of copper(II) sulphate and sulphuric acid
- Anode: Impure copper
- Cathode: A strip of pure copper

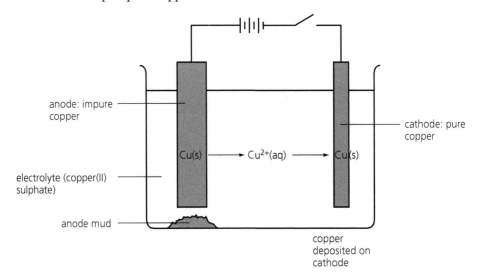

anode: impure copper

cathode: pure copper

Cu(s) → Cu²⁺(aq) → Cu(s)

electrolyte (copper(II) sulphate)

anode mud

copper deposited on cathode

Figure 13.21
Electrorefining of copper ▶

Anode half-reaction:

$$Cu^0(s) - 2e^- \rightarrow Cu^{2+}(aq)$$

Copper atoms leave the anode and enter the solution as copper ions.
Cathode half-reaction:

$$Cu^{2+}(aq) + 2e^- \rightarrow Cu^0(s)$$

Copper ions are discharged at, and deposited on, the cathode. Overall, copper leaves the anode and is deposited on the cathode.

The main impurities include zinc and gold. Zinc atoms in the impure copper anode behave like copper atoms, i.e. they lose electrons and pass into solution as zinc ions. The gold and other unreactive impurities collect at the bottom of the anode as a solid mixture, which is referred to as anode mud.

Rusting and its prevention

Many metals corrode.

- Corrosion of metals can sometimes be attractive, such as the patina on brass objects or the green colour of a copper roof.
- As we have seen with aluminium, corrosion can be protective.

- The corrosion of iron, commonly known as rusting, however, is very destructive and costly.

Figure 13.22
Rusting ▶

Rust is hydrated iron(III) oxide – $Fe_2O_3.xH_2O$. Rust does not stick to the iron underneath, instead it flakes off and a fresh surface is exposed for further attack. It is this continuous process that weakens the metal.

As much as 10–25% of the iron and steel manufactured each year is lost through rusting. To replace it costs huge sums of money. Rusting can also lead to loss of life; examples are when railway lines rust through and trains become derailed or when boilers and machinery malfunction in factories.

The rusting of iron is significant since iron and steel are widely used in construction (steel is an alloy of iron).

How does rusting take place?

Rusting is an oxidation process (see Section 12.7). Iron loses electrons according to the equation:

$$Fe(s) - 2e^- \rightarrow Fe^{2+}(aq)$$

This reaction occurs in parts of the iron that are not exposed to air.

In parts of the iron where there is a good supply of oxygen and water, the following reaction occurs:

$$O_2(g) + 2H_2O(l) + 4e^- \rightarrow 4OH^-(aq)$$

The Fe^{2+} ions react with the OH^- ions to form iron(II) hydroxide:

$$Fe^{2+} + 2OH^- \rightarrow Fe(OH)_2(s)$$

Oxygen from the air oxidises the iron(II) hydroxide to hydrated iron(III) oxide, which is rust:

$$Fe(OH)_2(s) + O_2(g) \rightarrow Fe_2O_3.xH_2O(s)$$
$$\text{rust}$$

Experiment 13.2 enables you to work out what the conditions for rusting are.

M&M
P&D

Experiment 13.2   Working out the conditions for rusting

Set up four test-tubes as shown in Figure 13.23. Look at each of the tubes for signs of rusting.

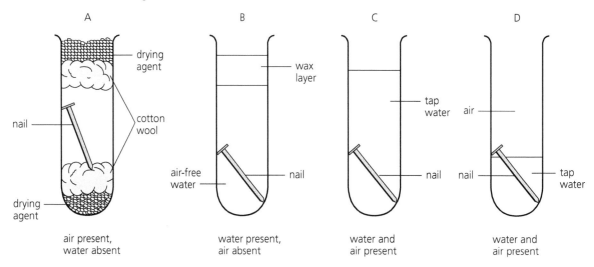

A
drying agent
nail
cotton wool
drying agent
air present, water absent

B
wax layer
air-free water
nail
water present, air absent

C
tap water
nail
water and air present

D
air
nail
tap water
water and air present

▲ **Figure 13.23**   Testing the conditions needed for rusting

Questions

Q1   In which of the test tubes did the nail rust?

Q2   Which variable(s) is/are kept constant in all four tubes?

Q3   A student concluded from the experiments that both water and air are required for rusting to take place. Are all four test-tubes necessary to draw this conclusion? Explain your answer.

How can rusting be controlled?

Coupling

Iron corrodes less readily when it is coupled with a more reactive metal. The iron and the more reactive metal, such as magnesium or zinc, must be in contact with each other and with an electrolyte. Such combinations of two metals (a couple) and an electrolyte are known as electrolytic cells.

The more reactive metal in the couple loses electrons (is oxidised) in preference to the iron. This is known as sacrificial protection; the magnesium or zinc corrodes (is sacrificed) in preference to the iron, and the iron is protected.

▲ **Figure 13.24**
Rusting was prevented here.

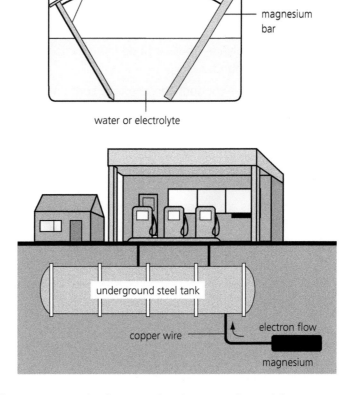

Figure 13.25
Iron does not rust here, even though it is in a salt solution. This is how you can show this in a school laboratory. ▶

Figure 13.26
It is cheaper to replace the bar of magnesium than the steel tank. ▶

Sometimes the iron is completely covered with a coat of zinc. This process is called galvanising. The zinc is oxidised and protects the iron from corrosion. Even if the surface becomes scratched, the zinc is preferentially oxidised to the iron. We often refer to galvanised sheets of iron as 'galvanise' or 'zinc' in the Caribbean. (Brightly coloured aluminium roofing is now very popular. Can you suggest some reasons why?)

There are other methods that can be used to try to prevent or control rusting.

- Tin plating – 'tins' are actually cans of steel coated with layers of tin. However, because tin is less reactive than iron, the iron corrodes when a 'tin' is scratched. Tin-coated cans can be safely used for food storage since tin salts are non-toxic. The small amounts of tin that may dissolve out into foods do not harm the consumer.
- Painting the object.
- Coating the object with organic plastics, rubbers and greases.
- Dipping the object into dilute solutions of oxidising agents such as sodium nitrite, potassium dichromate(VI) and sulphuric acid. These result in the formation of a tough oxide layer.

Summary

- Substances that conduct an electric current are:
  - metals and graphite;
  - electrolytes – these include ionic compounds and some polar covalent compounds.
- Electrolytes conduct because they contain ions that are free to move when these compounds are molten or in aqueous solution.
- When electrolytes conduct:
  - they are decomposed by the current;
  - the anions migrate to the anode where they are oxidised;
  - the cations migrate to the cathode where they are reduced.
- Electrons are removed from the cathode and given up to the anode.
- Generally, if two ions are competing at a given electrode:
  - the cation that is lower in the reactivity series is preferentially discharged;
  - the anion that is lower in the reactivity series is usually discharged.
- The transfer of 1 mole of electrons at each electrode corresponds to the passage of 96 500 C of electricity through the circuit.
- Industrial applications of electrolysis include:
  - the extraction of metals and non-metals from their compounds;
  - the anodising of aluminium;
  - chrome and nickel plating;
  - electrorefining (e.g. the refining of copper).
- The corrosion of iron (rusting) is an electrochemical process that requires the presence of water and oxygen.
- A more reactive metal in contact with iron protects it from corrosion, but a less reactive metal speeds up the rusting.

End-of-chapter questions

1 Compounds that conduct electricity and are decomposed by it are called:

  A Metals

  B Graphite

  C Insulators

  D Electrolytes

2 What quantity of electricity passes through the electrolyte when a current of 50 A flows for 5 minutes?

  A 15 000 C

  B 250 C

  C 100 C

  D 25 000 C

3 The charge of 1 mole of electrons (96 500 C) is known as:

  A An Ohm

  B Faraday's constant

  C Coulomb's law

  D Avogadro's number

4 Where does the following reaction take place during the electrolysis of copper sulphate:
$Cu^{2+}(aq) + 2e^- \; Cu(s)$?

  A At the anode

  B At the cathode

  C At the platinum electrode

  D At the graphite electrode

5 During the electrolysis of concentrated sodium chloride, which ion is preferentially discharged at the anode?

A OH
B Na+
C Cl
D H+

6 What are the active copper electrodes in the electrolysis of copper sulphate for purifying copper used for?

A Anodising
B Electroplating
C Coupling
D Electrorefining

7 A current of 2.5 A was passed for 386 s through a solution of $CuSO_4$ using inert electrodes. What was the mass of Cu deposited on the cathode?

A 0.32 g
B 0.64 g
C 0.5 g
D 0.16 g

---

8 a Comment on the economic importance of corrosion.
b Identify two different situations in which corrosion may lead to loss of life.
c Describe three different methods that can be used to slow down or prevent the rusting of iron.
d Explain why tin-plated articles rust rapidly once they have been scratched, whereas scratched galvanised objects usually do not rust.

9 Electrolysis involves 'redox reactions'. Explain this statement with the use of suitable examples.

10 Sodium hydride (NaH) is an ionic compound. Molten NaH was electrolysed using a current of 0.25 A for 10 minutes.
a What are the products of this electrolysis?
b Write ionic equations for the cathode and anode reactions.
c What mass of product is liberated at the cathode? Show all steps in the calculation.

11 1 920 C of electricity were passed through a dilute solution of sodium chloride, using inert electrodes.
a Draw a suitable diagram of the apparatus for this electrolysis.
b Give the name(s) of the gases evolved at each electrode.
c Give the equation for the reaction occurring at each electrode.
d Determine the volume of gas, at room temperature and pressure, produced at each electrode.
e If a steady current of 3.0 A was used during this electrolysis, calculate the time for which the electrolysis took place.

12 A current of 0.25 A was passed through molten lead chloride for 300 s, using inert electrodes.
a What element is liberated at the anode?
b What element is liberated at the cathode?
c Give equations for the reactions occurring at each of the electrodes.
d Determine the quantity of electricity passed during this experiment.
e Determine the volume of gas produced at room temperature and atmospheric pressure.

13 A current of 2 A is maintained through a copper sulphate solution (using copper electrodes) for 2 hours.
a Calculate the quantity of electricity passed.
b Describe the changes at each electrode.
c Give equations for the reactions occurring at each electrode.
d Calculate the theoretical change in the mass of each electrode.
e Identify one practical use to which this reaction can be put.

14 a Describe how you would obtain a sample of lead by electrolysis.
b How is pure copper obtained from blister (crude) copper?
c Why is sodium not deposited at the cathode during the electrolysis of aqueous sodium chloride?

14 The rates of chemical reactions

In this chapter, you will study the following:
- what is meant by reaction rate;
- factors that affect reaction rate;
- effect of factors on the reaction rate;
- how to interpret graphical representations of rates of reaction data.

This chapter covers
Objectives 10.1–10.4 of Section A of the CSEC Chemistry Syllabus.

'Rate', as you know, refers to speed. We use the idea of rates all the time in our daily lives. We talk about how fast someone drives, for example, in terms of miles per hour, or the run rate in cricket in terms of runs per over, or computer speed in terms of millions of cycles per second. We can also describe and measure the rates of chemical reactions.

Some chemical reactions occur almost instantaneously – think of the speed of an explosion. Others, such as the weathering of rocks, occur extremely slowly.

(a) Weathering of rocks.

(b) Bananas ripening.

(c) A fireworks display.

▲ **Figure 14.1** These reactions proceed at widely differing rates.

Between these two extremes, there are reactions that take place at rates which we can measure in the laboratory. This allows us to study factors that affect reaction rate. This is particularly important in industry so that optimum conditions can be applied to ensure that productivity is increased.

The rate of a chemical reaction is a measure of the increase in product against the time taken. Or, it can be the decrease in reactant against time taken. The rate of a reaction can be increased by changing any of the factors that affect the conditions required for the reactions.

This includes:
- particle size;
- concentration;
- temperature;
- light;
- catalyst.

14.1 Measuring the rates of chemical reactions

The **rate of a reaction** is the change in concentration of reactants or products in unit time.

During the course of a chemical reaction, the concentrations of the reactants decrease, while the concentrations of the products increase.

Figure 14.2
The blue colour of the solution fades as the concentration of the reactant, blue copper(II) ions, decreases while the concentration of the product, colourless zinc(II) ions, in the solution increases. The quantity of the solid zinc reactant decreases, while the quantity of copper, the brown solid product, increases. ▶

We can therefore measure the rate of a reaction by monitoring changes in the concentration of reactants or products.

The rate of reaction is given by:

$$\frac{\text{change (decrease) in the concentration of reactants}}{\text{time taken for change}}$$

or

$$\frac{\text{change (increase) in the concentration of reactants}}{\text{time taken for change}}$$

Though it is often difficult to measure changes in concentration, it becomes quite easy if one of the products is in a different physical state from the reactants. We may, for example, be able to measure:

- the change in volume of a gas;
- the rate of formation of a precipitate;
- a change in pH, where the pH of reactants and products differ.

Here are a couple of examples:

- Consider the reaction between a solid carbonate and hydrochloric acid. One of the products is carbon dioxide gas. The volume of gas produced as the reaction proceeds can be used to measure the rate of the reaction. A typical set of apparatus used is shown in Figure 14.3.

- The reaction between sodium thiosulphate solution and dilute acid forms a precipitate of sulphur. We can measure the time it takes for the reaction mixture to become opaque. (See Experiment 14.1.)

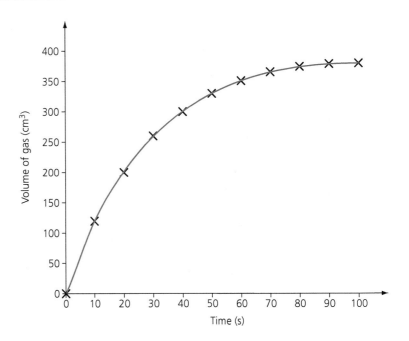

Figure 14.3
Measuring the
volume of carbon
dioxide produced in a
reaction ▶

14.2 Calculating the rate of reaction

A possible set of results obtained from the reaction between a carbonate and an acid can be presented as in Table 14.1 or plotted on a graph as in Figure 14.4.

▼ **Table 14.1** Experimental results.

Time (s)	0	10	20	30	40	50	60	70	80	90	100
Volume of carbon dioxide (cm³)	0	120	200	260	290	330	350	367	375	379	380

Figure 14.4
Graph of
experimental results ▶

To calculate the rate of the reaction over any period of time, you can use the formula:

$$\text{rate} = \frac{\text{change in concentration of product}}{\text{time taken for the change}}$$

From Figure 14.4, the change in volume of gas in the first minute (60 s) is 350 cm³. Therefore, the rate over this period is:

$$\frac{350 \text{ cm}^3}{60 \text{ s}} = 5.8 \text{ cm}^3 \text{ s}^{-1}$$

and the rate between 20 seconds and 40 seconds is:

$$\frac{100 \text{ cm}^3}{20 \text{ s}} = 5 \text{ cm}^3 \text{ s}^{-1}$$

Each of these rates is really the *average* rate during the specific time interval. Note that the rate of the reaction changes as the reaction proceeds. The steepness of the curve indicates the rate of the reaction. The steeper the curve, the faster the rate. The reaction is fastest at the beginning, and where the curve flattens, no more gas is produced. This is a typical graph for chemical reactions.

Understand it better

When the West Indies Cricket team is scoring at a run rate of 6 runs per over in a one-day match, it doesn't mean that they score exactly 6 runs in each and every over. Six runs per over is their average rate of scoring. Sometimes they would have scored at a faster rate and sometimes at a slower rate.

In a typical chemical reaction, the reaction is usually faster at the beginning, when there is a high concentration of reactants, and then it gets slower until no more product is produced. The shape of the graph in Figure 14.4 shows this clearly.

We can get the average rate over any period of time as shown above. We can tell the rate at a specific time by finding the gradient of the curve at that time.

There are specific mathematical skills needed here. Ask your teacher if you need some extra help.

To determine the gradient, we need to draw a tangent to the curve at the specified time.

▲ **Figure 14.5** How would you calculate the average run rate?

$$gradient = \frac{change\ in\ vertical\ distance}{change\ in\ horizontal\ distance}$$

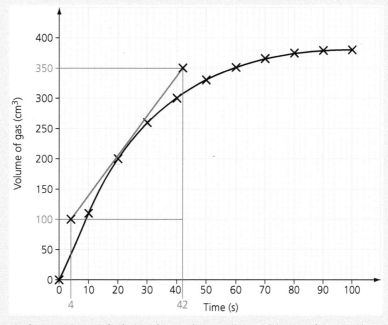

▲ **Figure 14.6** Calculating the gradient at time = 20 seconds.

Looking at a time of 20 s:
- The change in vertical distance shown in Figure 14.3 can be represented by:
  $350 \text{ cm}^3 - 100 \text{ cm}^3 = 250 \text{ cm}^3$
- The change in horizontal distance can be represented by:
  $42 \text{ s} - 4 \text{ s} = 38 \text{ s}$

The rate at 20 s $= \dfrac{250 \text{ cm}^3}{38 \text{ s}} = 6.6 \text{ cm}^3 \text{ s}^{-1}$

Use the graph in Figure 14.6 to answer these questions.

Q1 Calculate the rate at 50 s and at 80 s.

Q2 The average rate in the first 40 s of the reaction.

14.3 Conditions required for chemical reactions to occur

Chemical reactions involve the breaking of bonds (in the reactants) and the formation of new bonds (in the products). Chemists *believe* that for this to happen, three conditions are necessary:

Activation energy is a certain minimum energy required for bonds within reactant molecules to break and for the particles to become sufficiently energised for products to be formed.

Collision theory states that there must be effective collisions in order for chemical reactions to take place.

1 Reactant molecules must collide.

2 Reactant molecules, on collision, must have energy equal to or greater than the necessary '**activation energy**'.

3 Reactant molecules must collide in the correct position (i.e. with the correct orientation). Collisions that are sufficiently energetic would be most effective if the colliding molecules approached each other in such a way that the energy released on collision can be passed on directly to the bonds to be broken. Collisions of particles in any other position will not result in the formation of products.

Effective collisions are those collisions that result in the formation of products. They require that all the reactants are correctly oriented <u>and</u> have the required activation energy (Figure 14.7).

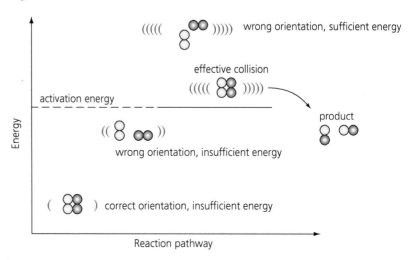

Figure 14.7
Formation of products via effective collisions. ▶

Factors that affect the rates of chemical reactions

The factors that can affect the rate of a chemical reaction are:

- concentration (or pressure for gaseous systems);
- temperature;
- catalysts;
- particle size/surface area;
- light, for some reactions.

The effect of concentration on the rates of reactions

The theory here is that an increase in concentration means there are more reactant molecules in a given volume. This increases the chances of more frequent effective collisions and thus leads to a faster reaction rate.

Figure 14.8
The effect of increasing the concentration of reactants ▶

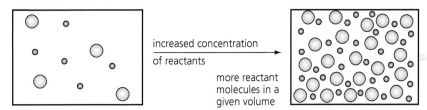

increased concentration of reactants

more reactant molecules in a given volume

Figure 14.9 shows the volume of hydrogen gas, $H_2(g)$, produced (a product) as the reaction between magnesium and excess hydrochloric acid (the reactants) proceeds. Different concentrations of hydrochloric acid have been used. The steepness of the curves indicates the rate of reaction in the initial stages.

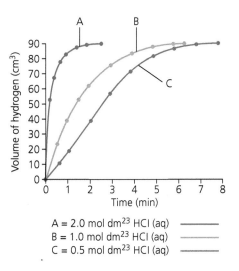

Figure 14.9
The reaction of 5-cm lengths of magnesium ribbon with three different concentrations of hydrochloric acid at 27 °C ▶

A = 2.0 mol dm^{23} HCl (aq) ——————
B = 1.0 mol dm^{23} HCl (aq) ——————
C = 0.5 mol dm^{23} HCl (aq) ——————

Curve A (2.0 mol dm^{-3}) has the steepest initial gradient and Curve C (0.5 mol dm^{-3}) has the least. The reaction rate, therefore, is fastest with 2.0 mol dm^{-3} hydrochloric acid.

In order to be able to make these comparisons, the experiment had to be carefully controlled. The following factors (variables) were kept identical (controlled) during the experiment:

- The amount of magnesium (use the same length)
- The temperature (27 °C)
- The surface area of the magnesium ribbon

The only variable that was changed (the 'manipulated' variable) was the concentration of the hydrochloric acid.

The responding, or 'dependent', variable was the rate of the reaction. It is this that is measured by the change (increase) in the volume of hydrogen over time.

The 'Understand it better' section below explains more about planning and designing experiments.

Changing concentrations in gaseous systems

In gaseous systems, increasing the pressure on the reactants will have the same effect as increasing the concentration. This is because, as you learnt before, the volume of a gas will decrease as the pressure increases (if the temperature remains constant). This is shown in Figure 14.10. This principle is applied in industrial processes involving gases.

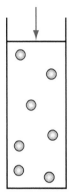 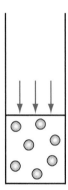

Figure 14.10
Increased pressure: The same number of molecules in a smaller volume, i.e. a higher concentration. ▶

Understand it better

Planning and designing experiments
A variable is any condition that can change in an experiment. Examples are temperature, pressure, concentration or light.

It is often possible to determine experimentally the effect of a particular variable (or changes in it) on a reaction.

In a good experiment, only one variable is deliberately altered or manipulated at a time. It is then possible to say that any effect observed is due to the change in this variable.

Here is an example of how to plan and design an experiment investigating the effect of temperature on the rate at which food spoils.

1 Identify as many variables as you can that may have an effect:
   • Type of food, amount of food, temperature, light, time, etc.
2 Identify the variable you wish to determine the effect of (this is the manipulated variable):
   • Temperature
3 Which values of this variable will you choose?
   • For example, 10 °C, 30 °C, 60 °C, 90 °C

4 Identify the change to be measured as you change the manipulated variable (this is the dependent or responding variable):
  • The rate at which the food spoils

5 Identify how you will measure this change:
  • Measuring the area of mould formed in a given time period

6 Keep all other variables listed in point 1 identical during the experiment (the controlled variables):
  • Type of food: cheese
  • Amount of food: a 4 cm cube
  • Light: place in identical black plastic bags, etc.

Experiment 14.1   The effect of concentration on reaction rate: the disappearing cross

Carry out the following experiment to help you understand the effect of concentration over time.

Theory

When acid is added to sodium thiosulphate solution, a fine colloidal suspension of sulphur is formed. This suspension scatters light. Viewed through the suspension, a cross on a sheet of paper disappears when enough sulphur particles are formed (see Figure 14.11).

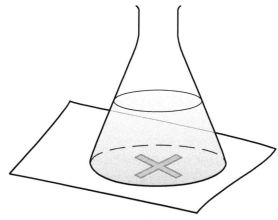

Figure 14.11
The 'disappearing cross' experiment. When a given quantity of sulphur is formed, the cross is no longer visible ▶

Procedure

1 Using a measuring cylinder, place 50 cm³ of 0.1 mol dm⁻³ sodium thiosulphate into a conical flask.

2 Add 50 cm³ of 0.5 mol dm⁻³ sulphuric or hydrochloric acid to this solution, noting the time that you do so.

3 Place the conical flask over the cross and record the time it takes for the cross to be no longer visible when viewed from above. Discard the mixture and wash the conical flask.

4 Repeat step 3, this time using the next thiosulphate/water mixture shown in Table 14.2.

▼ **Table 14.2** Results table for 'disappearing cross' experiment.

Experiment number	Volume of acid (cm³)	Volume of thiosulphate (cm³)	Volume of water (cm³)	Time, t (s)	$\frac{1}{t}$ (s⁻¹)
1	50	50	0		
2	50	45	5		
3	50	40	10		
4	50	35	15		
5	50	30	20		
6	50	25	25		
7	50	20	30		

Questions

Q1 $\frac{1}{t}$ is a measure of the rate of the reaction. Plot a graph of $\frac{1}{t}$ against volume of sodium thiosulphate used.

Q2 The volume of sodium thiosulphate is directly related to its concentration under the conditions of the experiment. How does the rate of this reaction depend on the concentration of the sodium thiosulphate?

Q3 State the variables controlled and the manipulated variable in this experiment.

Q4 What was the responding variable?

Q5 What steps could have been taken to keep the temperature constant?

Q6 Design an experiment, using the same apparatus and materials, to find out how the rate of this reaction depends on the concentration of the acid.

The effect of temperature on reaction rates

The rate of a chemical reaction increases as the temperature of the reactants increases. This is why:

- Particles move faster as their temperature increases (see Figure 14.12). This links to your studies in physics.
- Particles collide more frequently.
- The collisions are more effective since more particles have energy equal to or more than the activation energy.

Figure 14.12
The higher the temperature, the faster the particles move. ▶

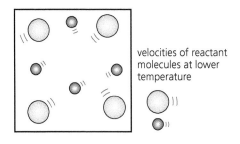

velocities of reactant molecules at lower temperature

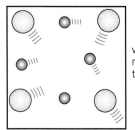

velocities of reactant molecules at higher temperature

For some chemical reactions (and for biological reactions up to about 40 °C) the reaction rate increases approximately two times for every 10 °C rise in temperature.

We store food at the lower temperature of the refrigerator to reduce the rate of spoilage reactions. We cook food in pressure cookers because water boils at about 120 °C in pressure cookers and the food cooks faster.

P&D

Experiment 14.2 The effect of temperature on the rate of reaction

The apparatus shown in Figure 14.13 can be used to investigate the effect of temperature on the rate of reaction between hydrochloric acid and magnesium ribbon.

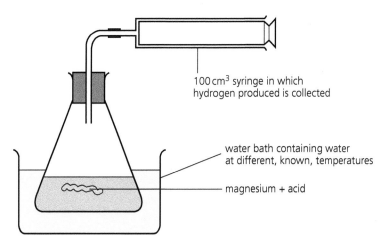

100 cm³ syringe in which hydrogen produced is collected

water bath containing water at different, known, temperatures

magnesium + acid

Figure 14.13
The apparatus to measure the relationship between reaction rate and temperature ▶

Procedure

1 Study the information on planning and designing experiments on page 258.

2 Plan and design an experiment to investigate the effect of temperature on the rate of reaction. Write up your plan under the following headings:
  • Problem statement
  • Statement of hypothesis
  • Aim
  • List of apparatus and materials
  • Procedure
  • Variables must be identified (manipulated, controlled, responding)
  • Data to be collected
  • How the data will be treated

3 Carry out your experiment.

4 Write up your results and discussion.
  • Construct a table to record your results.
  • Construct graphs to display your results.
  • Interpret your results.
  • Discussion (include chemical principles, limitations, sources of error).

The effect of catalysts on the rates of reaction

A **catalyst** is a substance that speeds up a chemical reaction, but is not itself used up in the reaction.

The rates of many reactions can be changed by the addition of small amounts of **catalysts**. Catalysts:

- do not appear in the chemical equation;
- take part in the reaction but are unchanged, chemically, at the end, though solid catalysts may undergo physical changes.

We do not fully understand exactly how catalysts work. It is believed that in the presence of a catalyst, a lower activation energy is needed (Figure 14.14).

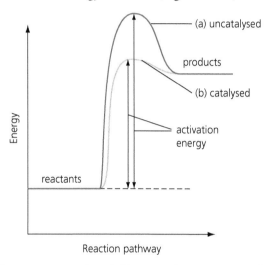

Figure 14.14
Energy profile diagrams for (a) an uncatalysed reaction and (b) a catalysed reaction. Note: The catalyst has lowered the activation energy required. ▶

This is because catalysts may provide a different and/or easier way for the reactants to form products.

- A catalyst in the solid state may catalyse reactions in the gaseous state or in aqueous solution by providing a surface for the reactants to react (Figure 14.15). An example is the use of manganese(IV) oxide as a catalyst in the decomposition of hydrogen peroxide.
- A catalyst in the same state as the reactants may provide an alternative (easier) route for the reaction to take place. An example is concentrated sulphuric acid, which acts as a catalyst in the reaction between ethanol and ethanoic acid to form ethyl ethanoate (see Section 19.2).

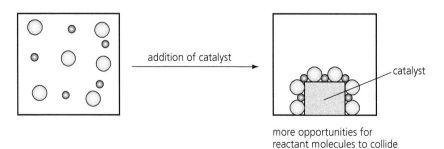

Figure 14.15
Catalysts may bring reacting molecules closer together. ▶

Experiment 14.3   Investigating the catalytic decomposition of hydrogen peroxide

Procedure

1   To 50 cm³ of '20 volume' hydrogen peroxide, add 45.0 cm³ of water, one drop of bench sodium hydroxide solution and 0.05 g of powdered manganese(IV) oxide.

2   Using the apparatus shown in Figure 14.16, collect the oxygen given off, noting the volumes at half-minute intervals, for 4 minutes.

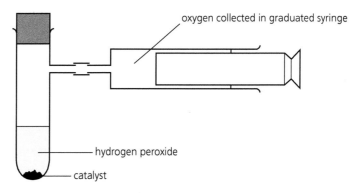

oxygen collected in graduated syringe

hydrogen peroxide

catalyst

Figure 14.16
Apparatus used in the catalytic decomposition of hydrogen peroxide ▶

3   Repeat steps 1 and 2, first using 0.10 g of powdered manganese(IV) oxide, then 0.2 g of powdered manganese(IV) oxide.

4   Inspect the catalyst – manganese(IV) oxide – after the reaction, recording any changes in appearance or texture.

5   Using a sheet of graph paper, plot volume of oxygen against time for all four experiments.

Questions

Q1   Make as many deductions as possible about the effect on reaction rate of the mass of catalyst used.

Q2   What further experiment needs to be carried out to ascertain that the reaction is in fact catalysed by the manganese(IV) oxide?

Extension

1   Repeat the above experiment, keeping all conditions the same but using different catalysts, such as copper(II) oxide and zinc oxide.

2   Plot, on the same axes, volume of oxygen gas evolved (y-axis) against time (x-axis) for the different catalysts used.

3   You will find that different catalysts increase the rate of oxygen evolution at different rates (Figure 14.17).

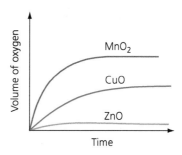

MnO₂

CuO

ZnO

Volume of oxygen

Time

Figure 14.17
Different catalysts have different effects on reaction rates. ▶

The effect of surface area on reaction rate

The rate of a chemical reaction involving a solid reactant is increased by increasing the state of subdivision (the surface area) of the solid, while decreasing the surface area of the solid has the opposite effect. In such reactions, collisions occur between moving molecules and the solid reactants. It follows that the smaller the particles of solid are, the greater surface area available for collisions, and the greater the reaction rate.

Finely divided solids catch fire more readily than the same mass of the solid in wire or sheet form. For example, a strip of heated iron wire is not affected when placed in a jar of oxygen, but heated steel-wool of the same mass spontaneously catches fire because of its greater surface area.

The action of the teeth helps to increase the surface area of food, by breaking it up into smaller pieces. This leads to more effective digestion by biological catalysts – enzymes.

 Understand it better

The more subdivided a surface, the greater the surface area exposed. This simple activity will help you to understand this.

1   Take four identical match boxes (or any four identical cuboid structures).

2   Use a rubber band to hold them together to make a single cuboid structure.

3   Calculate the surface area of the total exposed surface by:
    a   finding the area of each exposed surface;
    b   adding these areas together.
    This is the surface area of the large structure (x).

4   Separate the large cuboid into the four separate matchboxes.

5   Calculate the surface area of one matchbox.

6   Find the total surface area of the four matchboxes. This is the total surface area that is now exposed (y).

7   Compare x and y.

▲ **Figure 14.18**   A common photochemical reaction is photosynthesis, for which light is required. This process is crucial to life on Earth.

The effect of light on reaction rates

Light does not affect the rates of the majority of chemical reactions. However, those reactions that are affected by light are among the most important reactions. Perhaps the most important such reaction is photosynthesis. Reactions that are initiated or accelerated by light are known as photochemical reactions.

Conventional photography depends on light-sensitive chemical reactions. The reaction by which finely divided silver is formed when light interacts with silver halides is the basis of photography. Early photographs were probably produced using similar material, but exposure times of 6–8 hours were needed. It was later discovered that when silver salts are mixed with gelatine, the resulting emulsion has superior photographic properties. The silver salt in black and white photographic films is silver bromide.

▲ **Figure 14.19**   Silver salts are light sensitive.

Like conventional cameras, digital cameras use lenses to focus light to create an image on a screen. However, instead of using film, a digital camera focuses the light onto a semiconductor. The semiconductor device is able to record the light in electronic form. Digital cameras have a built-in computer. The image is represented in bits and bytes – that is, in the language of computers.

ORR

A&I

Experiment 14.4   The darkening of silver salts

Many silver salts are light-sensitive. For this reason, silver nitrate and other silver salts are stored in dark coloured bottles.

Procedure

1   Obtain four strips (8 cm × 2 cm) of filter paper.

2   Dip the strips in silver nitrate solution and leave them there until they are saturated. Set aside one strip to serve as a reference strip (a control).

3   Immerse one strip in aqueous potassium chloride, another in aqueous potassium bromide and the third in aqueous potassium iodide.

4   Record the initial appearance of each of the strips, including the reference strip.

5   Place each treated strip on a dark card, then place a coin at the centre of each strip. Expose all the strips to the light from a lamp.

6   Observe each strip over a period of about 20 minutes, recording all observations.

7   Switch off the lamp.

Results and observations

1 Remove the coins and comment on the contrast between the covered and uncovered parts of the strips. Also, compare the treated strips with the control (untreated) strip.

2 Make as many deductions as possible.

14.4 Catalysts in industry

In industrial processes, chemists try to choose the most favourable conditions so that they can get a good yield of product in a reasonable time. Table 14.3 shows how chemists apply some of the principles discussed in this chapter.

▶ **Table 14.3** The use of catalysts in industry.

Industrial process	Use of catalyst	Use of increased pressures
nitric acid production	platinum and rhodium	8 atmospheres
ammonia production	iron	250–1 000 atmospheres
sulphur(VI) oxide	vanadium(V) compounds	2 atmospheres
hardening of oils	nickel	5 atmospheres
cracking oil	aluminium oxide	

Practice

This section will help you to practise interpreting data collected from experiments involving rates of reaction.

The graph in Figure 14.20 contains some results of experiments carried out to determine the effect of temperature on rate of reaction.

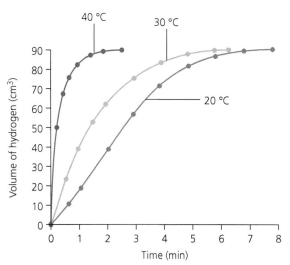

Figure 14.20
The results of the reaction between 5 cm lengths of magnesium ribbon and 1.0 mol dm^{-3} hydrochloric acid at different temperatures. ▶

1 The variable manipulated in this experiment was temperature. What variables were controlled?

2 What is the responding variable?

3 What is the final volume of hydrogen produced at each temperature? Give an explanation for your answer.

4 Imagine that a catalyst was added to the reaction mixture at 30 °C:
  a What would be the final volume of gas produced?
  b How would the shape of the graph change at that temperature?

Oxygen is given off when potassium chlorate(V) is heated. The reaction is catalysed by manganese(IV) oxide (MnO_2):

$$2KClO_3(s) \xrightarrow{\text{MnO}_2} 2KCl(s) + 3O_2(g)$$

In an experiment to study the rate of thermal decomposition of potassium chlorate(V), separate 5.0 g amounts of $KClO_3$ were heated with different masses of manganese(IV) oxide and the times taken for the reaction to be completed (i.e. to reach the final volume of oxygen) noted. Sample results are given in Table 14.4.

▼ **Table 14.4**

Mass of catalyst/5 g of KClO₃ (g)	0.00	0.10	0.20	0.40	0.75	1.00
Time to reach final volume (s)	480	360	300	240	180	150

5 Explain how the results support the view that the reaction is catalysed by manganese(IV) oxide.

6 Plot a graph of the mass of manganese(IV) oxide against time for completion of the reaction.

7 Use the graph you have plotted to show that the rate of this reaction depends on the mass of catalyst used.

This investigation looked at the effect of surface area of the solid on the rate of the reaction. The reaction is between excess 0.5 mol dm⁻³ hydrochloric acid and magnesium. The results are shown in Figure 14.21. **Note:** The same mass of magnesium is used in both experiments.

8 Comment on the significance of the results.

9 Use the graph to determine the times for the reaction to be half completed in each experiment.

10 Copy Figure 14.21 and add the curve you would expect to get if the same mass of magnesium ribbon were used with excess 0.5 mol dm⁻³ hydrochloric acid. (Magnesium ribbon provides larger pieces than magnesium turnings.)

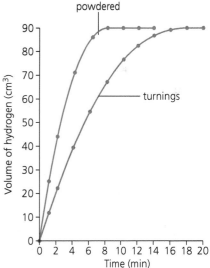

▲ **Figure 14.21** Investigating the state of subdivision of solid reactant on the rate of a reaction. The smaller the reacting particles, the faster the reaction rate.

11 What percentage of the mass of the magnesium is used after 4 minutes, for each of the experiments?

12 What test would you carry out to show that hydrogen is the gas evolved in both cases?

13 Draw a diagram of the apparatus you would use to carry out this investigation.

Summary

- Chemical reactions vary considerably in their rates.
- Chemical reactions can occur if the reactant molecules make effective collisions.
- Only a fraction of all the collisions of the right kind of molecules lead to reaction.
- An increase in the concentration of the reactants increases the rate of reaction.
- An increase in the pressure of gaseous reactants increases the rate of reaction.
- An increase in the surface area (or a decrease in particle size) of a solid increases the rate of reaction.
- The addition of a suitable catalyst increases the rate of reaction.
- Transition metals and their ions often display catalytic activity. The greater the surface area, the more effective the catalyst.
- Catalysts are not chemically changed by the reaction; however, they may undergo physical changes.
- Catalysts change the rate but not the extent of a chemical reaction.
- Some reactions e.g. 'photography' and photosynthesis take place when the system is exposed to light.

End-of-chapter questions

1 Which of the following has the slowest reaction rate?

   A Iron nails rusting in water

   B Fireworks

   C Metals reacting with acids

   D Catalyst supported reactions

2 Which one of the following is not a factor that affects the rate of chemical reactions?

   A Light

   B Collision of particles

   C Temperature

   D Catalysts

In a reaction between sodium carbonate and excess hydrochloric acid, a gas was produced. The volume of gas produced at different times was recorded as follows:

Time (s)	30	60	90	120	150	160
Volume (cm³)	10	18	25	30	32	32

3 What is the name of the gas produced in the reaction?

   A Carbon dioxide

   B Chlorine

   C Hydrogen

   D Oxygen

4 During which interval is the reaction the fastest?

   A 0–30 s

   B 60–90 s

   C 30–60 s

   D 90–120 s

5 Which two of the following will not affect the rate of the reaction?

   I   Increasing the temperature of the reaction

   II Increasing the concentration of the acid

   III Increasing the volume of acid

   IV Increasing the pressure

   A I and II

   B III and IV

   C II and III

   D I and III

6 **a** State the conditions necessary for chemical reactions to occur.
  b List the factors that affect the rate of a chemical reaction.
  c Explain how an increase in (i) temperature and (ii) concentration increases reaction rates.

7 **a** What are the units of rate?
  b Describe, giving a diagram of the apparatus, how you would study the rate of the reaction between calcium carbonate (marble chips) and hydrochloric acid. The equation for this reaction is:

$$CaCO_3(s) + 2HCl(aq) \rightarrow CaCl_2(aq) + H_2O(l) + CO_2(g)$$

8 **a** What is meant by the term 'the activation energy' for a reaction?
  b How do catalysts work?
  c It is known that solutions of sodium chlorate(I) (NaClO) slowly decompose according to the ionic equation:

$$2ClO^-(aq) \rightarrow 2Cl^-(aq) + O_2(g)$$

  Experiments show that the reaction is catalysed by an aqueous solution of cobalt nitrate. Explain how you would investigate whether it is the cobalt(II) ions or the nitrate ions that were responsible for the catalytic action.

9 Explain the following:
  a The label on bottles of powdered metals reads 'explosion possible'.
  b Unrefrigerated food spoils faster than refrigerated food.
  c Carbonated soft drinks are likely to explode if stored in a warm, enclosed cupboard.
  d It is difficult to hard-boil eggs at the top of a high mountain such as Mont Blanc (4 807 m) or Mount Kilimanjaro (5 895 m).

10 The results in Table 14.5 were obtained in an experiment on the catalytic decomposition of hydrogen peroxide.

▼ **Table 14.5**

Volume of O_2 (cm³)	0	39	70	91	106	115	120	120
Time (min)	1	2	3	4	5	6	7	8

  a What volume of oxygen is evolved in (i) the first minute (ii) the sixth minute? Account for the difference in these two values.
  b Plot a graph of volume of oxygen evolved against time and use the graph to determine:
   (i) the maximum volume of oxygen evolved;
   (ii) the time at which 65 cm³ of oxygen is evolved;
   (iii) the volume of oxygen evolved after 3.25 minutes;
   (iv) the initial rate of decomposition (you may use volume as a measure of concentration decrease).
  c Determine the number of moles of oxygen evolved in this reaction, which was carried out at room temperature and atmospheric pressure.
  d How many moles of hydrogen peroxide decomposed during this reaction?

In this chapter, you will study the following:

- exothermic and endothermic reactions;
- how to draw energy profile diagrams;
- measuring and calculating energy changes.

This chapter covers
Objectives 11.1–11.3 of Section A of the CSEC Chemistry Syllabus.

Chemical reactions always involve energy changes. Some of the energy changes you can observe are the production of heat and light. The use of some chemicals as fuels, for example, depends on their ability to produce vast amounts of heat and light when they burn. Other energy changes are more subtle. You may have noticed, for example, that when you carried out some reactions in the laboratory, the reaction vessel either became warm or cold to the touch. Chemical reactions can also produce other forms of energy, such as electricity. You will know that it is chemical reactions in dry cells that produce electricity.

Figure 15.1
Solar energy is converted to electricity via solar panels in St Lucia for powering water heaters in many hotels. ▶

Figure 15.2
Some energy changes are quite obvious, giving out much heat and light. Others, such as the production of electricity, are more subtle. ▶

15.1 Energy changes during chemical reactions

Experiment 15.1 Classifying reactions as exothermic or endothermic

Procedure

1 Half-fill a test-tube with hydrochloric acid, and then take the temperature of the acid. Add a few granules of zinc metal. Take the temperature again.

2 Half-fill a test-tube with water, and then take the temperature of the water. Add a few spatulas-full of ammonium chloride. Stir to dissolve the solid. Feel the test-tube and take the temperature again.

3 Half-fill a test-tube with sodium carbonate solution and half-fill another test-tube with a solution of magnesium chloride. Take the temperature of both solutions. Mix the two solutions in a beaker and take the temperature again.

4 Half-fill a test-tube with sodium hydroxide solution and half-fill another test-tube with dilute sulphuric acid. Take the temperature of both solutions. Mix the two solutions in a beaker and take the temperature again.

Analysis

1 Which reaction(s) led to an increase in the temperature of the system?

2 Which reaction(s) led to a decrease in the temperature of the system?

3 Use the information in the text below to classify reactions 1–4 as exothermic or endothermic.

Often the energy given off or absorbed in chemical reactions is in the form of heat.

Sometimes the reaction container and the contents become hotter, indicating that the reaction produces heat. Such changes are described as **exothermic**. Sometimes the reaction container and its contents become colder, in which case heat is absorbed. Such changes are described as **endothermic**.

> **Exothermic reactions** are chemical changes that result in an increase in the temperature of the surroundings.

> **Endothermic reactions** are chemical changes that result in a fall in the temperature of the surroundings. Endothermic reactions are much rarer than exothermic reactions.

Examples of exothermic changes include:

- the neutralisation reaction between aqueous sodium hydroxide and hydrochloric acid;
- dissolving solid sodium hydroxide in water;
- burning any fuel;
- respiration – a reaction occurring in living systems in which energy stored in the chemical bonds of food is released;
- dissolving soap powder in water.

Examples of endothermic changes include:

- the dissolving of some substances in water, e.g. ammonium nitrate, potassium iodide, urea and sodium thiosulphate;
- photosynthesis;
- the reaction between steam and carbon;
- using sodium acetate in a heat pad for relieving pain in sore muscles.

Chemicals have energy of three main types.

1 Energy in the nucleus of atoms (see Section 2.6) does not change in the course of ordinary chemical reactions, and does not concern us here.

2 The particles in chemicals have kinetic energy since they are always moving.

3 Most importantly for chemical reactions, chemicals also store energy in their bonds.

The kinetic energy and the chemical energy together make up the energy of a chemical (also called its heat content). This energy is represented by the symbol H.

When the **energy change** (ΔH) occurs at constant pressure, it is also described as the enthalpy change.

Energy changes occur during the course of chemical reactions, as reactants form new products. The energy change during a reaction is represented by ΔH.

ΔH is the difference between the energy of the products ($H_{products}$) and the energy of the reactants ($H_{reactants}$):

$$\Delta H = H_{products} - H_{reactants} \text{ (abbreviated to } H_p - H_r)$$

Enthalpy change is equal to the total heat content of the products minus the total heat content of the reactants.

- If H_p is less than H_r, then ΔH is negative and the reaction is exothermic (Figure 15.3a).

- If H_p is greater than H_r, then ΔH is positive and the reaction is endothermic (Figure 15.3b).

An energy level diagram shows the energy of the reactants and the products of a reaction.

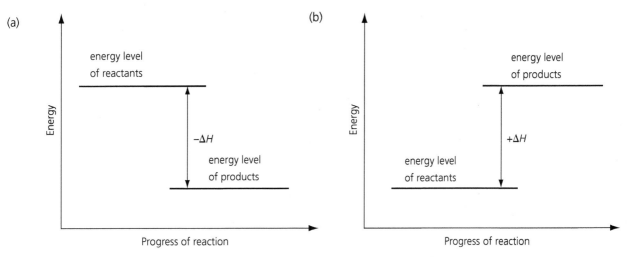

▲ **Figure 15.3** (a) Energy level diagram for an exothermic reaction and (b) energy level diagram for an endothermic reaction

For reversible reactions, if the forward reaction is exothermic, then the reverse reaction is endothermic, and vice versa.

$$\text{exothermic}$$
$$N_2(g) + 3H_2(g) \rightleftharpoons 2NH_3(g)$$
$$\text{endothermic}$$

The unit used for the change in enthalpy (ΔH) is joules per mole (J mol^{-1}) or kilojoules per mole (kJ mol^{-1}) and the energy change is stated for 1 mole of the substance under investigation.

Equations to show changes are written as in the example below:

$$C(s) + O_2(g) \rightarrow CO_2(g); \qquad \Delta H = -386 \text{ kJ mol}^{-1}$$

This means that when 1 mol of carbon burns completely in oxygen to form carbon dioxide gas, 386 kJ of heat are given off.

$$N_2(g) + O_2(g) \rightarrow 2NO(g); \qquad \Delta H = +180.4 \text{ kJ mol}^{-1}$$

This means that when 1 mol of nitrogen gas combines with oxygen to produce nitrogen monoxide (nitrogen(II) oxide), 180.4 kJ of heat are absorbed.

Practice

1 Draw energy level diagrams for the following reactions:

    a $C(s) + 2H_2(g) \rightarrow CH_4(g)$ $\Delta H = -74.8 \text{ kJ}$

    b $C(s) + H_2O(g) \rightarrow CO(g) + H_2(g)$ $\Delta H = +137 \text{ kJ}$

2 Write the equations and draw the energy level diagrams for this reaction:

    When 1 mol of liquid ethanol (C_2H_5OH) combines with 3 mol of oxygen gas to form 2 mol of carbon dioxide gas and 3 mol of water, the energy change is $-1\,367.3 \text{ kJ}$.

15.2   A closer look

Bond breaking and bond making

During chemical reactions, bonds are broken and formed. It takes energy to break chemical bonds and energy is released when bonds are formed.

Consider what happens, for example, when a substance A—B reacts with a substance C—D to form A—C and B—D. Overall, the reaction is:

$$A—B + C—D \rightarrow A—C + B—D$$

This reaction can be considered in more detail:

A—B	→	A and B	→	A—C
C—D	→	C and D	→	B—D
bonds intact in A—B and C—D	energy supplied (absorbed)	bonds broken to produce A, B, C, D	energy released	new bonds formed

The energy changes that occur as reactions proceed can be represented on graphs referred to as 'energy profile diagrams'.

Energy profile diagrams

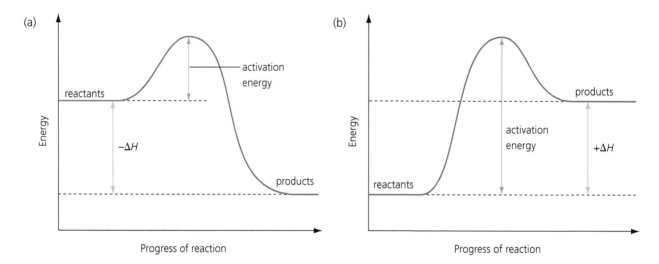

▲ **Figure 15.4**   Energy profile diagrams for (a) an exothermic reaction and (b) an endothermic reaction

You may notice several important features of the graphs in Figure 15.4. What actually happens is more complicated than in the energy level diagrams (Figure 15.3).

> The minimum energy that must be supplied before reaction proceeds is known as the **activation energy** of the reaction.

- **Activation energy** – in all chemical reactions old bonds must be broken before new ones can be formed. For this reason, reactants must be supplied with energy. Catalysts work by providing a pathway with lower activation energy (see Chapter 14).
- Energy is released during the formation of products.
- More energy is released than is supplied for exothermic reactions.
- More energy is supplied than is released for endothermic reactions.
- For exothermic reactions, the energy of the products (H_p) is less than the energy of the reactants (H_r) and ΔH is therefore negative.
- For endothermic reactions, H_p is greater than H_r and ΔH is positive.

Practice

3 Draw energy profile diagrams for the formation of carbon dioxide and the formation of nitrogen(II) oxide, using the equations shown on page 273.

4 The reaction between nitrogen and hydrogen to form ammonia (page 273) is exothermic and is catalysed by finely divided iron. Draw energy profile diagrams for this reaction (a) in the absence of the catalyst and (b) in the presence of the catalyst.

15.3 Measuring heat (enthalpy) changes in the laboratory

In measuring heat changes during chemical reactions, it is important to note that we never measure the actual heat content of chemicals. We simply measure the heat changes (q) that occur during the reaction. We can measure heat changes indirectly by measuring changes in temperature.

The **heat of neutralisation** is the energy change per mole of water formed during neutralisation of an acid by a base.

The **heat of solution** is the energy change when one mole of solute dissolves in a particular volume of solvent to form a very dilute solution.

We can measure the heat changes associated with specific chemical reactions: examples are the **heat of neutralisation** and the **heat of solution**.

When carrying out experiments to measure heat changes in the laboratory, the following items of apparatus are required:

- An insulated container to serve as a calorimeter (the apparatus to measure heat evolved or absorbed in a chemical reaction)

- A thermometer

- A balance

- Volumetric apparatus such as pipette and/or burette or measuring cylinder

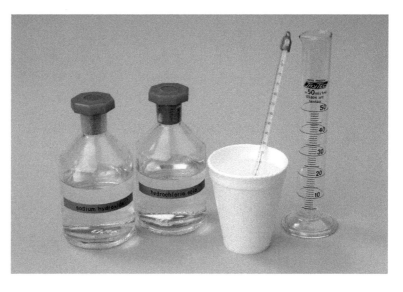

Figure 15.5
This apparatus is used in the laboratory to determine the heat of neutralisation of an acid/alkali reaction. ▶

Some general steps in the procedure are:

- allowing a known mass or volume of reactants to reach the steady temperature of the surroundings – this temperature is then recorded;

- thoroughly mixing the reactants and recording the highest or lowest temperature reached;

- determining the temperature change ($\Delta\theta$) for the reaction;

- calculating the heat evolved or absorbed in the experiment;

• calculating the enthalpy change for the reaction.

For a given mass (m kg) of reacting substance(s), the heat released or absorbed can be calculated from the relationship:

heat evolved or absorbed (in kJ) = mass (in kg) × specific heat capacity
(kJ kg^{-1} K^{-1}) × temperature change

$$= m \times c \times \Delta\theta$$

Specific heat capacity is the quantity of heat in joules required to raise the temperature of unit mass or unit volume of a substance by 1 °C or 1 K.

The heat evolved or absorbed per mole of the substance under investigation (ΔH, enthalpy) can then be calculated.

For aqueous solutions, it may be assumed that 1 cm^3 has a mass of 1 g, and that for dilute solutions, the **specific heat capacity** is 4.2 kJ kg^{-1} K^{-1}, i.e. a dilute aqueous solution has the same specific heat capacity as water.

Understand it better

How can we calculate heat change from temperature change?

In these experiments, we will calculate the heat change (quantity of heat absorbed or released) in a reaction by measuring the temperature change during the reaction and using the specific heat capacity of a dilute aqueous solution.

It is important to realise that 'temperature' is not the same as 'heat':

• Heat is a form of energy and the units of heat are joules (J) or kilojoules (kJ).
• Temperature is a measure of how much the particles in a substance are moving. The higher the temperature, the greater the average movement of the particles. You measure temperature with a thermometer and its units are °C.

When substances absorb heat, the movement of the particles in the substance may increase. The amount of heat that must be absorbed to increase temperature varies from one substance to another. For example, it takes 4 200 J of heat to raise the temperature of 1 kg water by 1 °C but 2 400 J of heat to raise the temperature of 1 kg of methylated spirit by 1 °C.

If we know the amount of heat required to raise the temperature of 1 kg of a dilute aqueous solution by 1 °C (that is, the specific heat capacity), and if we know the mass of the solution and the temperature change, then we can calculate the heat of reaction. We use this formula:

heat evolved or absorbed (in kJ) = mass (in kg) × specific heat capacity (c)
× temperature change ($\Delta\theta$)

Q1 How much heat is lost when 0.25 kg of water cools from 50 °C to 20 °C?

Q2 How much heat is gained if 0.64 kg of a liquid of specific heat capacity 2.5 kJ kg^{-1} K^{-1} heats up from 35 °C to 48 °C?

The following experiments show how to determine the enthalpy changes for three reactions.

Experiment 15.2   Measuring $\Delta H_{\text{neutralisation}}$

Aim

To measure the enthalpy change for the reaction between 2.0 mol dm^{-3} aqueous sodium hydroxide and 2.0 mol dm^{-3} hydrochloric acid.

This reaction is effectively the reaction between OH$^-$(aq) and H$^+$(aq). The reaction is exothermic.

⚠ The acid and alkali used in this experiment are concentrated. Handle with care.

Procedure

1 Run exactly 50 cm^3 of the hydrochloric acid from a burette into a plastic calorimeter, stir and record the steady temperature of the room (θ_1).

2 Run exactly 50 cm^3 of sodium hydroxide from a second burette into a plastic calorimeter, stir and set aside to allow to reach room temperature (θ_1).

3 Add the sodium hydroxide to the hydrochloric acid in the plastic cup, all at once. Stir, and note the maximum temperature reached (θ_2).

4 Calculate the temperature change: $\Delta\theta = \theta_2 - \theta_1$.

5 Calculate the quantity of heat released on mixing the sodium hydroxide and hydrochloric acid.

Calculations

heat released (in experiment) = 0.1 kg $\times$ 4.2 kJ kg^{-1} K^{-1} $\times$ $\Delta\theta$

0.1 kg corresponds to a volume of 100 cm^3 (i.e. 50 cm^3 of HCl + 50 cm^3 of NaOH).

50 cm^3 of 2.0 mol dm^{-3} sodium hydroxide acid solution contains 0.1 mol of sodium hydroxide.

50 cm^3 of 2.0 mol dm^{-3} hydrochloric acid solution contains 0.1 mol of hydrochloric acid.

In the reaction as carried out, 0.1 mol of sodium hydroxide reacted with 0.1 mol of hydrochloric acid to form 0.1 mol water.

Since enthalpy changes (neutralisation) are defined for 1 mol of water formed, the enthalpy change for this reaction is:

$$\Delta H \text{ (kJ mol}^{-1}) = \frac{-0.1 \times 0\ 4.2 \times \Delta\theta}{0.1}$$

Note: The − sign indicates that the change is exothermic.

Experiment 15.3   Measuring $\Delta H_{\text{solution}}$

Aim

To determine the enthalpy change when 1 mole of ammonium nitrate dissolves in water.

Many salts dissolve endothermically in water, i.e. the temperature of the water falls as the salts dissolve. The fall in temperature can be used to determine the enthalpy change of solution.

Procedure

1 Using a burette, place 50 cm³ of water into a plastic calorimeter.

2 Note and record the steady temperature (θ_1).

3 Weigh out 8.0 g of ammonium nitrate.

4 Add the ammonium nitrate crystals all at once, stirring to dissolve them.

5 Record the lowest temperature attained by the solution (θ_2) and calculate the value of $\Delta\theta = \theta_2 - \theta_1$.

6 Calculate the heat change when 8 g of ammonium nitrate dissolves in 50 cm³ of water.

Calculations

The enthalpy changes are displayed in Figure 15.6.

Figure 15.6
A diagrammatic representation of the enthalpy changes occurring when ammonium nitrate is dissolved in water. ▶

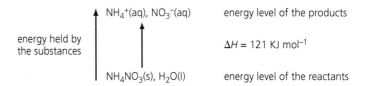

heat absorbed when 8 g of ammonium nitrate dissolves in 50 cm³ of water (approx. 0.05 kg) (in kJ) $= +0.05 \times 4.2 \times \Delta\theta$

Note: The $+$ sign indicates that the change is endothermic.

8.0 g of ammonium nitrate crystals $= 0.1$ mol

Therefore, the enthalpy change when 1 mol of ammonium nitrate dissolves completely in water $= \dfrac{+0.05 \times 4.2 \times \Delta\theta}{0.1}$ (kJ mol⁻¹).

M&M

A&I

Experiment 15.4   To determine heat change for a displacement reaction

Aim

To determine the heat evolved when zinc displaces copper from copper(II) sulphate solution. The apparatus used is illustrated in Figure 15.7.

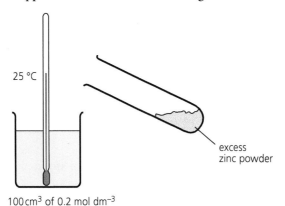

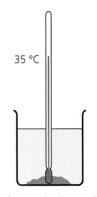

Figure 15.7
Determining the heat change for a displacement reaction ▶

100 cm³ of 0.2 mol dm⁻³ copper(II) sulphate solution

after the reaction is complete the temperature rose by 10 °C

The equation for the reaction is:

$$Zn(s) + CuSO_4(aq) \rightarrow Cu(s) + ZnSO_4(aq)$$

Question

Q1 Describe the steps that should be taken in carrying out this experiment.

Calculations

When this experiment was carried out by a student, the temperature increase was measured to be 10 °C.

We can now calculate q, the amount of heat produced when excess powdered zinc reacts with 100 cm³ of 0.2 mol dm⁻³ copper(II) sulphate solution.

$$q = m \times c \times \Delta\theta$$
$$m = 0.1 \text{ kg}$$
$$c = 4.2 \text{ kJ kg}^{-1} \text{ K}^{-1}$$
$$\Delta\theta = 10 \text{ °C}$$
$$q = -0.1 \times 4.2 \times 10$$
$$= -4.2 \text{ kJ}$$

100 cm³ of 0.20 mol dm⁻³ copper(II) sulphate contain 0.02 mol of $CuSO_4$.

Heat released when 1 mol of copper(II) sulphate solution reacts with excess zinc is:

$$\Delta H = \frac{-4.2}{0.02}$$
$$= -210 \text{ kJ mol}^{-1}$$

Further considerations

There are three sources of error in this experiment:

1 Some energy is lost to the surroundings before the maximum temperature is reached. This means that the temperature rise could have been a little more than 10 °C.

2 Some heat (energy) is transferred to the thermometer and container.

3 The specific heat capacity of the copper(II) sulphate solution was taken to be 4.2 kJ kg⁻¹ K⁻¹. This is only an approximate value for the specific heat capacity.

The reaction is exothermic, since the products are at a lower energy level than the reactants (Figure 15.9).

▲ **Figure 15.8**   Methane is made into methanol at the Methanol Holdings plant in Trinidad.

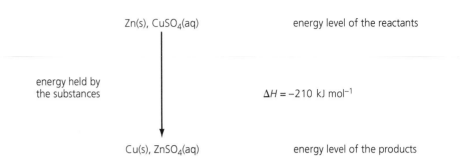

Zn(s), CuSO$_4$(aq)   energy level of the reactants

energy held by the substances $\Delta H = -210$ kJ mol^{-1}

Figure 15.9
The reaction when powdered zinc is added to copper(II) sulphate solution ▶

Cu(s), ZnSO$_4$(aq)   energy level of the products

Find out more

How do we use chemical energy?

Chemical energy, that is, energy stored in chemical bonds, is of vital significance to us. Millions of years ago, photosynthesis produced the plants that were changed into fossil fuels – oil, coal and natural gas. In a sense, the energy of the sun was trapped by photosynthesis, then stored in fossil fuels. When we burn the fuel, we release the energy again as heat and light.

- Coal is the most important solid fuel although it is difficult to mine and its bulk means it requires a lot of storage space.

- Liquid fuels include petrol and paraffin. They are obtained from crude oil.

- Methane (CH$_4$) is the most important gaseous fuel. It is found as natural gas, either alone or associated with crude oil.

Fossil fuels are examples of 'non-renewable' sources of energy – they cannot be replaced once they are used up. Also, pollutants are produced when fossil fuels are burnt.

▲ **Figure 15.10**   A pitch lake in Trinidad

The Caribbean depends heavily on oil as fuel, but only Trinidad and Tobago is a significant producer of fossil fuels, mainly oil and natural gas . These crude oil reserves are expected to last only another 10 years, but the natural gas production has increased significantly over the past few years.

Chemical energy in fuels is converted to many other forms of energy, such as electricity.

Q1 There is a great pitch lake in the south of Trinidad. It is believed to have been formed when petroleum seeped to the surface and the sun vaporised parts of the petroleum, leaving behind a vast lake of pitch.

Find out the names of two other Caribbean countries that are known to have oil and/or natural gas reserves.

Q2 Give examples of the conversion of chemical energy to:
  a mechanical energy;
  b heat energy;
  c electrical energy
  d light energy.

Q3 Find examples of renewable sources of energy used in your country.

Practice

The energy released when 1 mol of a substance burns completely in oxygen is known as the enthalpy of combustion. It is written as ΔH_c.

A student used the apparatus shown in Figure 15.11 to find a value of ΔH_c for ethanol.

He obtained the following results:
  initial mass of spirit lamp and ethanol = 65.20 g
  final mass of spirit lamp and ethanol = 64.28 g
  final temperature of water in can = 47.1 °C
  initial temperature of water in can = 28.5 °C
  mass of water = 0.30 kg

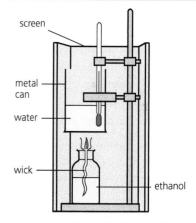

▲ **Figure 15.11** Finding the heat of combustion of ethanol.

5 What are the products of complete combustion of ethanol?

6 What mass of ethanol was burnt in this experiment? How many moles of ethanol is this? (M_r of ethanol = 46)

7 What quantity of heat was transferred to the 0.30 kg of water during the experiment?

8 Use your answers to questions 6 and 7 to determine a value for ΔH_c of ethanol.

9 Identify three possible sources of error in this experiment.

10 Do you consider ethanol to be a good fuel? Explain your answer.

Summary

- Chemical changes involve rearrangements of the atoms or groups within substances.
- Chemical reactions may be endothermic or exothermic.
- In endothermic reactions, heat energy is absorbed from the surroundings. The enthalpy change for endothermic reactions is positive.
- In exothermic reactions heat is released to the surroundings. The enthalpy change for exothermic reactions is negative.
- Energy is stored in chemical bonds.
- Energy must be supplied to break bonds.
- Energy is released when bonds are made.
- Energy profile diagrams show how the energy of substances changes during the course of a reaction.
- The activation energy is the minimum energy which reactant molecules must possess for a reaction to take place.
- Energy can be converted from one form to another. Chemical energy can be converted to heat, light, electricity and mechanical energy.

End-of-chapter questions

1 Using an alkali to neutralise an acid is:
  A Explosive
  B Endothermic
  C Exothermic
  D A displacement reaction

2 Endothermic reactions:
  A Produce heat
  B Take very long
  C Are more common than exothermic reactions
  D Result in a fall in temperature in the surroundings

3 Activation energy is the:
  A Minimum energy needed to start a reaction
  B Energy changes that occur during a neutralisation activity
  C Heat evolved at the start of a reaction
  D Heat of solution

4 A calorimeter is used in experiments that:
  A Measure pH
  B Measure energy changes
  C Measure the change in concentration of a solution
  D Only involve acids

5 3.25 g of zinc were added to 50 cm³ of 4.0 mol dm⁻³ hydrochloric acid. When the reaction was complete, the temperature had risen by 28 °C.
  a All the zinc dissolved in this reaction. How many moles of zinc reacted?
  b How many moles of acid are there in 50 cm³ of 4.0 mol dm⁻³ hydrochloric acid?
  c Write a balanced equation for the reaction between zinc and hydrochloric acid.
  d Which of zinc or hydrochloric acid is used in excess? Explain your answer.
  e Is the reaction between zinc and hydrochloric acid endothermic or exothermic?
  f Calculate the heat change when 3.25 g of zinc reacts with 50 cm³ of 4.0 mol dm⁻³ hydrochloric acid.
  g Calculate the enthalpy change for the reaction between zinc and hydrochloric acid. State all the assumptions you have made in your calculations.
  h Draw an energy level diagram for this reaction.

6 The heat given out when 1 mol (12 g) of carbon burns completely in oxygen is −386 kJ. The heat evolved when 1 mol (60 g) of propanol is completely burnt is −2 020 kJ.

  a Which of carbon or propanol has the higher energy value per gram?

  b Which of charcoal (carbon) and propanol (a liquid) is easier to store? Explain.

  c Assuming no heat losses, calculate the maximum temperature change when the heat released from burning 1 g of charcoal is used to heat 0.25 kg of water.

7 When 100 cm³ of 1.0 mol dm⁻³ nitric acid was added to 100 cm³ of 1.0 mol dm⁻³ sodium hydroxide in a well-lagged cup, the temperature rose from 27 °C to 33.9 °C.

  a How many moles of sodium hydroxide are there in 100 cm³ of a 1.0 mol dm⁻³ solution?

  b How many moles of nitric acid are there in 100 cm³ of a 1.0 mol dm⁻³ solution?

  c How many moles of sodium nitrate were produced in the above reaction?

  d Calculate the heat given out in this reaction.

  e What is the heat produced when 1 mol of nitric acid reacts with 1 mol of sodium hydroxide?

  (1 cm³ of a dilute solution has a mass of 1 g. Take the specific heat capacity of this solution to be 4.0 kJ kg⁻¹ K⁻¹.)

8 The heat of combustion of methanol is given as 640 kJ mol⁻¹.

  a Determine the heat released when 8 g of methanol are completely burnt in oxygen. (CH_3OH = 32.0)

  b If the heat released in part (a) is used to raise the temperature of 1 kg of water, how much would the temperature of the water rise? (The specific heat capacity of water can be taken as 4.0 kJ kg⁻¹ K⁻¹.)

9 10.1 g of potassium nitrate was dissolved in 50 cm³ of water. The temperature fell by 16 °C. Calculate:

  a the number of moles of KNO_3 dissolved in 50 cm³ of water;

  b the concentration of the resulting solution;

  c the heat change that would be produced if 1 mol of KNO_3 dissolves completely in 1 dm³ of water.

  (Take the specific heat capacity of the solution as 4.0 kJ kg⁻¹ K⁻¹ and heat change = mass × specific heat capacity × temperature rise.)

10 Draw suitable energy level diagrams to illustrate the changes taking place in the following reactions:

  a 50 cm³ of 2.0 mol dm⁻³ NaOH mixed with 50 cm³ of 2.0 mol dm⁻³ HCl gave a temperature rise of 13.6 °C.

  b Dissolving 20 g of ammonium nitrate in 250 cm³ of water leads to a drop in temperature of 5.0 °C.

  (4.2 kJ of energy is required to heat 1 kg of a dilute solution through 1 °C.)

Multiple-choice questions for Chapters 11–15

You will need to use this data in these questions:
A_r of C = 12, H = 1

1 A weak base:
  A Has a pH greater than 12
  B Does not turn litmus blue
  C Is a very dilute solution of a base in water
  D Produces only a few ions when dissolved in water

2 Which two terms can be used to describe the electrolysis of fused potassium chloride?
  A Redox and neutralisation
  B Substitution and thermal decomposition
  C Redox and decomposition
  D Ionic precipitation and substitution

3 In a reaction between potassium manganate(VII) and another substance, the following change took place:

$$MnO_4^-(aq) \rightarrow Mn^{2+}(aq)$$
  purple   colourless

Which of the following is true about the reaction, moving from left to right?
  A The oxidation number (state) of manganese (Mn) increased.
  B The oxidation number of manganese (Mn) remained the same.
  C The manganese in MnO_4^- is reduced.
  D The manganese in MnO_4^- has lost electrons.

4 In which of the following electrolytic processes is the electrode listed made the anode of the electrolytic cell?

Process	Electrode
I anodising of aluminium	aluminium
II electroplating of a steel object with chromium	pure chromium
III electrorefining of impure copper	pure copper

  A I only
  B I and II
  C II and III
  D I, II and III

Use this information to answer questions 5–7.
In a reaction between sodium carbonate and excess hydrochloric acid, a gas was produced. The volume of gas produced at different times was recorded as follows:

Time (s)	30	60	90	120	150	160
Volume (cm³)	10	18	25	30	32	32

5 What is the name of the gas produced in the reaction?
  A Carbon dioxide
  B Chlorine
  C Hydrogen
  D Oxygen

6 During which interval is the reaction the fastest?
  A 0–30 s
  B 60–90 s
  C 30–60 s
  D 90–120 s

7 Which **TWO** of the following will **NOT** affect the rate of the reaction?
  I   Increasing the temperature of the reaction
  II   Increasing the concentration of the acid
  III Increasing the volume of the acid
  IV   Increasing the pressure
  A I and II
  B III and IV
  C II and III
  D I and III

8 The heat of neutralisation is the heat change:
  A Per mole of acid used during neutralisation
  B Per mole of base used during neutralisation
  C Per mole of salt formed during neutralisation
  D Per mole of water formed during neutralisation

9 6 g ethane (C_2H_6) burn completely in oxygen, with the evolution of 156 kJ of heat. What is the molar heat of combustion of ethane?
  A 26 kJ mol^{-1}
  B 156 kJ mol^{-1}
  C 780 kJ mol^{-1}
  D 1 560 kJ mol^{-1}

10 Which statement is true of acid salts?
  A They are formed from monobasic acids.
  B They contain anions that can produce hydrogen ions.
  C They are made from weak acids only.
  D They are all insoluble.

11 Which method is most suitable for the preparation of copper(II) sulphate in the laboratory?
  A Add copper(II) oxide to dilute sulphuric acid
  B Add copper(II) nitrate to dilute sulphuric acid
  C Add copper metal to dilute sulphuric acid
  D Add copper(II) carbonate to sodium sulphate

12 In which of the following are all the salts soluble?
  A KCl, $NaBr$, $MgSO_4$, $CaSO_4$
  B $Ba(NO_3)_2$, $PbCl_2$, $KMnO_4$, Na_2CO_3
  C Na_2SO_4, K_2CO_3, $BaCl_2$, $AgNO_3$
  D $Al_2(SO_4)_3$, $BaSO_4$, K_2SO_4, $CaCl_2$

Use the following information for questions 13 and 14.
A student heated hydrated zinc nitrate, a colourless crystalline salt, in a hard glass test-tube. It was heated gently at first, and then more strongly.

13 Which of the following changes will the student observe?
  I Blue litmus held in the mouth of the tube turned red.
  II A glowing splint was rekindled when plunged into the tube.
  III The residue was orange whilst hot but changed to yellow on cooling.
  IV Droplets of a colourless liquid were seen on the upper walls of the tube.
  A I and II only
  B II and III only
  C I, II and IV only
  D I, II, III, and IV

14 The type of chemical reaction occurring is:
  A Dehydration
  B Oxidation
  C Neutralisation
  D Decomposition

15 A substance:
  • releases oxygen on heating;
  • turns acidified potassium dichromate(VI) from orange to green;
  • turns a solution of iron(II) sulphate from green to pale yellow.

Which of the following statements is true about the substance?
  A The substance oxidised potassium dichromate(VI).
  B The substance reduced iron(II) sulphate.
  C The substance has both oxidising and reducing properties.
  D The substance is not decomposed by heating.

16 Which of the following statements is/are true about a catalyst?
  I it is chemically changed in the reaction
  II it may be physically changed in the reaction
  III it alters the enthalpy change for the reaction
  IV it alters the energy of activation for the reaction
  A I only
  B II and IV only
  C III and IV only
  D IV only

17 Which statement is **TRUE** of electrolytes <u>and</u> metallic conductors?
  A They conduct electricity with decomposition.
  B They contain free moving charged particles.
  C They conduct electricity in the solid state.
  D They are the only substances that conduct electricity.

18 In which of the following electrolyses, using platinum electrodes, will the products at the cathodes be the same?
  I Aqueous sodium hydroxide
  II Molten sodium chloride
  III Dilute aqueous sodium chloride
  IV Concentrated aqueous sodium chloride
  A I, III and IV only
  B II and III only
  C III and IV only
  D II and IV only

19 Between which of the following reactants will an ionic precipitation reaction occur?

 A Calcium(II) chloride and sodium hydroxide
 B Barium nitrate and hydrochloric acid
 C Ammonium chloride and sodium hydroxide
 D Sodium hydroxide and hydrochloric acid

20 Which one of the following equations represents a displacement reaction?

 A $3Mg + 2FeCl_3 \rightarrow 2Fe + 3MgCl_2$
 B $BaCl_2 + Na_2SO_4 \rightarrow BaSO_4 + 2NaCl$
 C $HCl + NaOH \rightarrow NaCl + H_2O$
 D $2FeCl_2 + Cl_2 \rightarrow 2FeCl_3$

→ ### In this chapter, you will study the following:

- that carbon atoms can form single and double bonds, branched and unbranched chains, and ring compounds;
- the formulae to represent simple organic compounds;
- the general characteristics of a homologous series;
- the general and molecular formulae for members of a given homologous series;
- the homologous series given the fully displayed and condensed formulae of compounds.

This chapter covers
Objectives 2.1–2.5 of Section B of the CSEC Chemistry Syllabus.

Organic chemistry is the study of organic compounds. An organic compound is one that contains carbon atoms bonded to one another and to other atoms. Most organic compounds contain carbon and hydrogen atoms. These organic compounds are called hydrocarbons. Other types of organic compounds may contain oxygen, nitrogen and sulphur atoms. Organic chemistry is therefore concerned with the chemistry of hydrocarbons and their derivatives.

Not all carbon-containing compounds are organic. Examples of inorganic carbon compounds include the oxides of carbon, such as carbon monoxide and carbon dioxide, and metallic carbonates, such as sodium carbonate and magnesium carbonate.

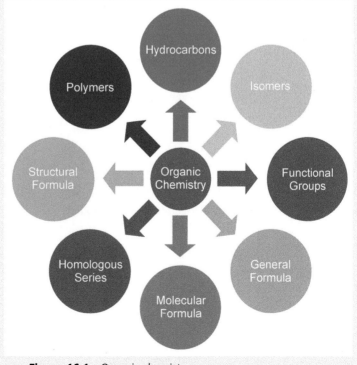

▲ **Figure 16.1** Organic chemistry

16.1 The bonding properties of carbon

The carbon atom has very unique bonding properties. One of these properties is that the carbon atom is tetravalent. This means that it has a valency of four and normally forms covalent bonds.

In addition, the carbon atom has the ability to catenate – form bonds between atoms of itself – and this allows it to form long branched and unbranched chains as well as ring structures with single, double and triple covalent bonds.

The structure of an organic compound may be based on a number of features.

Classification based on structure

Organic compounds may have a structure that is a:

*   straight chain structure;
*   branched chain structure;
*   cyclic structure.

Straight (unbranched) chain compounds

Straight-chain molecules are unbranched. They may contain single, double or triple bonds.

*   A straight-chain molecule containing carbon to carbon and carbon to hydrogen single bonds (see Chapter 17)

butane, C_4H_{10}

*   A straight-chain molecule containing carbon to carbon, carbon to hydrogen, carbon to oxygen and oxygen to hydrogen single bonds (see Chapter 18)

propan-1-ol, C_3H_7OH

*   A straight-chain molecule containing a carbon to halogen (bromine) single bond (see Chapter 17)

1-bromobutane, C_4H_9Br

- A straight-chain molecule containing a carbon to carbon double bond (see Chapter 17)

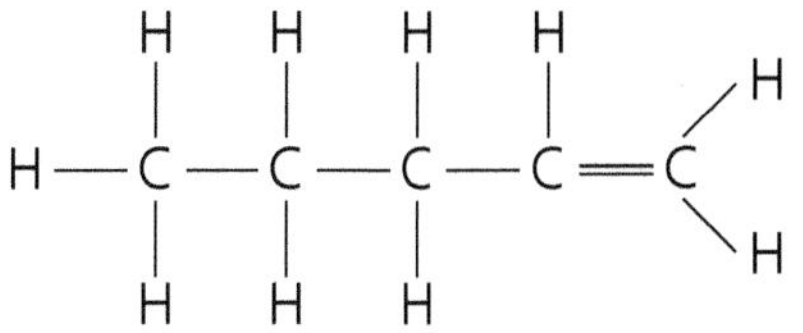

pent-1-ene
C_5H_{10}

- A straight-chain molecule containing a carbon to oxygen double bond, a carbon to oxygen single bond and an oxygen to hydrogen single bond (see Chapter 19)

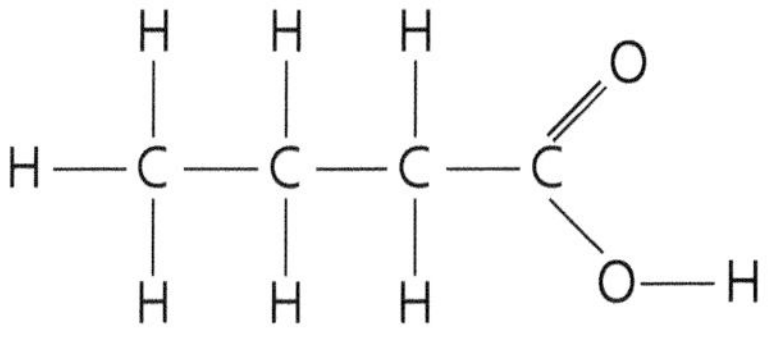

butanoic acid
C_3H_7COOH ($C_4H_8O_2$)

- These are straight-chain molecules containing carbon to nitrogen single and triple bonds respectively.

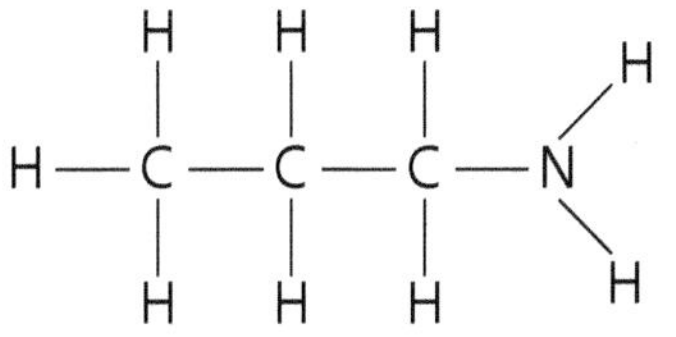

$C_3H_7NH_2$

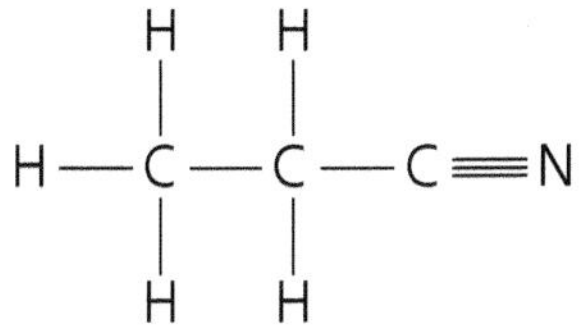

C_2H_5CN

Branched-chain compounds

Look at the structures in Figure 16.2. These are all examples of branched-chain compounds. Each compound has a main chain that contains five carbon atoms.

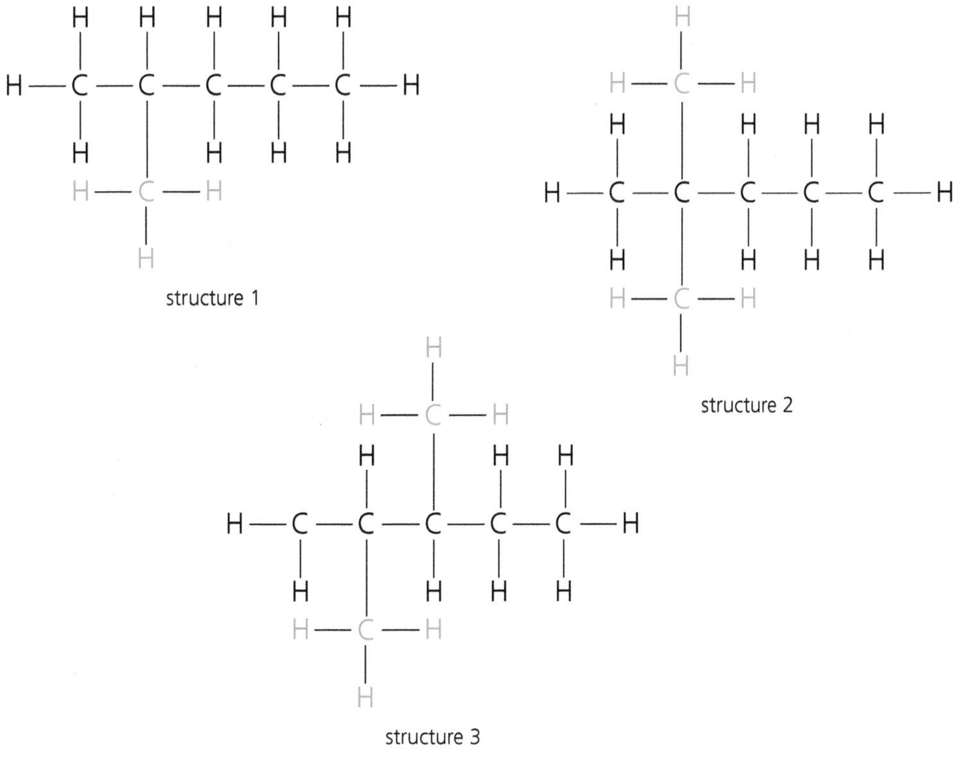

structure 1

structure 2

structure 3

Figure 16.2
Branched-chain compounds. Note: Some of the bonds are drawn longer for clarity. ▶

- Structure 1: This molecule has one branch, which is on the second carbon of the main chain.
- Structure 2: There are two branches on the same carbon atom of the main chain.
- Structure 3: There are two branches on different carbons of the main chain.

Note that every carbon atom in all the compounds has a total of four bonds (see Section 4.4).

Classification based on saturation

The saturation of a carbon atom suggests that it has four neighbouring atoms, i.e. four atoms directly bonded to a carbon atom. Chains can either be saturated (with carbon-carbon single bonds only) or unsaturated (with carbon-carbon double or triple bonds).

When all the carbon to carbon bonds within an organic compound are single bonds, we say that the organic compound is **saturated**.

When there is at least one carbon to carbon double or triple bond within an organic compound, we say that the compound is **unsaturated**.

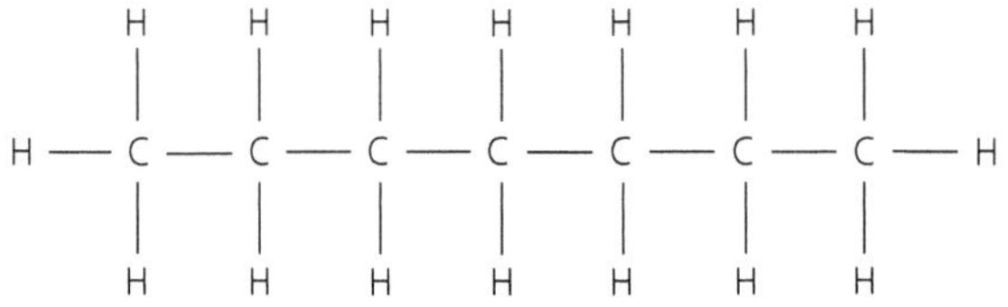

heptane
saturated (single bond)

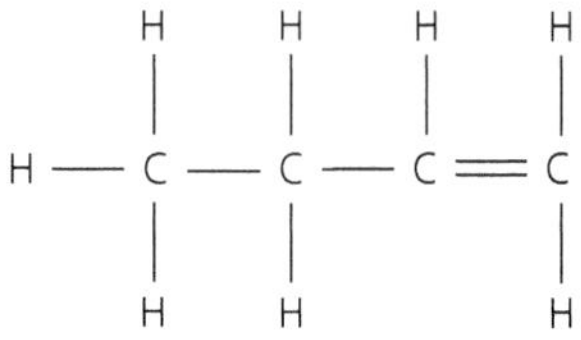

but-1-ene
unsaturated (double bond)

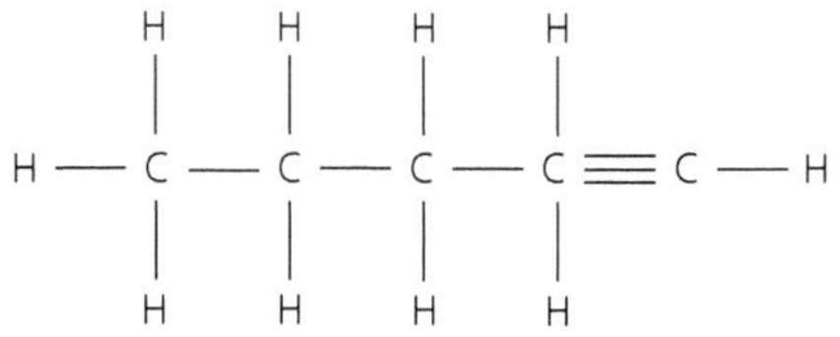

pent-1-yne
unsaturated (triple bond)

▲ **Figure 16.3** Saturated and unsaturated organic compounds

16.2 Formulae of organic compounds

There are five types of formulae that can be used to describe and identify organic compounds:

- Molecular formula
- Empirical formula
- Full structural or displayed formula
- Condensed structural formula
- General formula

We have discussed molecular and empirical formulae in Sections 7.1 and 8.2. These formulae do not give any information on how the atoms are arranged with respect to one another (see Section 7.1).

Molecular formula

The molecular formula of a chemical compound indicates the type of atoms and the number of each type of each atom in a molecule of a compound, for example, $C_8H_4O_2$.

Empirical formula

The empirical formula is the highest common factor of the number of atoms present in the compound. For example, in the compound $C_8H_4O_2$, the highest common factor of the number of atoms is 2. Therefore, the empirical formula would be C_4H_2O, which is the molecular formula divided by 2.

Structural formula

The structural formula of a chemical compound shows how the atoms in a molecule of a compound are arranged. There are two types of structural formulae: displayed and condensed.

Displayed structural formula

The displayed formula of a chemical compound is a diagram of the molecule that shows how the atoms are arranged. Each atom is shown as many times as it occurs and lines between them are used to show how they are bonded.

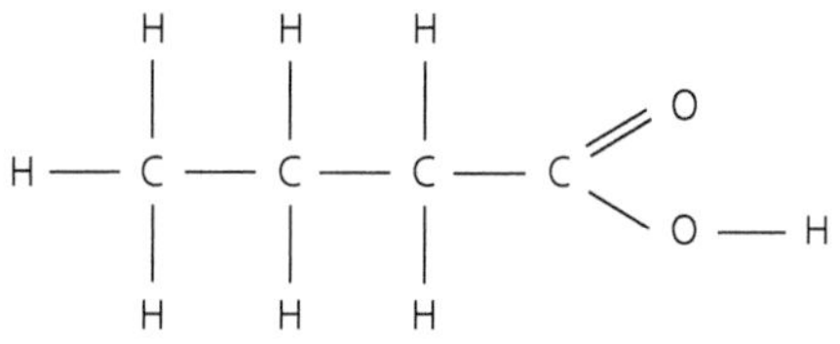

Figure 16.4
Butanoic acid ▶

The compound shown in Figure 16.4 is a displayed structural formula. Note the following:

- All bonds in the molecule are shown when writing full structural formulae.

- Full structural formulae are shown only in two dimensions – full structural formulae are flat and do not indicate the actual shapes of the molecules, nor do they show how the atoms are positioned in space relative to one another.

Condensed structural formula

The condensed structural formula of a compound gives less detail than the full structural formula. Figure 16.6 shows some full structural formulae and the corresponding condensed structural formulae. When you are interpreting condensed structural formulae, you need to remember that each carbon atom always forms a total of four covalent bonds in organic compounds.

A condensed structural formula indicates how atoms are grouped based on their structural arrangement without the bonds. For example, the condensed formula for butanoic acid in Figure 16.5 is $CH_3CH_2CH_2COOH$ or $CH_3(CH_2)_2COOH$. The $-CH_2$ group in parentheses is attached to the carbon written before it.

As another example, we write the condensed structural formula of the branched compound 2,2,3,3–tetramethylpentane as $CH_3CH_2C(CH_3)_2$ $(CH_3)_2CH_3$, and again the group in the brackets is attached to the atom written before it. The two red methyl ($-CH_3$) groups are attached to the blue carbon before it, and the two purple methyl ($-CH_3$) groups are attached to the yellow carbon before it.

The chemical compounds in Figure 16.6 are drawn with both structural displayed and condensed formulae (written in purple) written beneath each compound. Their IUPAC (International Union of Pure and Applied Chemistry) name is written in red.

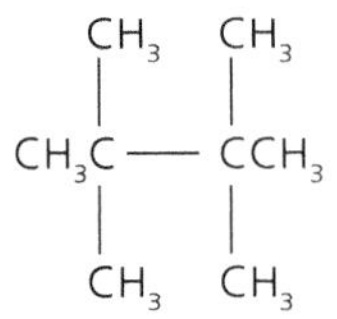

2,2,3,3-tetramethylbutane
$CH_3C(CH_3)_2C(CH_3)_3$

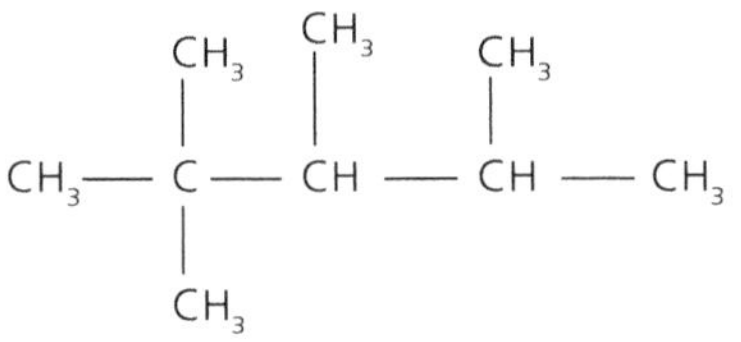

2,2,3,4-tetramethylpentane
$CH_3C(CH_3)_2CH(CH_3)CH(CH_3)_2$

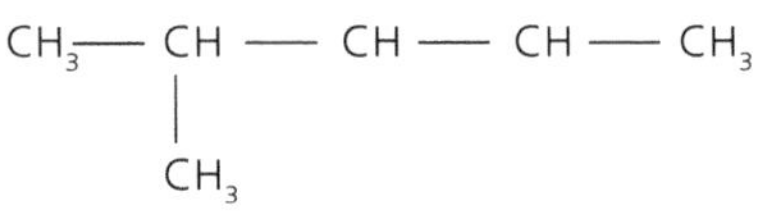

4-methylpent-2-ene
$CH_3CH(CH_3)CHCHCH_3$

▲ **Figure 16.6** Structural displayed and condensed formulae of some organic compounds

Practice

1 Determine the empirical formulae for each of these compounds.
 a $C_6H_{12}O_2$
 b C_7H_{14}
 c C_5H_{12}
 d C_6H_8O

2 Draw the structural displayed formula for each of these chemical compounds.
 a CH_3CHCH_2
 b $C(CH_3)_3C(CH_3)C(CH_3)C(CH_3)_2$
 c $C(CH_3)_2CH(CH_3)C(Br)(CH_3)CH_2CH_3$
 d $C(CH_3)_2C(OH)(CH_3)CH(CH_3)_2$

3 Write the condensed formula for each of these chemical compounds.
 a

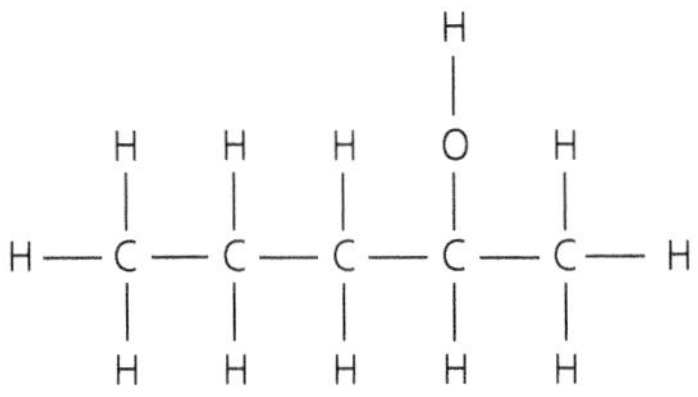

 b

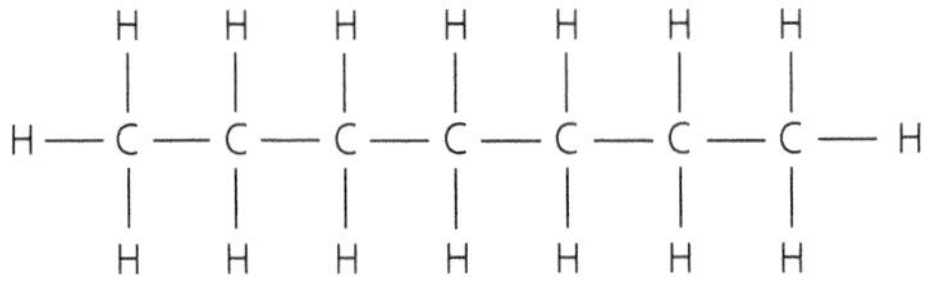

General formula

The general formula of a chemical compound is the simplest algebraic formula for a member of the homologous series. For example, the general formula for the alkanoic acid homologous series is $C_nH_{2n+1}COOH$.

Practice

4   Using the general formulae in Table 16.1, write the molecular formulae for:

   a   alkanes with values of $n = 1, 3, 4, 7$;

   b   alkenes with values of $n = 2, 3, 4, 10$;

   c   alkanols with values of $n = 1, 2, 3$;

   d   alkanoic acids with values of $n = 0, 1, 2, 3, 6$.

5   Use the following molecular formulae to determine the general formula of the homologous series to which each of these compounds belong:

   a C_3H_8

   b CH_3OH

   c C_2H_5COOH

   d C_4H_8

6   What happens if you make $n = 0$ for alkanes, alkenes and alkanols?

16.3   Homologous series and their functional groups

The functional group is an atom or group of atoms that determines the specific chemical properties of an organic compound in a homologous series. It is usually a reactive group such as the hydroxyl group (–OH) of alcohols, the amino group (–NH_2) of the amines, or the carboxyl group (–COOH) of organic acids.

A homologous series is a family of organic compounds that have similar general formulae. Compounds in a homologous series have similar chemical properties because they have the same **functional group** (i.e. the suffix, the end of the name, is the same), and show a gradation in physical properties because their molecular size and molecular mass increases through the series. For example, the alkanes all have the general formula C_nH_{2n+2}. Members of this homologous series include:

- methane (CH_4);
- ethane (C_2H_6);
- propane (C_3H_8);
- butane (C_4H_{10});
- pentane (C_5H_{12}).

The characteristics of a homologous series are the following:

- All members of a series have the same general formula.
- The molecular mass of each member of the homologous series differs from the membered positioned before or after it (its nearest neighbour) by a –CH2 unit, which corresponds to a molecular mass of 14.
- Members of a homologous series have similar chemical properties.
- For a given homologous series, physical properties (e.g. boiling point, density and enthalpy of combustion) change in a regular way as the molecular mass or carbon chain length of the compound increases.
- All members of a particular homologous series may be prepared by the same general series.

The most reactive functional group in a compound determines the suffix in the name. The prefix (beginning) part of the compound is called the **alkyl group** (see Table 16.1 below), which is determined by the number of carbons in the compound. For example, in the compound propanol C_3H_7OH, the C_3H_7 is the alkyl radical and the –OH is the functional group.

A functional group approach to the study of organic chemistry, greatly reduces the number of specific facts that you have to memorise. With this approach, you can predict the chemical properties of organic compounds from the functional groups they possess.

▼ **Table 16.1** Naming homologous series.

Homologous series	General formula	Functional group	Prefix + suffix	Examples
alkanes	C_nH_{2n+2} $n \geq 1$	carbon–carbon single bond	prefix + ane	
alkenes	C_nH_{2n} $n \geq 2$	carbon–carbon double bond	prefix + ene	
alkynes	C_nH_{2n-2} $n \geq 2$	carbon–carbon triple bond	Prefix + yne	
alkanols (alcohols)	$CnH_{2n-1}OH$ $n \geq 1$	hydroxyl group −OH	prefix + anol	
alkanoic (carboxylic) acids	$CnH_{2n-1}COOH$ $n \geq 0$	carboxylic acid where R represents an alkyl group	prefix + anoic acid	pentanoic acid
esters	$CnH_{2n-1}COOC_mH_{2m+1}$ $n \geq 0$ and $m \geq 1$	ester linkage	prefix + yl prefix + anoate	
aldehydes	$CnH_{2n-1}CHO$ $n \geq 0$		prefix + anal	

| ketones | $CnH_{2n-1}CHO$
$n \geq 2$ | 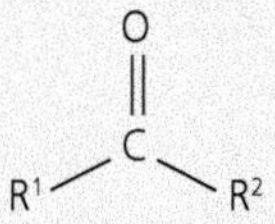

R¹ and R² may
be the same or
different alkyl
groups | prefix + anone | 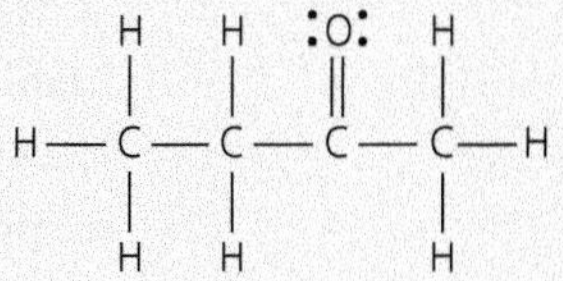 |
| amine | $CnH_{2n-1}NH_2$
$n \geq 1$ | 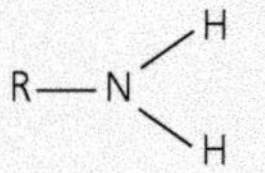 | prefix + yl amine/
amino | 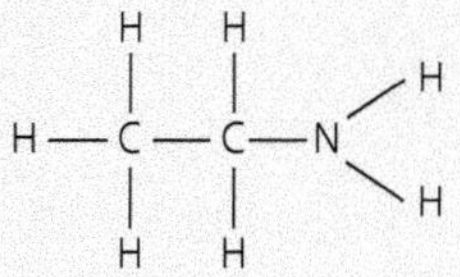 |

Practice

7 Draw diagrams to show the structural formulae of compounds with the following molecular formulae:

   a C_2H_6
   b C_2H_2
   c C_2H_5OH
   d C_4H_9COOH
   e CH_3NH_2
   f C_4H_8
   g $C_3H_7CONH_2$

8 Examine the structures shown below and answer the questions that follow.

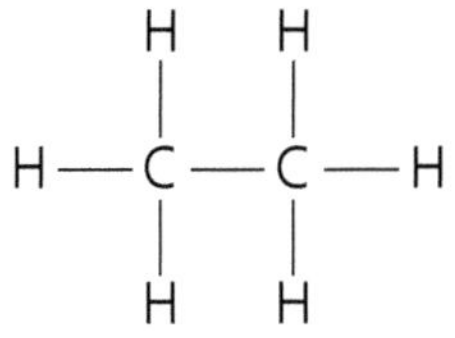

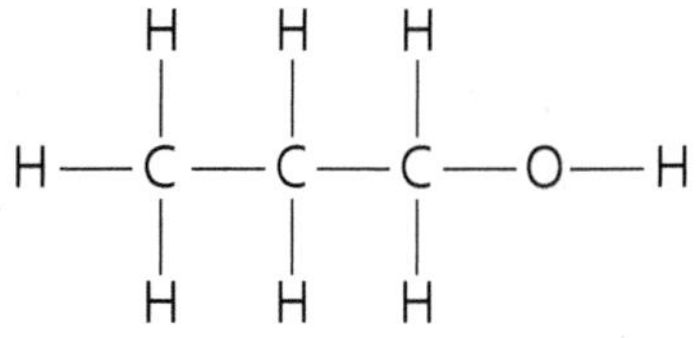

   a Identify two compounds belonging to the same homologous series.
   b What is the functional group in these compounds?
   c Separate the compounds A to E into saturated and unsaturated compounds.

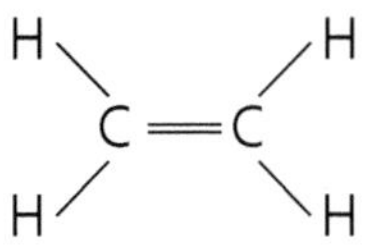

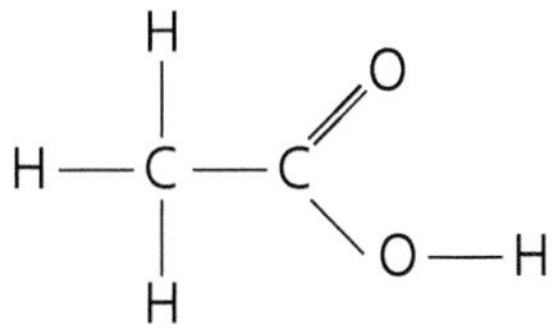

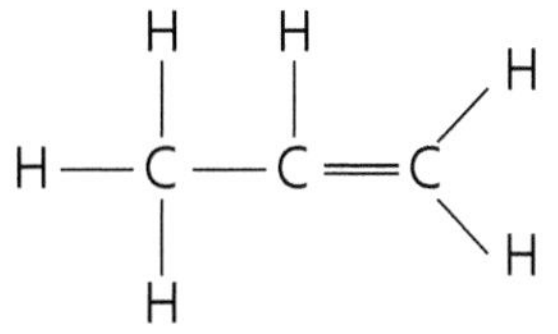

9 a What type of formulae is shown in the diagrams?
   b Write a different type of formula for any two of the compounds shown.
   c Name the type of formula you have written in each case.

16.4 The IUPAC system of naming organic compounds

The International Union of Pure and Applied Chemistry (IUPAC) derived a system for naming compounds. Through the IUPAC naming system, there are two parts to the name of an organic compound: the prefix and the suffix. The prefix is the hydrogen radical and the suffix is the functional group. The prefix is derived from the number of carbons present within the compound:

- **Meth** – one carbon atom
- **Eth** – two carbon atoms
- **Prop** – three carbon atoms
- **But** – four carbon atoms
- **Pent** – five carbon atoms

The remaining alkyl groups are named **hex-, hept-, oct-, non-** and **dec-** for six, seven, eight, nine and ten carbon atoms respectively.

Table 16.2 gives some examples of the names of different compounds and the homologous series to which they belong.

Naming straight-chain alkanes and alkenes

A straight-chain compound may look vertical, horizontal or even a zig-zag. As we stated earlier, the name of an organic compound has two parts. The prefix is based on the number of carbon atoms in the compound and the suffix depends on the functional group or the homologous series to which it belongs.

- The molecular formula allows you to determine the total number of carbon atoms in one molecule of the compound.
- The structural displayed formula allows you to determine a compound's functional group, and hence its homologous series.
    - If the compound has **only** carbon–carbon single bonds, the functional group is alkane and the suffix is '**–ane**'.
    - If the molecule has a carbon–carbon double bond, the functional group is alkene with the suffix '**–ene**'.
- Combine the prefix and the suffix to get the IUPAC name.

Hint: Always draw out the structural displayed formula for a given molecular formula. Never assume that the molecule is saturated or unsaturated.

▼ **Table 16.2** Naming straight-chain alkanes and alkenes.

Homologous series	Molecular formula	prefix based on number of carbon atoms	Structural displayed formula	Suffix based on homologous series	Name (combined prefix and suffix)
alkane	C_5H_{12}	pent–	H H H H H $\vert$ $\vert$ $\vert$ $\vert$ $\vert$ H—C—C—C—C—C—H $\vert$ $\vert$ $\vert$ $\vert$ $\vert$ H H H H H	–ane	pentane

▼ **Table 16.2** *continued*

alkene	C_7H_{14}	hept−		−ene	heptene

Naming straight-chain alkanols and alkanoic acids

Unbranched (straight-chain) alkanols and alkanoic acids are named as follows:

- Count the number of carbons in the molecular formula, which will give you the prefix.
- Draw out the structural displayed formula and identify the functional group.
  - If the structural displayed formula has a hydroxyl (–OH) group that is not attached to a carbon that is double bonded to an oxygen atom, it belongs to the alkanol homologous series and the suffix is '–**anol**'.
  - If the structural displayed formula has a carboxylic acid, it belongs to the alkanoic acid homologous series and the suffix is '–**anoic acid**'.
- Combine the prefix and the suffix to get the IUPAC name.

▼ **Table 16.3** Naming straight-chain alkanols and alkanoic acids.

Homologous series	Molecular formula	Prefix based on number of carbon atoms	Structural displayed formula	Suffix based on homologous series	Name (combined prefix and suffix)
alkanol	$C_8H_{17}OH$	oct−		−anol	octanol
alkanoic acid	$C_6H_{13}COOH$	hex−		−anoic acid	hexanoic acid

Naming branched-chain alkanes and alkenes

When naming a branched-chain chemical compound, it is always important to draw out the structural displayed formula. If you are given the molecular or structural condensed formula, draw out the structural displayed formula. If you do this, you reduce your chance of making a mistake when identifying the homologous series.

A branched-chain chemical compound has two parts:

- Parent chain: The parent chain is the longest continuous carbon atom chain. This chain can be vertical, horizontal or even a zig-zag.
- Substituent (alkyl) group: The substituent (alkyl) group of a branched chain compound is a group that has fewer carbons than the parent chain. A parent chain may have more than one alkyl group attached to it.

- When there are two of the same alkyl groups attached to a parent chain, '**di**' is written before the alkyl group.
- When there are three of the same alkyl groups attached to the parent chain, '**tri**' is written before the alkyl group.
- When there are four of the same alkyl groups attached to the parent chain, '**tetra**' is written before the alkyl groups.

Alkyl groups cannot exist on their own. They are usually linked to a hydrogen atom, as in alkanes, or to some functional group, for example, –O–H in alkanols and –COOH in alkanoic acids. We use the formula R–H to represent an alkane, R–O–H to represent an alkanol and R–COOH to represent an alkanoic acid.

We get the formula of each alkyl group by inserting the value of 'n' into the general formula (see Table 16.4). The prefix of the name of each alkyl group depends on the number of carbon atoms in the group and then the suffix '-yl' is added.

▼ **Table 16.4** Naming branched-chain alkanes and alkenes

Value of n	General formula of alkyl group (C_nH_{2n+1})	Formula of alkyl group	Name of alkyl group
1	$C_1H_{2(1)+1}$	CH_3	methyl–
2	$C_2H_{2(2)+1}$	C_2H_5	ethyl–
3	$C_3H_{2(3)+1}$	C_3H_7	propyl–
4	$C_4H_{2(4)+1}$	C_4H_9	butyl–
5	$C_5H_{2(5)+1}$	C_5H_{11}	pentyl–
6	$C_6H_{2(6)+1}$	C_6H_{13}	hexyl–

The following steps describe how to name a chemical compound using the IUPAC system of nomenclature.

Step 1: Identify and name the parent chain as above. This is written last.

Step 2: Number the parent chain starting with the end nearest a (branch) substituent.

- Start numbering from the end nearest a substituent in such a way that the numbers of the substituent carbons are as low as possible. We call these numbers **locant** numbers.
- If the parent chain is an alkene, write the lower of the two locant numbers for the carbon atom that holds the double bond. For example, if the double falls between carbon atoms three and four, write the three to specify the position of the double bond.

Step 3: Circle and name the substituent (alkyl) group(s) that is/are attached to the parent chain.

- If the alkyl group has the formula –CH_3, the name of the branch is methyl.
- If there are two alkyl groups, then write 'dimethyl'
- If the parent chain has different types of alkyl groups attached, place them in alphabetical order.

Step 4: Write the locant number of the carbon atom that has the alkyl (substituent) group attached to it.

- If there are two alkyl groups attached to the same carbon atom on the parent chain, write the locant number twice.
- If there are two or more different alkyl groups on the parent chain, each will need its own locant number. You must write this locant number before the attached alkyl group that it identifies, for example, 2-ethyl-4-methylheptane.
- Use a hyphen to separate numerals and words.
- Use commas to separate numerals.

Step 5: Write the substituent groups first followed by the parent name.

Worked example 1

What is the IUPAC name for the chemical compound given below?

(structural formula of 2,3-dimethylbutane)

Solution

Step 1: Identify the parent chain. The longest chain has four carbon atoms, hence the prefix is **but–**. As there are only carbon–carbon single bonds in the compound, the suffix is **–ane** as it belongs to the alkane homologous series.

Step 2: Number the parent chain to get the locant numbers of the substituent groups. The substituent groups are two methyl groups, and so we have **dimethyl**. They are on two different carbon atoms, and so the locant numbers are 2 and 3. Thus the name of the compound is 2,3-dimethylbutane.

Practice

10 Name the compounds represented by the following structural displayed formulae:

a   CH₃CH₂CH₂CH —— CH —— CH —— CH₂
                       | | | |
                      CH₃   CH₂   CH₃   CH₂
                                  | |
                                CH₂   CH₂
                                  | |
                                CH₃   CH₃

b

<pre>
                    H
                    |
        H | H
        | | |
   H —— C —— C —— C —— H
        | | |
        H | H
                    |
              H —— C —— H
                    |
                    H
</pre>

11 Name the compounds with the following molecular formulae:
a C_3H_8
b HCOOH
c C_4H_9OH
d C_5H_{10}

12 For each of the compounds in question 11, write:
a the empirical formula;
b the full structural formula.

13 Name the compounds represented by the following full structural formulae:

a

<pre>
        H   H   H   H
        | | | |
   H —— C — C — C — C —— H
        | | | |
        H | H   H
            |
      H —— C —— H
            |
      H —— C —— H
            |
            H
</pre>

b

<pre>
              H
              | H
              | /
   H —— C —— C == C
              | \
              H   H
              |
        H —— C —— H
              |
              H
</pre>

Naming organic compounds with more than one functional group

So far, we have discussed compounds with only one functional group. However, organic compounds may have more than one functional group, and these groups may be the same or different.

To name such compounds, we need to indicate:
- the type or types of functional groups;
- the number of functional groups;
- the positions of the functional groups that the compound contains.

We will now work out names for these compounds:

- Count the number of carbon atoms to get the main part of the name. Compound H has two carbon atoms, so its name is based on ethane. Compound I has six carbon atoms, so its name is based on hexane.
- Both molecules contain the same functional group, which is a hydroxyl group, so we need to have '-ol' in the names of the compounds.
- The compounds are diols because there are two alkanol functional groups.
- We need to use numbers to indicate the positions of the functional group in the two molecules. The numbers we need are '1,2-' for H and '1,6-' for I.

H is called ethane-1,2-diol. I is called hexane-1,6-diol.

Sometimes molecules contain two or more different functional groups. Here are two examples:

2, 3-dihydroxybutanoic acid
compound H

2-aminoethanoic acid
compound I

Practice

14 Draw full structural formulae for the following named compounds:
   a 2-chlorobutane
   b 2-aminopropanoic acid
   c 3-ethylpentane
   d 1,1,1-trichloroethane
   e 2,3-dimethylbut-2-ene

15 Name the compounds with the following structures:
   a

b

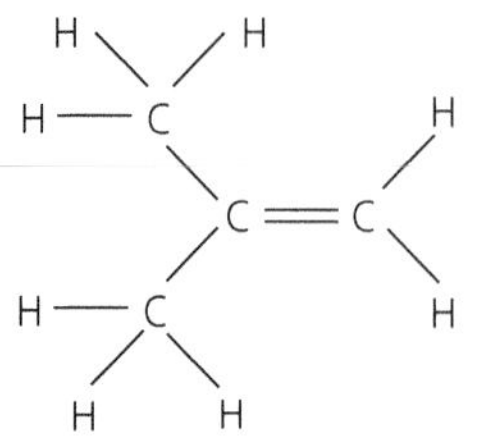

c

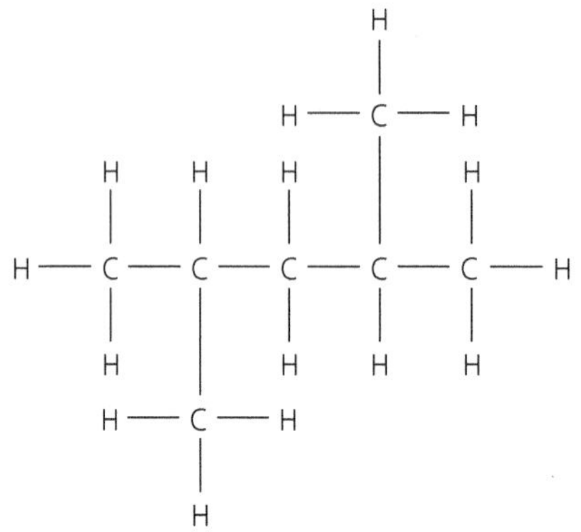

d

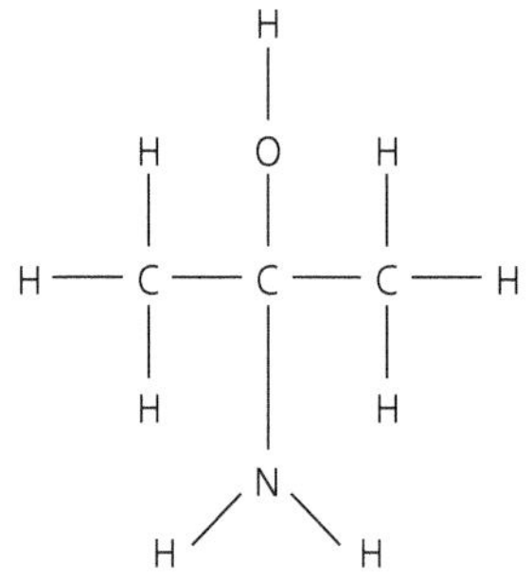

16.5 Deducing the homologous series from the structural formulae of organic compounds

We determine the homologous series of an organic compound from the functional group in the compound. An organic compound may contain more than one functional group. For example, amino acids contain an amine and a carboxylic acid functional group.

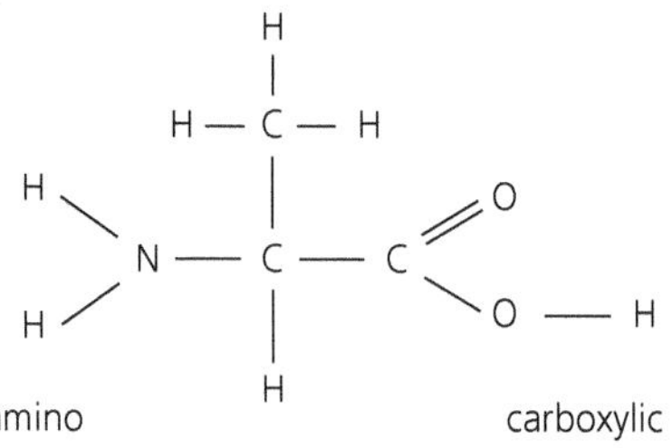

amino   carboxylic
                         acid

Figure 16.7
Amino acids contain an amine and carboxylic functional group. ▶

The functional group in an organic compound determines the homologous series to which the organic compound belongs (see the tables in the previous section).

16.6 Determining the homologous series when given the molecular formula

- When given the molecular formula for an organic compound, draw out the structural displayed formula. When drawing the structural displayed formula, first write down all the carbon atoms in the longest chain and attach the functional group. If the functional group belongs to the alkene homologous series, use a double line between the two carbon atoms based on the position given in the molecular formula.

- Attach all the other groups according to their frequencies and positions. Finally, insert the hydrogen atoms. Only use the same number of atoms that you are given in the molecular formula of the organic compound.

- Check each carbon to ensure that it has four bonds, i.e. four lines attached to it.

- Circle the functional groups.

- Name the homologous series based on the functional groups.
  - Carbon–carbon single bond (—C—C—) only: Alkane
  - Carbon–carbon double bond(—C=C—): Alkene
  - Oxygen–hydrogen single bond(—OH—): Alkanol (alcohol)
  - Carbon double bonded to oxygen also bonded to a hydroxyl group Alkanoic (carboxylic) acid

$$\begin{array}{c} O \\ \parallel \\ C \\ \diagup \quad \diagdown \\ \qquad OH \end{array} \; :$$

Worked example 2

Determine the homologous series to which each of the following organic compounds belongs:

a C_5H_{10}
b C_4H_{10}
c C_3H_7OH
d HCOOH
e $CH_2CH(CH_2)_3CH_3$

Solution

a Draw out the structural displayed formula, then circle the functional group. Note that all the carbons are tetravalent.

$$\begin{array}{c} \quad\;\; H \quad\;\; H \quad\;\; H \quad\;\; H \\ \quad\;\; | \quad\;\; | \quad\;\; | \quad\;\; | \\ H\diagdown \\ \quad\; C = C \; - \; C \; - \; C \; - \; C \; - \; H \\ H\diagup \\ \quad\;\; | \quad\;\; | \quad\;\; | \\ \quad\;\; H \quad\;\; H \quad\;\; H \end{array}$$

Therefore, C_5H_{10} belongs to the **alkene** homologous series, as it has a carbon–carbon double bond.

b C_4H_{10}

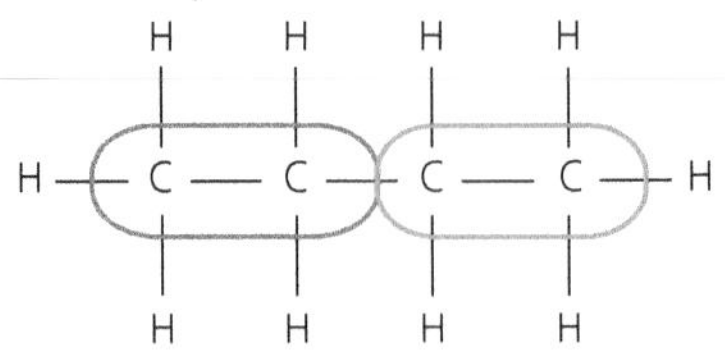

Note that there are only carbon–carbon single bonds, and so C_4H_{10} is an **alkane.**

c C_3H_7OH

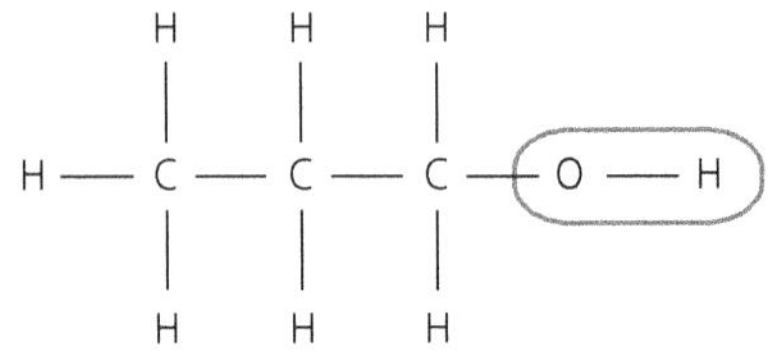

The compound has a hydroxyl (oxygen bonded to a hydrogen) group, and so it belongs to the **alkanol (alcohol)** homologous series.

d HCOOH

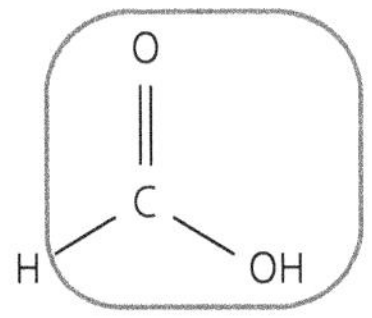

The compound contains –COOH, and so it belongs to the **alkanoic (carboxylic) acid** homologous series.

e $CH_2CH(CH_2)_3CH_3$

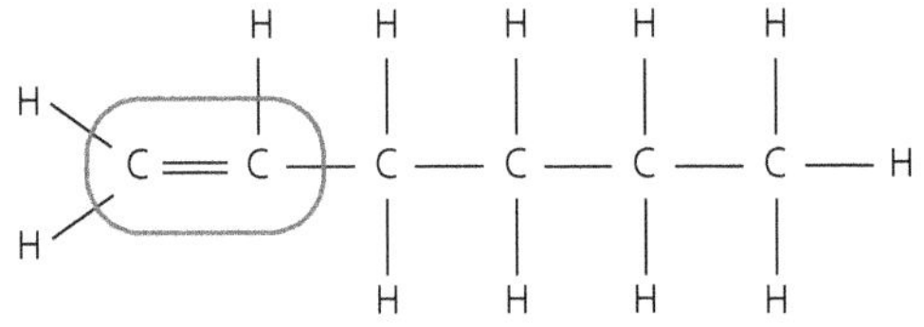

The compound contains a carbon–carbon double bond, and therefore it is an **alkene.**

Practice

16 Determine the homologous series to which each of the following compounds belongs:

a $CH_3(CH_2)_3CH_3$
b CH_3COOH
c CH_2CH_2
d $CH_3C(CH_3)_2C(CH_3)CH_2$

17 Tick the alkene, streak the alkanol and place a circle over the alkanoic acid functional groups that are present in this structural formula.

16.7 Distinguishing an alkane and an alkene using a general formula approach

You can determine the homologous series of a hydrocarbon (alkane or alkene) from its general formula using algebra. The general formula for an alkene is C_nH_{2n}. This means that the number of hydrogen atoms in one molecule of that compound is twice (double) the number of carbon atoms. So, if you divide the number of hydrogen atoms by the number of carbon atoms, the compound is an **alkene** if the answer is 2.

The general formula of an alkane is C_nH_{2n+2}. This means that the number of hydrogen atoms in one molecule of that compound is twice the number of carbon atoms plus two. Hence, if you divide the number of hydrogen atoms by the number of carbon atoms, the compound is an **alkane** if the answer is 2 with a remainder of 2.

Worked example 3

Determine whether each of the molecular formulae given below is an alkane or alkene using the general formula approach:

a C_5H_{10}
b C_6H_{14}

Solution

a The number of carbon atoms is 5 and the number of hydrogen atoms is 10.

$$\frac{10}{5} = 2$$

Since the answer is 2 with no remainder, the compound C_5H_{10} belongs to the alkene homologous series.

b The number of carbon atoms is 6 and the number of hydrogen atoms is 14.

$$\frac{14}{6} = 2 \text{ remainder } 2$$

Since the answer is 2 with a remainder of 2, the compound C_6H_{14} belongs to the alkane homologous series.

Practice

18 Using the general formula approach, determine whether each of the molecular formulae given below is an alkane or an alkene:

   a $C_{15}H_{30}$

   b $C_{12}H_{26}$

   c $C_{10}H_{22}$

   d C_8H_{16}

Summary

- In organic compounds, the carbon atoms are bonded to hydrogen and sometimes other atoms including oxygen, nitrogen and sulphur. Hydrocarbons are organic compounds that have only carbon and hydrogen atoms.

- Hydrocarbons are classified as alkane, alkene, alkynes and arenes, depending on their structure.

- When drawing the structural displayed formula, first write down all the carbon atoms in the longest chain and attach the functional group. If the functional group belongs to the alkene homologous series, use a double line between the two carbon atoms based on the position given in the molecular formula.

- The word 'organic' means present in, or derived from, plants and animals.

- All organic compounds contain carbon covalently bonded to one or more atoms of hydrogen, oxygen, nitrogen and the halogens.

- The large number and wide range of organic compounds are due to the unique bonding properties of the carbon atom.

- A homologous series is a set of organic compounds having the same general formula, similar chemical and physical properties, which change in a regular way with the number of carbon atoms and whose molecular masses differ by whole multiples of 14 (corresponding to $-CH2-$ units).

- A general formula is a mathematical representation of the formula of any member of a homologous series.

- The functional group is that atom or group of atoms in an organic compound whose reactions determine certain chemical properties.

- Some organic compounds have more than one functional group.

- The IUPAC (International Union of Pure and Applied Chemistry) system is used to name organic compounds.

End-of-chapter questions

1 Why is it possible for carbon to form chain and ring structures with single, double or triple covalent bonds?

  A It has the ability to catenate.
  B It has the ability to share electrons.
  C It is in group four.
  D It is not very reactive.

2 For the compound given below, identify the homologous series to which it belongs.

  A Alkanes
  B Alkenes
  C Alkanols
  D Alkanoic acids

3 Which of the following is an unsaturated compound?

  A $CH_3CH_2CH_2CH_3$
  B CH_2CHCH_3
  C $CH_3CH_2(OH)$
  D $CH_3CH(OH)CH_3$

4 Determine the general formula for the homologous series to which the compound below belongs.

  A CnH_{2n} $(n \geq 2)$
  B CnH_{2n+2} $(n \geq 1)$
  C $CnH_{2n+1}OH$ $(n \geq 1)$
  D $CnH_{2n+1}COOH$ $(n \geq 0)$

5 Caprylic (octanoic) acid is naturally found in palm and coconut oil, and in the milk of humans and bovines. Nutritionists often recommend caprylic acid to treat candidiasis (yeast infections) and bacterial infections. The structure for caprylic acid is given below.

  a Which of the following condensed formula is correct for caprylic acid?
    A $CH_3(CH_2)_6COOH$
    B $CH_3(CH_2)_5COOH$
    C $CH_3(CH_2)_4COOH$
    D $CH_3(CH_2)_3COOH$
  b What is the general formula for caprylic acid?
    A CnH_{2n} $(n \geq 2)$
    B CnH_{2n+2} $(n \geq 1)$
    C $CnH_{2n+1}OH$ $(n \geq 1)$
    D $CnH_{2n+1}COOH$ $(n \geq 0)$
  c What is the IUPAC name for caprylic acid?
    A Pentanoic acid
    B Hexanoic acid
    C Heptanoic acid
    D Octanoic acid
  d How many carbons are there in the chemical compound that comes after caprylic acid?
    A 7
    B 8
    C 9
    D 10

6 What feature of an organic compound determines its chemical properties?
    A Homologous series
    B Structural formula
    C Functional group
    D Active site

7 Which of the following is true of a homologous series?

I   Gradation in physical properties

II Each consecutive members differs by $-CH_2$

III Same general formula

IV Different methods of preparation

  A I and II only

  B II and IV only

  C I, II and III only

  D I, II and IV only

8 Which of the following lists of chemical compounds belong to the same homologous series?

  A C_3H_6, C_3H_8, C_3H_7OH, C_3H_7COOH

  B C_4H_{10}, C_5H_{12}, C_8H_{18}, $C_{10}H_{22}$

  C CH_3OH, C_2H_5COOH, C_3H_7OH, $HCOOH$

  D C_3H_8, C_6H_{12}, C_5H_{10}, C_2H_4

9 A student is employed in a chemistry laboratory for the summer. She is given six chemical compounds with their structural condensed formulae and is asked to categorise them as alkanes, alkenes, alkanols, alkanoic acids and esters. Copy and complete the table below that will help her to categorise these chemicals:

- CH_3CH_2COOH
- $CH_3CH_2CH_2CH_2CH_3$
- CH_2CHCH_3
- $CH_3C(CH_3)_2CH_2CH_2OH$
- $CH_3CH_2COOCH_3$
- $CH_3CH(OH)CH_3$

Alkanes	Alkenes	Alkanols	Alkanoic acids	Esters

10 The structural formula below has five different functional groups. Copy the diagram and circle each functional groups.

11 Use the compounds below, to answer the questions that follow.

I $CH_3(CH_2)_3CH_3$

II

III

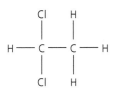

a What type of formula describes all of these compounds?
b To which homologous series do the compounds belong?
c Which of the compounds is unsaturated?
d What is the general formula for compound III?
e Draw the structural displayed formula for the compound that appears before compound I in its homologous series.

12 The carbon atoms in ethene (shown below) are unsaturated. Explain what this means in terms of the structural displayed formula of ethene.

$$H \diagdown \quad \diagup H$$
$$C = C$$
$$H \diagup \quad \diagdown H$$

13 Ethene is made by cracking hydrocarbons obtained from petroleum. Ethene is naturally found in plants and is used to accelerate the ripening of fruits. Ethene is a member of homologous series with the general formula C_nH_{2n}.
  a Name the homologous series to which ethene belongs.
  b Draw and name the structural displayed formula for the third member of this series.
  c Ethene decolourises bromine solution from reddish brown to colourless. What is responsible for this?

14 List four reasons why carbon forms so many compounds.

15 Using examples, distinguish between a molecular formula and a structural formula.

16 What is meant by the term 'homologous series'?

17 An organic compound has the molecular formula $C_4H_8O_2$. Write possible structural formula corresponding to this molecular formula.

18 Write structural formulae corresponding to each of the following names:
  a Butanoic acid
  b Dichloromethane
  c 1,2-dibromoethane
  d 2,2-dimethylpropane

19 Give the names of the compounds with the following structural formulae:
  a

$$\begin{array}{ccc} Cl & & H \\ | & & | \\ H - C & - & C - H \\ | & & | \\ Cl & & H \end{array}$$

b

```
       H   H
       |   |
   H — C — C — Br
       |   |
       H   H
```

c

```
       H   H   H
       |   |   |
   H — C — C — C — O — H
       |   |   |
       H   H   H
```

20 This compound is a constituent of sour milk:

```
       H   H       O
       |   |      //
   H — C — C — C
       |   |      \
       H   O       O — H
           |
           H
```

    a What is the molecular formula of this compound?
    b Give the names of two functional groups present.
    c Give the structural formula of a possible isomer of this compound.

21 The active ingredient in rubbing alcohol is:

```
       H   H   H
       |   |   |
   H — C — C — C — H
       |   |   |
       H   O   H
           |
           H
```

    a Name this compound.
    b Write the structural formula of a possible isomer. Name the isomer.

22 2-methylpropan-2-ol is a gasoline additive.
    a Write the structural formula of this compound.
    b Write the structural formula of an isomer of this compound. Name the isomer.

23 A compound contains 85.7% carbon and 14.3% hydrogen. Its relative molecular mass is 70.
    a Determine the empirical formula of the compound.
    b What is the molecular formula of the compound?
    c Is the compound an alkane or an alkene?
    d Write structural formulae of three possible isomers with the molecular formula obtained in part (b).
    e Name each isomer from your answer to part (d).
    C = 12.0, H = 1.00

17 Hydrocarbons: Alkanes and alkenes, and their derivatives

In this chapter, you will study the following:

- natural gas and petroleum as natural sources of hydrocarbons;
- the main uses of at least three fractions obtained from the fractional distillation of petroleum;
- the cracking of petroleum fractions;
- fully displayed structures and names of branched and unbranched alkanes, and unbranched alkenes, alcohols and alkanoic acid;
- isomerism fully displayed structures of isomers;
- the reactions of alkanes and alkenes;
- the characteristic reactions of alkanes and alkenes;
- how to distinguish between alkanes and alkenes;
- how to relate the properties of hydrocarbons to their uses.

This chapter covers
Objectives 2.6–2.8 and 3.1–3.4 of Section B of the CSEC Chemistry Syllabus.

Hydrocarbons are organic compounds that contain only carbon and hydrogen atoms, which are covalently bonded together. We will discuss the hydrocarbons alkanes and alkenes in this chapter.

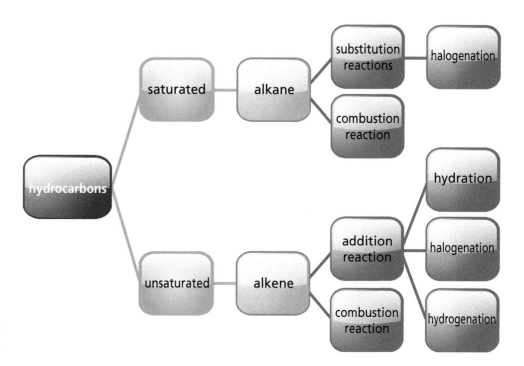

Figure 17.1
The chemistry of hydrocarbons ▶

17.1 Natural sources of hydrocarbons

Fossil fuels are fuels that have been made by the anaerobic decomposition of buried dead plants and animals. These remains have been exposed to high temperatures and pressure for millions of years. Fossil fuels include coal, peat, natural gas and petroleum (crude oil).

Coal and peat are formed from the remains of plants, while the other fossil fuels are formed from both aquatic plants and animals.

Natural gas and petroleum occur together in the Earth's crust and are mined using special drills called rigs. After being mined, the petroleum and natural gas are sent to oil refineries to be separated and purified.

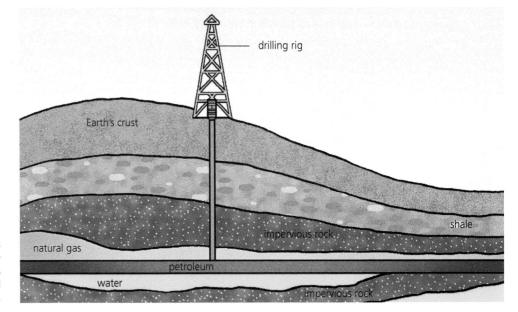

Figure 17.2
Hydrocarbons are mined from the Earth's crust using drills called rigs. ▶

Petroleum is the general term for all naturally occurring hydrocarbons, and includes gasoline, diesel and fuel oils such as kerosene oil. Petroleum is a thick, flammable, yellow-to-black mixture of solid, liquid and gaseous hydrocarbons that occurs naturally beneath the Earth's surface.

- **Solid hydrocarbons**: Asphalt
- **Liquid hydrocarbons**: Crude oil
- **Gaseous hydrocarbons**: Natural gases, for example, methane, propane and butane

17.2 Fractional distillation of crude oil

Fractional distillation (discussed in Chapter 5) is how we separate a liquid mixture into its separate fractions (components). This is achieved by increasing the temperature of the mixture and collecting the fractions that boil off over different ranges of temperature.

We can use fractional distillation to separate crude oil into its different fractions: petrol and diesel. Each fraction of crude oil is a mixture of hydrocarbon chains of similar length. The products of crude oil are miscible and have different boiling points. This is why we can use fractional distillation to separate it. We use fractional distillation to separate miscible liquids with different boiling points.

Table 17.1 lists the products of crude oil with their respective boiling points.

▼ **Table 17.1**   Products of crude oil and their boiling points

Fractions (products)	Boiling Point (K)
gases	310
petrol	310–450
naphtha	400–490
kerosene	430–523
gas oil	590–620
fuel oil and wax	above 620

Crude oil is separated into its fractions as follows:

- The crude oil is first heated in a furnace, which is treated to remove sulphur to reduce pollution.

- A mixture of liquid and vapour passes into a fractionating tower that is cooler at the top than at the bottom.

- The vapour passes up the fractionating tower via a series of trays containing bubble caps, until it arrives at a tray that is sufficiently cool (i.e. has a lower temperature than the vapour's boiling point). The vapour condenses to a liquid here.

- The condensed liquid on each tray is collected.

- The hydrocarbons with the shorter chains condense in the trays nearer to the top of the tower. This is because it is cooler here and they have lower boiling points than the longer chain hydrocarbons.

- Bitumen, which is a thick layer, is collected at the base of the tower. We use bitumen or tar for road surfacing, but it can be further broken down to provide more valuable products.

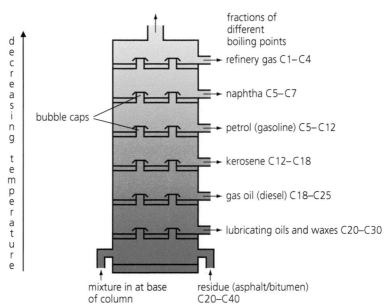

Figure 17.3
An industrial fractional distillation column ▶

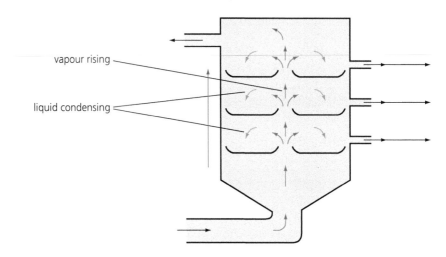

Figure 17.4
The processes occurring in an industrial fractionating column ▶

vapour rising

liquid condensing

Figure 17.5
An oil refinery ▶

▼ **Table 17.2** The fractions from petroleum

Fractions	Number of carbons in chain	Approximate boiling range (K)	Percentage present	Uses
petroleum gases	1–5	310	2	fuels for heating, lighting
petrol (gasoline)	5–10	310–450	8	fuels for small vehicles
naphtha	8–12	400–490	10	feedstock for producing high octane gasoline; diluent in bitumen mining
kerosene	11–16	430–523	14	fuels for stoves, jets
gas oil (diesel)	16–24	590–620	21	fuels for trucks, boilers
heavy oil and wax	2– 50	above 620	45	lubrication; cracking; candles; petroleum jelly; polish
residue (bitumen)	> 70	non-distillable		surfacing roads and roofs

Cracking of petroleum fractions

We can break up the fractions obtained from crude oil distillation into smaller ones. This is important, because fractional distillation of crude oil produces more large hydrocarbons than small hydrocarbons. Smaller hydrocarbons are more useful in today's world.

Small hydrocarbons, such as ethane, are precursors for many organic compounds such as ethanol. Hydrocarbons are broken up into smaller hydrocarbons in a process called **cracking**. Cracking always results in the formation of at least one **alkene**, and is therefore a major source of alkenes.

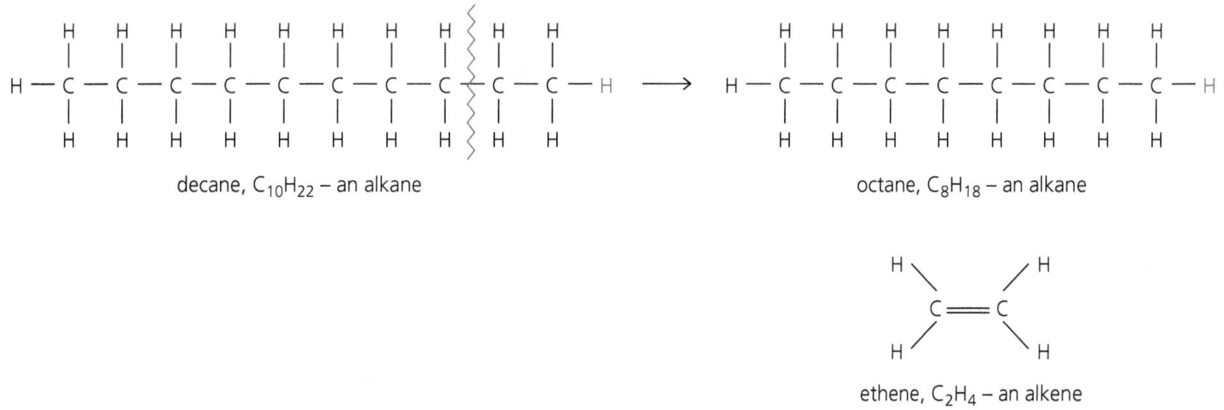

decane, $C_{10}H_{22}$ – an alkane

octane, C_8H_{18} – an alkane

ethene, C_2H_4 – an alkene

▲ **Figure 17.6** The cracking of decane.

Table 17.3 shows the approximate composition of the vapour leaving the cracker when **naptha** is cracked.

Naptha is the fraction that contains hydrocarbons with 6–7 carbon atoms.

▼ **Table 17.3** Products of the cracking of naphtha

Component	Amount (%)
ethene	23
propene	15
methane	12
butadiene	4
other gases	12
gasoline	34

We can crack hydrocarbons by using only heat (thermal cracking), or we can use a combination of heat and a catalyst (catalytic cracking). Heat makes the hydrocarbon molecules vibrate so much that the carbon–carbon bonds break at certain points.

- Thermal cracking: Thermal cracking (also called steam cracking) uses heat to break up large hydrocarbons into smaller ones. Thermal cracking is used to manufacture ethylene. During thermal cracking, alkanes are heated up to between 700 K and 1 200 K and under high pressure (7 000 kPa). At this temperature and pressure, the carbon–carbon single bonds break to form radicals. The radicals are formed because one electron from the pair in a covalent bond goes to each carbon atom. Catalytic cracking, however, forms ions instead of radicals. These radicals are highly reactive.

- Catalytic cracking: Catalytic cracking takes place at a lower temperature (773 K) and pressure than thermal cracking. Catalytic cracking is used mainly to produce motor fuel. The products are branched alkanes, cycloalkanes and aromatic compounds. In catalytic cracking, the hydrocarbons are mixed with a very fine catalyst powder. Modern catalysts are called zeolites (complex aluminosilicates), which are more efficient than older mixtures of aluminium oxide and silicon dioxide. The mixture is then propelled like a liquid through a reaction chamber at a temperature of approximately 773 K. Because the mixture behaves like a liquid, we call this type of cracking **fluid catalytic cracking**.

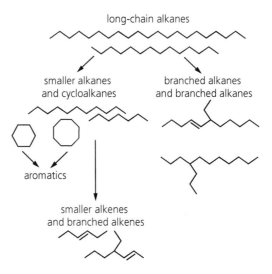

Figure 17.7
An example of the catalytic cracking of petroleum hydrocarbons. ▶

Reforming

The process of **reforming** is similar to cracking – straight-chain hydrocarbons are broken up and then reassembled into branched-chain hydrocarbons.

In **reforming**, low-grade gasoline (with a low octane rating) is converted to high-grade gasoline (with a high octane rating).

 Find out more

Gasoline with a high proportion of straight-chain alkanes has a 'low octane' rating because it causes knocking in the engine. Knocking occurs when the fuel in the engine's cylinders explodes too early. Knocking results in overheating of the engine and a loss of power, as well as causing the 'knocking' noise and possibly damaging parts of the engine. The straight-chain alkane heptane is assigned an 'octane rating' of 0. The branched-chain alkane 2,2,4-trimethylpentane, on the other hand, is assigned an octane rating of 100.

$CH_3–CH_2–CH_2–CH_2–CH_2–CH_2–CH_3$
heptane; octane rating = 0

$$CH_3—\underset{\underset{CH_3}{|}}{\overset{\overset{CH_3}{|}}{C}}—CH_2—\underset{\underset{CH_3}{|}}{\overset{\overset{H}{|}}{C}}—CH_3$$

2,2,4-trimethylpentane; octane rating = 100

Next time you visit a petrol station, look at the octane rating of the fuel you use.

17.3 Structural isomerism and nomenclature

Structural isomerism is the existence of different compounds that have the same molecular formula but have different structural formulae. Structural isomers can belong to the same homologous series or to a different homologous series.

Structural isomers found within a homologous series (i.e. that have the same functional group) have different physical properties, but similar chemical properties. We call this type of structural **chain isomers**. Structural isomers can also belong to different homologous series. This means they have different functional groups, which means they have different physical and chemical properties. We call this type of isomer **functional group isomers**.

There are three types of structural isomers:

- Chain isomers
- Functional group isomers
- Positional isomers

Chain isomers

Chain isomers are structural isomers in which the carbon skeletons are arranged differently. They have the same functional group and belong to the same homologous series. Pentane, 2-methylbutane and 2,2-dimethylpropane all have the same molecular formula of C_5H_{12}.

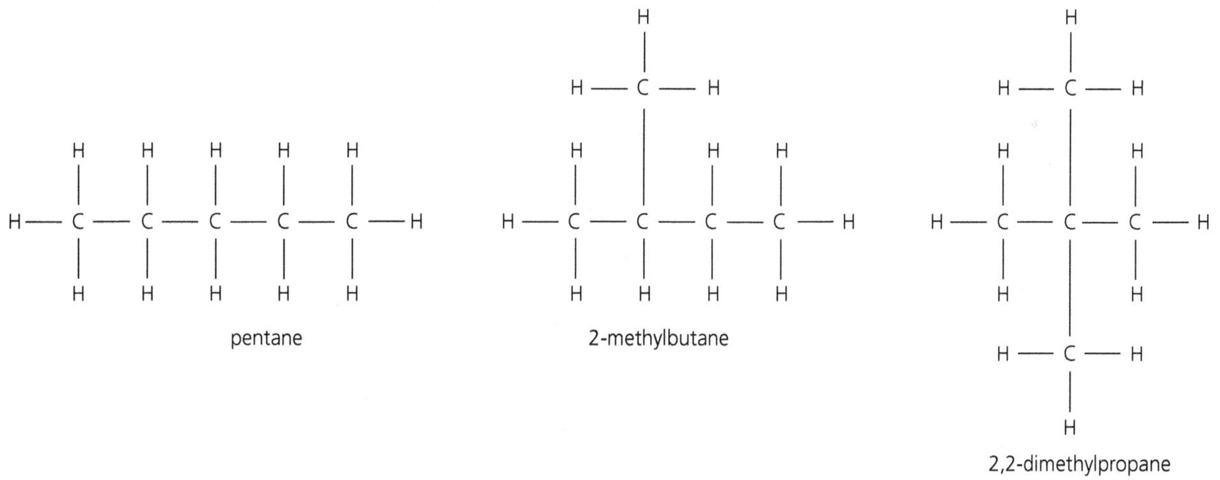

Figure 17.8 ▲ Examples of chain isomers

Chain isomers have different physical properties because the different shapes change the strength of the **dispersion forces**. Pentane boils at 36 °C, 2-methylbutane at 28 °C, and 2,2-dimethylpropane at 10 °C. However, they have the same chemical properties, because they all belong to the same homologous series – alkanes.

The **London dispersion force** is a weak, temporary attractive force that forms between non-polar molecules when the two electrons in two adjacent atoms occupy positions such that the atoms form temporary dipoles.

Functional group isomers

Functional group isomers are structural isomers that have different functional groups and belong to different homologous series. They have different physical properties as well as different chemical properties, because the functional group is different. Figure 17.9 shows some examples.

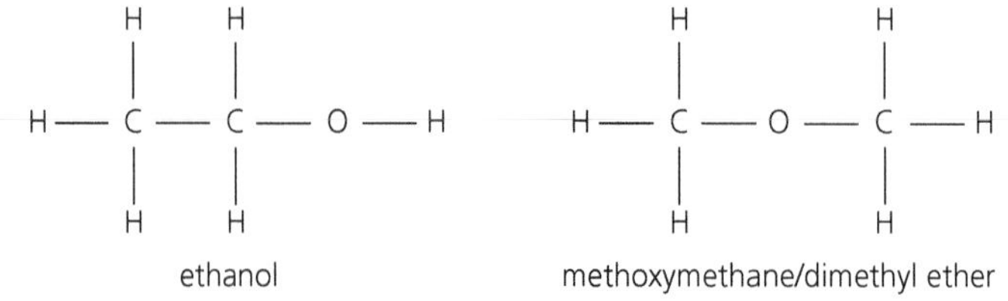

ethanol   **methoxymethane/dimethyl ether**

(a) An alcohol and an ether with molecular formula C_2H_6O.

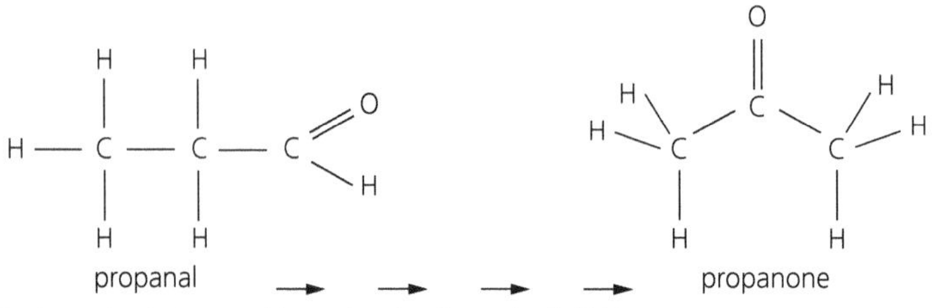

propanal   **propanone**

(b) An aldehyde and a ketone with molecular formula C_3H_6O.

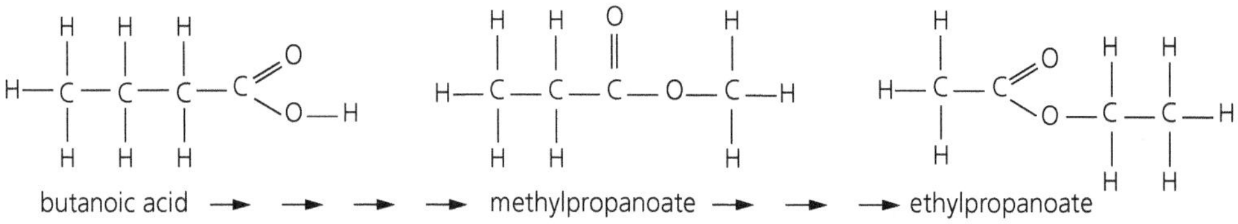

Figure 17.9
Examples of
functional group
isomers. ▶

butanoic acid → → → → **methylpropanoate** → → → **ethylpropanoate**

(c) A carboxylic acid and two esters with molecular formula C_4H_8O.

Positional isomers

Positional isomers are structural isomers that have a functional group or substituent group in different positions on the same carbon skeleton (parent chain). In positional isomers, the parent chains are the same. Positional isomers are chemically similar because they possess the same functional group. Note the position of the functional groups in Figure 17.10.

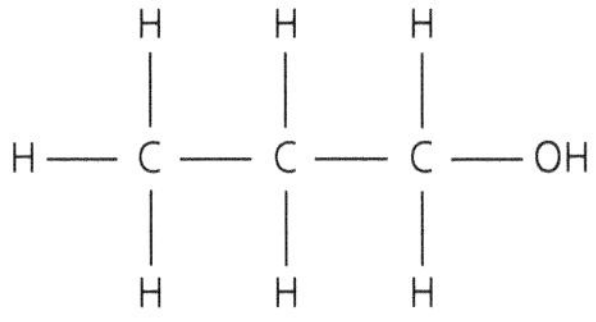

(a) Propan-1-ol and Propan-2-ol

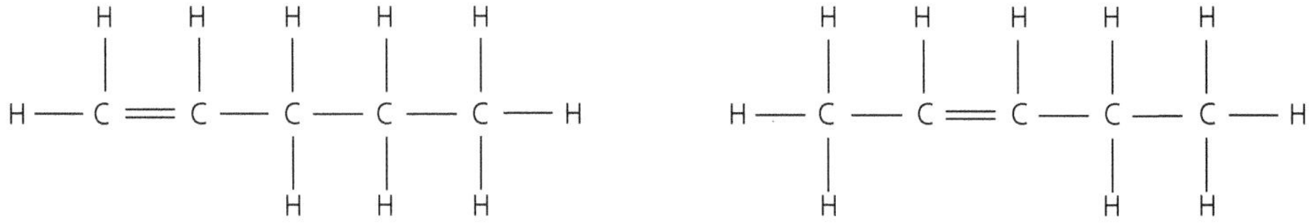

(b) Pent-1-ene and pent-2-ene

Ortho, *meta* and *para*
must be italicised.

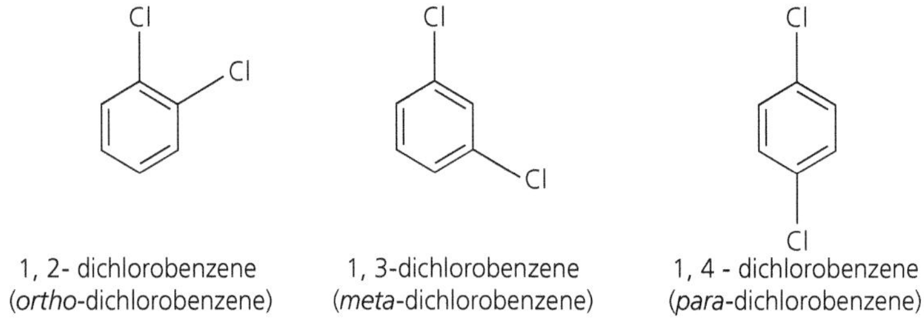

Figure 17.10
Examples of
positional isomers ▶

1, 2- dichlorobenzene   **1, 3-dichlorobenzene**   **1, 4 - dichlorobenzene**
(*ortho*-dichlorobenzene)   (*meta*-dichlorobenzene)   (*para*-dichlorobenzene)

(c) Benzene derivatives

Structural formulae of alkenes

We can only draw one structure for alkenes that contain two or three carbon atoms:

ethene

propene

Alkenes with four carbon atoms show isomerism:

but-1-ene   but-2-ene   2-methylprop-1-ene

straight-chain
structures

branched-chain
structure

When you study these structures, you should notice the following:

- The straight-chain structures have the double bond in two different positions. The third isomer is a branched-chain compound.

- When naming alkenes, the position of the double bond in the molecule must be indicated.

The position of the double bond is given by the lower number of the two carbon atoms that are linked by the double bond. The general rules for naming organic compounds also apply.

Nomenclature of branched and unbranched organic compounds

Organic compounds can exist as branched, unbranched and ring structures. This chapter focuses on naming branched and unbranched alkanes, and unbranched alkenes, alkanols, and alkanoic acids. The following steps are involved in naming organic compounds using the IUPAC nomenclature system:

Step 1: Identify the longest chain and number the carbon atoms from the end nearest the functional group. Any carbon left without a number is a branch.

Step 2: Name the longest chain using the Greek prefix plus the suffix of the homologous series to which it belongs.

Step 3: Name the attached groups (branches or substituents) based on their positions and prefixes, and then write them in alphabetical order.

Step 4: If there is more than one particular group, add **di, tri-** or **tetra-** to the prefixes, for example, dibromo– (2 × Br) and tetrachloro– (4 × Cl).

Step 5: Place a **comma** between two adjacent numbers and a **hyphen** between a number and a word.

These steps apply to alkanes, alkenes, alkanols and alkanoic acid homologous series.

Drawing structural formulae for organic compounds

Step 1: Write down all the carbon atoms in the parent chain that have single bonds.

Step 2: Place the functional group to the end or as specified.

Step 3: Attach all the other groups.

Step 4: Attach the hydrogen atoms so that each carbon atom has four bonds.

Step 5: If any carbon atom still needs two hydrogen atoms, place a double bond between two of the carbon atoms.

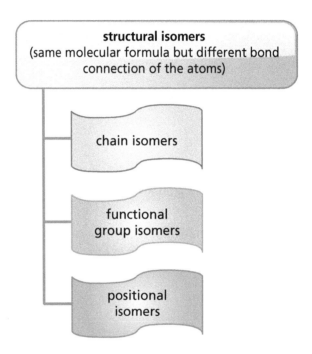

Figure 17.11
Structural isomers ▶

Practice

1 Draw full structural formulae for the alkanes containing three and four carbon atoms. Hint: There is one C_3 alkane and two C_4 alkanes.

Look at the structures that you drew for the alkanes with four carbon atoms. The atoms in compounds with a molecular formula of C_4H_{10} can be arranged in two ways, either as:

butane
(straight-chain molecule)

or as:

2-methyl propane
(branched-chain molecule)

Both these structural formulae correspond to a molecular formula of C_4H_{10}. This means that the two compounds are isomers. As discussed previously, isomers differ in their structure. They cannot be converted to one another by turning the molecule or by rotation about carbon–carbon bonds.

The physical properties of isomers such as butane and 2-methylpropane differ. However, their chemical properties are very similar because the two compounds contain the same functional groups.

Understand it better

It would help if you used atomic models to make structures as you work through this section.

Many students think that the following structures are different and so represent different compounds:

If you trace the sequence of carbon atoms with a pencil, without lifting the pencil, or if you manipulate the models you have made, you will see that the four carbon atoms always form a continuous chain.

Now look at this structure:

Here the longest chain that you can trace in any direction without lifting your pencil has only three carbon atoms. In whichever direction you trace, one of the carbon atoms is always left out of the continuous chain. This compound has the same molecular fomula but is different from the three compounds above and is therefore an isomer.

Note that in a branched-chain isomer of a straight-chain hydrocarbon:
• one of the carbon atoms must be attached to at least three carbon atoms;
• the branch must be positioned at least one carbon atom from the end of the main chain.

Practice

2 Draw and name the structural chain isomers of the compound $C_4H_8O_2$.

3 Draw and name three structural positional isomers of the compound $C_3H_6Cl_2O$.

4 Draw and name three structural positional isomers of the compound C_4H_6BrCl.

5 Write the formulae of three structural isomers with the formula C_3H_8O.

6 a Draw structural formulae for the alkenes that have five carbon atoms.

   b Name all the structures that you drew in part (a)

7 Draw the isomers for compounds with the molecular formula C_5H_{12}.

8 a Three of the following structures represent the same compound. Identify them and name the compound.

   b The other structure represents an isomer of the compound in part (a). Name this isomer.

   c Draw the structure of another isomer with the molecular formula C_6H_{14}.

17.4 The alkanes

The alkanes are the homologous series of saturated hydrocarbons that are obtained mainly from crude oil and natural gas. Alkanes have the general formula C_nH_{2n+2} ($n \geq 1$), and we can use this fact to determine the molecular formula of the members of the alkane series.

Alkanes can be gases, liquids or solids at room temperature. The physical states of alkanes at room temperature are as follows:

- 1–4 carbon atoms are **gases**.
- 5–17 carbon atoms are **liquids**.
- 18+ carbon atoms are **solids**.

The molecular formulae, names and physical states of the first ten members of the alkane homologous series are shown in the table below.

▼ **Table 17.4** Molecular formulae, names and physical states of the first ten members of the alkane homologous series.

Value of n in the general formula C_nH_{2n+2}	Molecular formula	IUPAC name	Physical state at room temperature
1	CH_4	methane	gas
2	C_2H_6	ethane	gas
3	C_3H_8	propane	gas
4	C_4H_{10}	butane	gas
5	C_5H_{12}	pentane	liquid
6	C_6H_{14}	hexane	liquid
7	C_7H_{16}	heptane	liquid
8	C_8H_{18}	octane	liquid
9	C_9H_{20}	nonane	liquid
10	$C_{10}H_{22}$	decane	liquid

Practice

9 Examine Table 17.4. Identify two trends that support the fact that alkanes form a homologous series.

10 Write the molecular formulae for the alkanes containing:
   a 8 carbon atoms;
   b 19 carbon atoms.

17.5 Physical properties of alkanes

The members of the alkane homologous series show a gradual change in physical state as the number of carbon atoms in the chain increases. This phenomenon occurs because of an increase in van der Waals forces between molecules of unbranched chains. The larger unbranched alkanes have greater van der Waals forces of attraction because their surface area is larger. A large molecule with a large surface area and van der Waals forces will have a high boiling point.

Branched alkanes that are more compact have a smaller surface area on which van der Waals forces act, and so they have a lower boiling temperature than an unbranched chain isomer with the same number of carbon atoms. Solid alkanes that are normally soft have low melting points. The solid alkanes are soft because within their crystalline solids, there are strong repulsive forces between electrons on neighbouring atoms. The strong repulsive forces counterbalance the weak van der Waals forces of attraction. The physical properties of alkanes, in summary, are as follows:

- Alkanes are colourless, non-polar compounds with distinctive odours.

- Alkanes are less dense than water.

- The first four alkanes are gases, the next 12 are liquids, and the rest are solids at room temperature and pressure.

- The boiling point increases with increasing mass.

Alkanes may be used to show that physical properties change in a regular way for a given homologous series. For example, boiling point, density and enthalpy of combustion all increase with an increase in molecular mass or length of the carbon chain. Figure 17.12 shows that an almost linear relationship exists between the boiling point of the alkanes and the number of carbon atoms in the chain. Can you explain this trend?

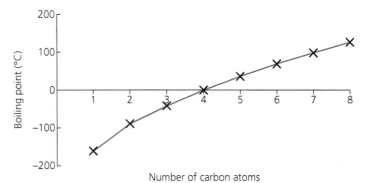

Figure 17.12
The relationship between boiling point and number of carbon atoms in alkanes ▶

17.6 Chemical properties of alkanes

Members of the alkane homologous series have similar chemical properties. Alkanes are less reactive than alkenes, which in turn are less reactive than alkynes. Alkanes are a saturated compounds, which is why they are less reactive than unsaturated alkenes and alkynes. In previous chapters, we explained that electrons determine the chemistry of atoms. The members of an alkane homologous series have only carbon–carbon single bonds. This characteristic means that there is only one pair of electrons that are shared between any carbon–carbon single bond.

Figure 17.13
Bonding in alkanes, alkenes and alkynes ▶

Alkenes and alkynes both have two and three pairs of electrons that are shared between their functional groups, and carbon–carbon double bonds and carbon–carbon triple bonds, respectively. The two and three pairs of electrons present in the functional groups of alkenes and alkynes respectively, make them more reactive than saturated alkanes. Consequently, alkanes generally undergo substitution reactions, whereas alkenes and alkynes undergo addition reactions. Other reactions of alkanes include cracking, reforming and combustion.

Combustion reaction of alkanes

When you ignite an alkane, it burns with a blue smokeless flame. Alkanes burn in excess air (or excess oxygen) to produce carbon dioxide, water and heat. This reaction is called **complete combustion**. If the supply of air or oxygen is insufficient, **incomplete combustion** occurs, producing carbon monoxide and smoke or soot (solid carbon particles). These are the equations for the combustion of ethane:

$$2C_2H_6(g) + 7O_2(g) \rightarrow 4CO_2(g) + 6H_2O(l) + \text{heat (\textbf{complete combustion})}$$

$$C_2H_6(g) + 3O_2(g) \rightarrow CO_2(g) + CO(g) + 3H_2O(l) + \text{heat (\textbf{incomplete combustion})}$$

Note that there is an excess of oxygen present in the complete combustion of ethane. The mole ratio of oxygen to ethane is 1:3.5. This means that for every 1 mole of ethane burnt, it will use 3.5 moles of oxygen. Incomplete combustion uses 3 moles of oxygen to burn 1 mole of ethane.

This is the general formula for the complete combustion of alkane:

$$nC_nH_{2n+2} + (3n + 1)O_2 \rightarrow 2nCO_2 + (2n + 2)H_2O$$

Practice

11 When the hydrocarbon C_2H_6 burns in air, possible products are carbon (C), carbon monoxide (CO), and carbon dioxide (CO_2).

  a Which of these products are formed in (i) complete combustion and (ii) in incomplete combustion?

  b Determine the oxidation number of carbon in:
    (i) C_2H_6;
    (ii) the element C;
    (iii) CO;
    (iv) CO_2.

  c Use your answers to parts (a) and (b) to compare the extent of the oxidation of carbon in complete and incomplete combustion.

Substitution reaction of alkanes

Alkanes undergo **substitution reactions** with the halogens (especially chlorine) in the presence of bright light, usually sunlight.

The substitution reaction gives a mixture of organic products and hydrogen chloride gas, as shown in the following equations:

$$CH_4(g) + Cl_2(g) \rightarrow CH_3Cl(g) + HCl(g)$$
monochloromethane

Further substitution can occur, as indicated by the following equations:

$$CH_3Cl(g) + Cl_2(g) \rightarrow CH_2Cl_2(g) + HCl(g)$$
dichloromethane

$$CH_2Cl_2(g) + Cl_2(g) \rightarrow CHCl_3(l) + HCl(g)$$
trichloromethane

$$CHCl_3(l) + Cl_2(g) \rightarrow CCl_4(l) + HCl(g)$$
$$\text{tetrachloromethane}$$

You can show that substitution is taking place by reacting the HCl(g) produced with concentrated ammonia. Dense white fumes of ammonium chloride will be formed:

$$HCl(g) + NH_3(g) \rightarrow NH_4Cl(s)$$

The organic products are known as halogenoalkanes. The number of hydrogen atoms replaced in the reaction depends on the relative quantities of the alkane and halogen used.

Find out more

Why is light needed for the reaction between alkanes and halogens?

The energy of the light, or other radiation, breaks the bond in the halogen molecule to produce reactive halogen atoms or free radicals (atoms or groups of atoms containing an unpaired electron). For example:

$$Cl_2 \xrightarrow{\text{light}} Cl\bullet + Cl\bullet$$
$$\text{free radicals}$$

The free radicals then rapidly attack the alkane molecule to produce organic compounds known as chloroalkanes and hydrogen chloride. The reaction occurs in stages.

Practice

12 Using the equations for the reaction of chlorine with methane as a guide, write equations for the reaction of bromine with ethane.

17.7 Uses of alkanes and their derivatives

Alkanes

Alkanes are used mainly as fuels – examples are LPG and natural gas. This is because alkanes produce a lot of heat when they burn. In technical terms, we say that their heat of combustion, ΔH_c, is large. Other properties of a good fuel are:
- they ignite easily but not spontaneously under normal conditions;
- they burn well, but not explosively;
- they have low smoke and ash content;
- they are inexpensive;
- they are safe to use;
- they are easy to store and transport.

Halogenoalkanes

Some halogenoalkanes are:
- solvents – examples are tetrachloromethane and trichloromethane.
- anaesthetics. Halothane, for example, is a halogenoalkane which is widely used in hospitals. It contains the atoms bromine and fluorine as well as chlorine. Chloromethane is used for local anaesthesia by spraying on the skin.
- pesticides –halogenated hydrocarbons are also very widely used as pesticides.

17.8 The alkenes

The alkenes have the general formula C_nH_{2n}. They are unsaturated hydrocarbons, containing at least one carbon–carbon double bond. The alkynes are also unsaturated hydrocarbons, and they contain at least one carbon–carbon triple bond. The carbon–carbon double bond is the functional group of alkenes. Unsaturated hydrocarbons have similar properties.

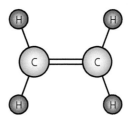

Figure 17.14
A computer model of the ethene molecule ▶

▼ **Table 17.5** The molecular formulae, names and physical states of the first four members of the alkene homologous series

Value of n in the general formula (C_nH_{2n})	Molecular formula	Name	Physical state at room temperature
2	C_2H_4	ethene	gas
3	C_3H_6	propene	gas
4	C_4H_8	butene	gas
5	C_5H_{10}	pentene	volatile liquid

Sources of alkenes

Cracking and dehydration of alcohols are the main sources of alkenes.
- Cracking: We have already discussed this earlier in this chapter.
- Dehydration of alcohols: Alcohols react with excess concentrated sulphuric acid at 180 °C to produce their corresponding alkenes. For example, when ethanol is dehydrated it produces ethene.

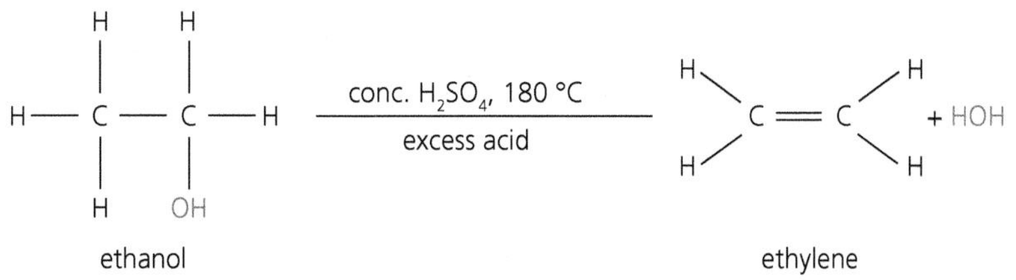

Figure 17.15
The dehydration of alcohols ▶

ethanol

$$\text{conc. H}_2\text{SO}_4\text{, 180 °C} \quad \text{excess acid}$$

+ HOH

ethylene

17.9 Physical properties of alkenes

The physical properties of alkenes are similar to those of alkanes. The first three alkenes (ethene, propene and butane) are colourless gases. Members of the alkene homologous series with between five and 15 carbon atoms are liquids, and chains with more than 15 carbon atoms are solids at room temperature and pressure.

The boiling points of alkenes are similar to those of alkanes with the same number of carbon atoms. The boiling points of alkenes depend on the chain length (molecular mass). The longer the chain length, the greater the boiling points. The intermolecular forces of alkenes become stronger with increasing molecular size.

17.10 Chemical properties of alkenes

Members of the alkene homologous series are defined primarily by their functional group: the carbon–carbon double bond. The presence of this carbon–carbon double bond functional group makes alkenes unsaturated. This double bond has two pairs of electrons that are shared between the two carbon atoms. This makes alkenes susceptible to addition reactions.

Because the alkene hydrocarbons are unsaturated, they can react with hydrogen atoms in the presence of a catalyst. The resulting compound, with no multiple bonds, is saturated.
17.11   Reactions of alkenes

Combustion of alkenes

Alkenes, like alkanes, burn in a plentiful supply of oxygen to produce carbon dioxide and steam. Again, the reaction is highly exothermic.

$$C_2H_4(g) + 3O_2(g) \rightarrow 2CO_2(g) + 2H_2O(g) + heat$$

However, alkenes are not used as fuels because:

* they burn with a smoky flame (this is in contrast to alkanes, which burn with a clean blue flame);

* they are prized starting materials for making other valuable organic compounds such as polymers (see Section 20.1).

$$C_nH_{2n} + (n + 1)O_2 \rightarrow nCO_2 + nH_2O + heat$$

Combustion experiments can be used to determine the molecular formulae of hydrocarbons. You need to follow the stages in this worked example and then answer the questions at the end.

Worked example 1

0.25 mol of a hydrocarbon, X, with a molar mass of 72 g mol^{-1}, was burnt completely in an excess of oxygen. The volume of carbon dioxide produced was 30 dm^3. What is the molecular formula of the hydrocarbon X?

Solution

The aim in a problem such as this is to use the information to find the number of atoms of carbon and hydrogen in 1 mole of the hydrocarbon.

Step 1: Find the mass of C in 1 mole of X.

X was burnt in excess oxygen, so we can deduce that complete combustion has taken place.

This means that all the carbon in X will be converted to CO_2. This means that the amount of carbon in 0.25 mol of X is the same as the amount of carbon in 30 dm³ of CO_2.

We know that the volume of 1 mole of gas is 24 dm³ at RTP.

30 dm³ is the volume of $\frac{30}{24} \times 1$ mol = 1.25 mol CO_2.

0.25 mol of X produced 1.25 mol CO_2.

1 mol of X produces $\frac{1.25}{0.25} \times 1$ mol of CO_2 = 5 mol of CO_2

5 mol of CO_2 contains 5 mol of C atoms, so there are 5×12 g of C in 1 mol X = 60 g.

Step 2: Find the mass of H in 1 mole of X.

We are told that the molar mass of X is 72 g mol⁻¹.

We have just calculated that 1 mole of X contains 60 g of C, so there must be 72 − 60 g = 12 g of H in 1 mole of X.

We know that the mass of 1 mol of hydrogen atoms is 1 g, so 12 g is equivalent to 12 moles of H.

Therefore 1 mole of X has 12 mol of H atoms and 5 mol of C atoms.

> The molecular formula of X = C_5H_{12}.

Q1 Is the hydrocarbon X an alkane or an alkene? Explain your answer.

Q2 Write a balanced equation for the combustion of X.

Q3 Calculate the number of moles of oxygen used in the experiment.

Addition reactions of alkenes

The typical reactions of the alkenes are **addition reactions**.

Reaction with hydrogen

Hydrogen adds to alkenes in the presence of finely divided nickel, which serves as a catalyst. This reaction, which results in the formation of alkanes, is commonly described as hydrogenation:

alkene + hydrogen → alkane

propene   finely divided nickel   propane

Note that the hydrogen molecule is added across the double bond. One hydrogen atom is added to each of the carbon atoms linked by the double bond.

Hydrogenation reactions with unsaturated alkene-type molecules are carried out in industry.

Find out more

How is margarine made?

The hydrogenation reaction is used in the edible oil industry to make vegetable oils more spreadable. When oils are hydrogenated – a process known as hardening – they are converted to margarine and solid cooking fats.

vegetable oils (low melting point)

↓ hydrogenation of a few double bonds

soft fats

↓ more extensive hydrogenation

hard fats (higher melting point)

Margarine is an emulsion. It consists of water droplets dispersed in an oil/fat mixture.

Vegetable oils consist of esters of long-chain carboxylic acids, which contain many double bonds. It has been found that while margarine made from vegetable oil contributes no dietary cholesterol, it contributes trans fatty acids, which may raise total cholesterol and LDL or 'bad cholesterol'.

▲ **Figure 17.16** What can you tell about the ingredients in this margarine?

The hydration of alkenes to produce alkanols

The equation for this hydration reaction is:

$$\underset{\text{ethene}}{\ce{H2C=CH2}} \quad + \quad \underset{\text{water}}{\ce{H2O}} \quad \longrightarrow \quad \underset{\text{ethanol}}{\ce{CH3CH2OH}}$$

Hydration can be achieved both indirectly and directly.

The indirect method uses sulphuric acid to form an intermediate. For example:

$$\underset{\text{ethene}}{\ce{H2C=CH2}} \xrightarrow[\text{heat (170 °C)}]{\text{excess conc. H}_2\text{SO}_4} \underset{\text{ethyl hydrogen sulphate}}{\ce{CH3CH2-SO3H}} \xrightarrow[\text{water}]{\text{warm with}} \underset{\text{ethanol}}{\ce{CH3CH2-O-H}}$$

The direct hydration of ethene is used to produce ethanol commercially. Ethene is compressed to about 70 atmospheres, mixed with water and passed through a furnace. The heated mixture is then passed over a phosphoric(V) acid catalyst at a temperature of 300 °C. Ethanol is formed and the mixture of gases, ethanol vapour plus uncombined ethene and steam, is then passed through a condenser where ethanol vapour and steam are condensed and drawn off. Unreacted ethene can then be recycled. The process is shown in Figure 17.17.

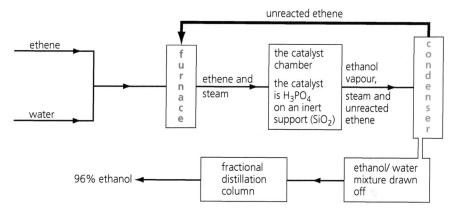

Figure 17.17
The commercial production of ethanol. ▶

17.11 Distinguishing between alkanes and alkenes

The differences between alkanes and alkenes are based on the facts that:

- alkanes are saturated compounds, so they undergo substitution reactions;
- alkenes are unsaturated compounds, so they undergo addition reactions.

Alkenes can therefore be told apart from alkanes by testing for the presence of unsaturation. Alkenes rapidly decolorise an acidified solution of potassium manganate(VII), $MnO_4^-/H^+(aq)$. The acidified potassium manganate(VII) reagent has no effect on alkanes.

Note that two hydroxyl groups are added across the double bond, forming a di-alcohol. The reaction is rapid and there is an obvious colour change from deep purple to very pale pink (see Table 12.5 for more about the MnO_4^-/H^+ reagent).

Alkenes also rapidly decolorise a solution of liquid bromine in 1,1,1-trichloroethane from red-brown to colourless. The reaction occurs 'in the dark', i.e. in the absence of sunlight.

335

Both these reactions are addition reactions.

Alkanes do not react with bromine or other halogens, except in the presence of bright light or ultraviolet radiation (e.g. sunlight). Even then, alkanes react at a slower rate than do alkenes.

Experiment 17.1   To distinguish between an alkane and an alkene

In this experiment, it is convenient to use cyclohexane and cyclohexene, as these are both liquid hydrocarbons. Ethane and ethene are gases, which makes them difficult to use in experiments, and they must be prepared prior to testing.

Procedure

Test 1: Add a few drops of acidified potassium manganate(VII) to 1 cm³ of cyclohexene, with shaking.

Test 2: Repeat test 1, using cyclohexane in place of cyclohexene.

Test 3: Add a few drops of aqueous bromine (bromine water) or bromine dissolved in tetrachloromethane to 1 cm³ of cyclohexene, with shaking.

Test 4: Repeat test 3, using cyclohexane in place of cyclohexene.
Record your results in a table.

Questions

Q1 Cyclohexane has the molecular formula C_6H_{12} whereas cyclohexene has the molecular formula C_6H_{10}. Compare these formulae with those of the straight-chain alkanes and alkenes containing six carbon atoms. Comment on the differences you find.

Q2 Write balanced chemical equations for any positive reactions observed in this experiment.

Q3 Which of the compounds you tested showed unsaturation? Explain your answer.

17.12   Uses of alkenes

Figure 17.18 summarises the major uses of a typical alkene, ethene. As can be seen, like other alkenes, ethene is a versatile compound. It is one of the major building blocks of the chemical industry.

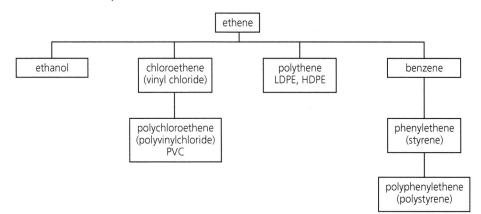

Figure 17.18
The uses of
ethene ▶

From crude oil to gasoline and ethene

Crude oil is a natural source of hydrocarbons. Fractional distillation (see Section 5.3) is used to separate the constituents of crude oil.

On an industrial scale (Figure 17.3), the fractionating column consists of perforated horizontal trays on which the rising vapour condenses and is then vaporised later by more rising vapour. Thus each tray functions both as a condenser and a boiling flask.

The components (fractions) of petroleum range from volatile liquids used as gasoline (boiling at 50 °C to 200 °C) and kerosene (boiling at 175 °C to 325 °C) to the heavier fractions that are used as diesel fuel (boiling above 275 °C). The residue that remains behind after fractional distillation of heavy oil is asphalt, which is used for surfacing highways.

The composition of crude oil from different sources varies considerably. Components that are in lesser demand can be changed into more useful ones by two important processes: **cracking** and **reforming**.

17.13   The alkanols

The alkanols are a homologous series with the general formula $C_nH_{2n+1}OH$ and they have the hydroxyl (–OH) functional group. They are saturated carbon atom chains like those of **ALKAN**es and alcoh**OL**s. Because of this similarity, alkanols are named using the prefix of the alkane with the –ol of the alcohol, for example ethane → ethanol and propane → propanol.

Alcohols with four or more carbon atoms can be either straight- or branched-chain isomers. Also, the position of the hydroxyl functional group can vary, which produces positional isomers. The chemical properties of alcohols are determined by the hydroxyl (–OH) group. Table 17.6 lists the first four alkanols.

▼ **Table 17.6**   The first four alkanols.

Value of n in general formula	Molecular formula	IUPAC name
1	methanol	CH_3OH
2	ethanol	C_2H_5OH
3	propanol	C_3H_7OH
4	butanol	C_4H_9OH

Sources of alkanols

We obtain alkanols from fermentation of suitable carbohydrates or hydration of the corresponding alkenes.

Fermentation is the conversion of a carbohydrate to alcohol using yeast. Yeast is mixed with the carbohydrate in an airtight container. The enzyme invertase breaks down the carbohydrate into glucose and fructose. When the yeast respires anaerobically (without oxygen), an enzyme zymase (in its cells) breaks down glucose and fructose into ethanol and carbon dioxide.

Figure 17.19
The process of
fermentation ▶

$$C_6H_{12}O_6 \text{ (aq)} \xrightarrow{\text{yeast}} 2CH_3CH_2OH \text{ (aq)} + 2CO_2 \text{ (g)}$$

Glucose   Ethanol   Carbon dioxide

Ethanol stops being produced when the mixture contains 14% ethanol, because the action of zymase is inhibited. The mixture must then be filtered and fractionally distilled to get proof rum (50% alcohol) or industrial alcohol (96% ethanol).

17.14 Physical properties of alkanols

- Alkanols are colourless and sweet-smelling.
- An alkanol is less volatile than an alkane or ester of similar mass, because the polar hydroxyl (–OH) group causes hydrogen bonding between the molecules. This hydrogen bonding adds to the van der Waals' forces of attraction between the hydrocarbon parts. Thus, the first 12 alkanols are liquids at room temperature and pressure.
- Due to the polar –OH group, alkanols are soluble in water. However, the solubility decreases as the hydrocarbon chain gets longer.
- The non-polar hydrocarbon chain allows alkanols to dissolve some non-polar compounds.

17.15 The alkanoic acids

The alkanoic acids are a homologous series that have the carboxyl (–COOH) functional group. The alkanoic acids have the general formula $C_nH_{2n+1}COOH$. Alkanoic acids with four or more carbon atoms can be straight- or branched-chain isomers. The properties of alkanoic acids are caused by the presence of the carboxyl (–COOH) functional group.

▼ **Table 17.7** The first four alkanoic acids

Value of n in general formula	Molecular formula	IUPAC name
0	methanoic acid	HCOOH
1	ethanoic acid	CH_3COOH
2	propanoic acid	C_2H_5COOH
3	butanoic acid	C_3H_7COOH

Sources of alkanoic acids

The main source of alkanoic acids is from the oxidation of the corresponding alkanols. The alkanol is heated under reflux with acidified potassium or sodium dichromate, for example, ethanol is oxidised to ethanoic acid.

Figure 17.20
Ethanol is oxidised to
ethanoic acid ▶

$$CH_3CH_2OH \xrightarrow[\text{H}^+]{K_2Cr_2O_7} CH_3CHO \xrightarrow[\text{H}^+]{K_2Cr_2O_7} CH_3COOH$$

Ethanol   Acetaldehyde   Acetic acid

17.16 Physical properties of alkanoic acids

- The alkanoic acids are colourless and have a vinegary smell.
- An alkanoic acid is less volatile than an alcohol or ester of similar mass because of the polar carboxyl (–COOH) group.
- Due to the polar –OH group, the alkanoic acids are soluble in water. However, the solubility decreases as the hydrocarbon chain increases.

Summary

Alkanes and alkenes are hydrocarbons obtained from crude oil.

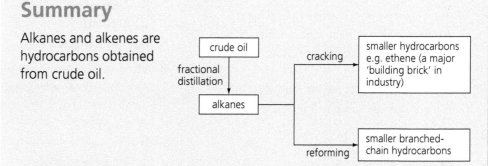

▼ **Table 17.8** Similarities and differences between alkanes and alkenes

	Alkanes	Alkenes
Similarities		
Isomerism	Members containing four or more carbon atoms show isomerism.	
Combustion	They undergo complete combustion reactions in excess oxygen to produce carbon dioxide and water and to release heat. In limited supply of oxygen or air the combustion is incomplete.	
Differences		
General formula	C_nH_{2n+2}	C_nH_{2n}
Saturation	Saturated hydrocarbons – they contain only carbon–carbon single bonds	Unsaturated hydrocarbons – they contain at least one carbon–carbon double bond
Reactions	Undergo substitution reactions, for example, with chlorine (and bromine) in the presence of sunlight or ultraviolet radiation	Undergo addition reactions: • with bromine to form dibromoalkanes • with hydrogen to produce alkanes • with water to form alcohols

End-of-chapter questions

1 Which of the following is a typical reaction of a saturated hydrocarbon?
   A Addition
   B Combustion
   C Substitution
   D Hydration

2 Why are alkenes more reactive than alkanes?
   A Alkenes are better hydrocarbons than alkanes.
   B Alkenes have two pairs of electrons in their functional groups.
   C Alkenes reacts more readily than alkanes.
   D Alkanes are saturated hydrocarbons.

3 How many isomers of butene exist?
   A 3
   B 4
   C 5
   D 6

4 Ethanol cannot be directly converted to:
   A Ethene
   B Ethane
   C Ethanoic acid
   D Carbon dioxide

5 Ethane may be used as a fuel because it:
   A Is readily available
   B Burns with a blue, smokeless flame
   C Can be converted to ethene
   D Burns with a yellow, smoky flame

6 Which of the following is the correct functional group for alcohol?
   A The hydroxyl group
   B The carbon-carbon double bond
   C The carbon-carbon single bond
   D The carboxyl group

7 Compounds A and B were reacted with acidified potassium manganate (VII). Compound A rapidly decolourised the solution and compound B had no effect on the solution. To what homologous series is compound A likely to belong?
   A Alkanes
   B Alkenes
   C Alkanols
   D Alkanoic acids

8 What is the name of the product formed when propene is hydrogenated in the presence of finely divided nickel at 5 atm and 100 °C?
   A Propane
   B Propanol
   C Propanoic acid
   D Propanal

9 Why do branched alkanes have a lower boiling point compared to unbranched alkanes with the same number of carbon atoms?
   A They are more compact and have a smaller surface area for van der Waals forces to act on.
   B They have a larger molecular mass than unbranched alkanes.
   C The bonds break more easily than unbranched alkanes.
   D They have a larger surface area than unbranched alkanes.

10 Which of the following gives the correct IUPAC name for isomers of butene?
   A But-1-ene, but-2-ene and 1-methylprop-2-ene
   B But-1-ene, but-2-ene and 1-methylprop-2-ene
   C But-1-ene, but-2-ene and 2-methylprop-1-ene
   D But-1-ene, but-2-ene and 2-methylprop-1-ene

11 Carefully study the information in the following reaction scheme and answer the questions that follow. The structure of A has been identified for you.

A CH₃CHCH₂ $\xrightarrow[\substack{\text{excess} \\ \text{conc. H}_2\text{SO}_4}]{\text{H}_2\text{O, 170 °C}}$ **B** $\xrightarrow[\text{heat}]{\text{H}^+ \text{ K}_2\text{Cr}_2\text{O}_7}$ **C** turns moist blue litmus paper red

D saturated compound

| Cl₂ uv light

E (a halogenoalkane)

a Draw the fully displayed chemical structures for the organic compounds B, C and D.

b Name the type of reaction involved in the conversion of:
(i) A to D;
(ii) B to C.

c Describe a chemical test to identify the functional group present in compound A. In your answer, indicate which reagent should be used and write a chemical equation for the reaction that occurs.

12 Consider the structures of compounds A–F below and answer the questions follow.

A

$$\begin{array}{ccccc} & H & H & H & H \\ & | & | & | & | \\ H- & C- & C- & C- & C- H \\ & | & | & | & | \\ & H & H & H & H \end{array}$$

B

$$\begin{array}{ccccc} & H & H & H & OH \\ & | & | & | & | \\ H- & C- & C- & C- & C- H \\ & | & | & | & | \\ & H & H & H & H \end{array}$$

C

$$\begin{array}{ccccc} & H & H & OH & H \\ & | & | & | & | \\ H- & C- & C- & C- & C- H \\ & | & | & | & | \\ & H & H & H & H \end{array}$$

D

$$\begin{array}{cccc} & H & H \\ & | & | \\ H- & C- & C- & C=C- H \\ & | & | & | \ | \\ & H & H & H \ H \end{array}$$

E

$$\begin{array}{cccc} & H & H & H & O \\ & | & | & | & || \\ H- & C- & C- & C- & C-OH \\ & | & | & | \\ & H & H & H \end{array}$$

F

$$\begin{array}{ccc} & H & H \\ & | & | \\ H- & C- & C=C \\ & | & | \\ & H & \end{array} \begin{array}{c} O \\ // \\ \backslash \\ O-H \end{array}$$

H—C—H
|
H

a Give the IUPAC name for the compounds A–F.

b Identify and name two pairs of compounds that belong to the same homologous series.

c Define the term structural isomerism.

d Which two compounds exhibit:

  (i) chain isomerism;

  (ii) positional isomerism.

e Draw and name a structural isomer of compound D.

f Which type of structural isomer did you draw for part (e)?

g Describe a simple chemical test that you could use to distinguish between compounds D and E. You should state the reagent to be used and any changes that occur. Write an equation for the reaction that occurs.

h Show, using an equation, how you could convert compound D into compound A. You should indicate the catalyst to be used.

i Which one of compounds B or D would you expect to be more soluble in water? Explain your answer.

13 The arrows in the diagram below show that compounds in a given homologous series may be converted to compounds in other homologous series. For each conversion, fill in:

a the type of reaction;

b the conditions of the reaction.

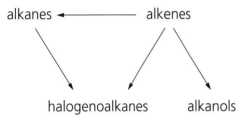

14 Test your knowledge of alkanes and alkenes by completing the table below. You should describe any positive reactions that occur, and where possible, write chemical equations for the reactions.

	Alkane	Alkene
Odour	no distinctive smell	somewhat sweetish smell
Colour	colourless	colourless
Reaction with bromine		
Reaction with KMnO$_4$		
What happens when it burns?		

15 a Identify three fractions produced by the fractional distillation of petroleum (crude oil).

b Suggest one use for each fraction identified.

c Why is fractional distillation rather than simple distillation used in separating the components of crude oil?

d Is fractional distillation a physical or chemical process? Justify your answer.

16 Cracking is considered to be a very important process.

a What do you understand by the term cracking?

b Why is the naptha fraction produced in the refining of petroleum cracked?

c Alkenes, such as ethene, are sometimes produced in cracking.

  (i) What is the primary use of ethene industrially?

  (ii) Write a balanced chemical equation for the cracking of the C20 alkane, given that ethene is also produced.

d What is the main difference between cracking and fractional distillation?

17 What products are formed when but-2-ene reacts with the following?

  a Hydrogen in the presence of a catalyst

  b Bromine dissolved in 1,1,1-trichloroethane

  c Steam in the presence of a catalyst

18 a Write a balanced equation for the complete combustion of butane.

  b What volume of carbon dioxide is formed when 7.25 g of butane is burnt completely in oxygen?

  c What volume of oxygen is needed for this combustion?

  d What volume change takes place when 7.25 g of butane is completely burnt in oxygen, all volumes being measured at room temperature and atmospheric pressure?

19 a How does propane react with chlorine?

  b Write chemical equations to support your answer in part (a).

20 a What is meant by the term 'hardening of fats'?

  b How is this achieved, industrially?

21 How are alkenes obtained commercially from alkanes? Give an equation to illustrate your answer.

22 Use the following information to identify the lettered compounds A–E.

  • 1 mole of hydrocarbon burns in oxygen to produce 3 moles of carbon dioxide.

  • Hydrocarbon B has an M_r of 42 and it decolourises bromine in 1,1,1-trichloroethane, forming a dense oily liquid, C.

  • B reacts with hydrogen and nickel to produce a compound D of M_r of 44.

  • D burns with a clean flame.

  • In the presence of sunlight, D combines with a controlled quantity of chlorine to form a compound E of M_r of 78.5.

23 The following are some of the processes that are used in industry:

  • Fractional distillation

  • Hydrogenation

  • Cracking

  • Reforming

Describe how each process is used in industry to obtain useful products.

In this chapter, you will study the following:

- the typical chemical reactions of alkanols;
- that the types of chemical reactions are determined by the functional groups;
- that physical properties are also linked to the presence of the functional groups;
- the industrial production of ethanol and alcoholic drinks;
- how to test for levels of alcohol consumption;
- the impact of alcohol consumption on individuals and society.

This chapter covers

Objectives 3.5–3.11 of Section B of the CSEC Chemistry Syllabus.

The alkanols form a homologous series with the general formula $C_nH_{2n-1}OH$, but can more conveniently be represented as R—OH, where R is an alkyl group such as —CH_3 or —C_2H_5 and the hydroxyl group, —OH, is the functional group. Ethanol, one of the most widely used organic compounds, is taken as being representative of the alkanols.

18.1 The reactions of alkanols

Ethanol and the other alkanols undergo a number of reactions based on the presence of the hydroxyl (—OH) functional group. In this chapter, we will look, primarily, at ethanol. However, the alkanols are a large group of chemicals with a wide range of uses.

Reactions with more reactive metals

Highly reactive metals, such as sodium, lithium and calcium, react with low molecular mass alkanols, such as methanol, ethanol and propanol. These reactions give hydrogen gas and salts of the alkanols, called alkoxides. The equation for the reaction of ethanol with sodium metal is:

$$2Na(s) + 2CH_3CH_2OH(l) \rightarrow 2CH_3CH_2O^-Na^+(s) + H_2(g)$$

sodium ethoxide

This reaction is similar to that of sodium and water, but it is less vigorous. We can think of ethanol as having the same functional group as water.

Practice

1 a Write the molecular formulae for the C_1 and C_2 alkanols and name them.

   b **(i)** Draw the displayed formulae for alkanols containing two and four carbon atoms.

      (ii) Give the name of all the molecules you have drawn.

      (iii) Which of the molecules do you consider to be isomers?

      (iv) Explain your answer to part (iii).

2 Compare the reaction between alkanols and reactive metals with the reaction between mineral acids (such as sulphuric acid) and reactive metals.

3 What does the evolution of hydrogen gas suggest about the alkanols?

4 Write an equation for the reaction of methanol with calcium and name the organic product formed.

Oxidation of alkanols

Powerful oxidising agents, such as acidified potassium dichromate(VI) or acidified potassium manganate(VII), convert alkanols to organic acids. In these reactions an intermediate alkanal (or aldehyde) is formed:

$$CH_3CH_2OH(l) + [O] \xrightarrow[\text{warm}]{KMnO_4/H^+} H_2O + CH_3CHO(l)$$

ethanol   ethanal
                                            (intermediate)

$$CH_3CHO(l) + [O] \xrightarrow[\text{warm}]{KMnO_4/H^+} CH_3COOH(aq)$$

ethanal   ethanoic
(intermediate)   acid

During this reaction, the reaction mixture changes colour from purple to colourless. The same reactions take place with potassium dichromate(VI), but the colour change is from orange to green.

Figure 18.1
The oxidation of alcohol with potassium permanganate ▶

Figure 18.2
Oxidation with potassium dichromate ▶

ORR
M&M
A&I

Experiment 18.1 Some typical reactions of alkanols

⚠ Alkanols are flammable.

Procedure

1 Add a mixture of potassium dichromate(VI) and sulphuric acid to some methanol in a test-tube. Heat the mixture in a water bath. Record all your observations.

2 Repeat step 1 with samples of ethanol and propanol.

3 React samples of methanol, ethanol and propanol with a *small* piece of sodium metal in separate test-tubes. Test to identify the gas evolved (see Section 25.3 for details of the tests you will need to use).

4 Record your observations in a table.

Questions

Q1 **a** What type of chemical reaction do alkanols undergo with acidified potassium dichromate(VI)?
 b What visible changes support your answer to part (a)?
 c What other chemical reagent can you use in the laboratory to bring about this change?

Q2 The reaction between alkanols and sodium metal can be classified as the same type of reaction as between alkanols and acidified potassium dichromate(VI), even though the products are different. Discuss by referring to oxidation number changes that occur as the reactants are changed to products in each reaction.

Q3 Use the results of the experiments, the information you have found and the information supplied, to show that methanol, ethanol and propanol are members of the same homologous series.

Find out more

How can you tell whether someone is driving under the influence of alcohol?

In some countries, people who are suspected of driving under the influence of alcohol (ethanol) are given a breathalyser test.

The breathalyser test is based on the fact that high levels of ethanol vapour in the breath reduce acidified orange potassium dichromate(VI) crystals, turning them green.

Q1 What is the legal limit for alcohol for drivers in your country?

Q2 Why should someone not drive if they have been drinking alcohol?

Dehydration is a chemical reaction in which a molecule of water is removed from one molecule of a compound to form a new compound. ⚠

Dehydration of alkanols to alkenes

Alkanols react with dehydrating agents, such as concentrated sulphuric acid or anhydrous aluminium oxide, to form alkenes. The reaction is described as a **dehydration**. The water is lost across two adjacent carbon atoms in the alkanol. This results in the formation of a double bond between the two carbon atoms.

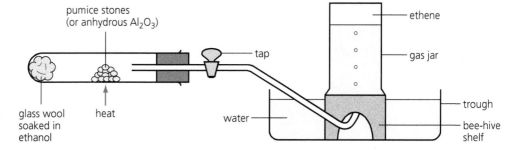

ethanol (a liquid)　　　　　　　ethene (a gas)　　　water

▲ **Figure 18.3**　The dehydration of ethanol

Dehydration can be achieved by:

* heating the ethanol with excess concentrated sulphuric acid at a temperature of 170 °C;

* passing ethanol vapour over activated alumina (Al_2O_3) heated to 450 °C.

ORR
A&I

Experiment 18.2　Preparation of ethene from ethanol (Teacher demonstration)

Procedure

1　Set up the apparatus as shown in Figure 18.4.

2　Heat the dehydrating agent.

3　Collect the gas evolved as ethanol vapours pass over the hot pumice stone.

pumice stones (or anhydrous Al_2O_3)

ethene

tap

gas jar

trough

water

bee-hive shelf

Figure 18.4
The apparatus used to prepare ethene ▶

glass wool soaked in ethanol

heat

Questions

Q1　What method is being used to collect ethene gas?

Q2　What is the function of the tap in the apparatus?

Q3　Describe two tests that you can carry out on the gas to prove that it is ethene.

Q4　Will ethanol react with either of the reagents you mention in question 3?

Q5　**a**　Can ethene be reconverted to ethanol?
　　　b　If your answer to part (a) is 'Yes', then give the conditions for the reaction.

Conversion of alkanols to esters

Another reaction of alkanols is their conversion to esters.

$$\text{alkanoic acid} + \text{alkanol} \underset{\text{as catalyst}}{\overset{\text{conc. } H_2SO_4}{\rightleftharpoons}} \text{ester} + \text{water}$$

This reaction is described more fully in Chapter 20. It is commonly called a 'condensation' reaction.

Combustion of alkanols

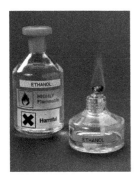

The combustion of alkanols is a highly exothermic reaction; hence the use of alkanols, especially methanol and ethanol, as fuels. Ethanol burns with a blue flame to yield carbon dioxide and water (see the 'Find out more' section below).

$$2CH_3OH(l) + 3O_2(g) \rightarrow 2CO_2(g) + 4H_2O(g) + \text{heat}$$
$$C_2H_5OH(l) + 3O_2(g) \rightarrow 2CO_2(g) + 3H_2O(g) + \text{heat}$$

These equations represent the complete combustion of the alkanols. In an insufficient supply of oxygen, carbon monoxide, carbon and even hydrogen may be formed. This is similar to the combustion of hydrocarbons (see Sections 17.6 and 17.11).

▲ **Figure 18.5** The combustion of ethanol

Find out more

Ethanol as a fuel
Ethanol is being increasingly used as a fuel alternative today. Brazil is one of the countries leading this initiative. It is estimated that more than 40% of Brazil's energy comes from 'green' sources.

Q1 What are Brazil's 'green' sources of energy?

Q2 Find out what other 'green' sources of energy are being developed in other countries. Which locally grown crop(s) do you think could be used in ethanol production?

Q3 Look at the criteria for a good fuel in Section 17.7. To what extent does ethanol fit these criteria?

Figure 18.6
The Jamaica Broilers Ethanol plant was commissioned into service on July 23, 2007. The plant dispatched a shipment of over 3.2 million gallons of ethanol to the USA three days later for fuel production. ▶

18.2   Properties and uses of alkanols

Like water, ethanol is a polar covalent compound. However, when compared with water, it has a higher molar mass and higher heat of combustion, but lower melting and boiling points. The boiling point of water is 100 °C, whereas the boiling point of ethanol is 78 °C. This suggests that the intermolecular forces are weaker in ethanol than in water. Ethanol is therefore a more volatile liquid than water.

These properties make ethanol useful as:

- a solvent;
- a fuel;
- a germicide;
- an antifreeze.

These uses of ethanol are discussed below.

- Ethanol is a solvent with a wide range of uses.
    - Ethanol dissolves a wide range of substances, both polar and non-polar.
    - Ethanol dissolves many compounds that do not readily dissolve in water, so it is a complementary solvent to water.
    - Many useful products are obtained by dissolving suitable substances in ethanol or ethanol-based solvents. Among these substances are paints, thinners and polish, inks, lacquers and varnishes, adhesives, deodorants and colognes, and a wide range of pharmaceutical preparations.
    - Ethanol's volatility makes its removal as a solvent easy.
    - Ethanol's miscibility with water allows it to act as a solvent in reactions between ionic and covalent substances.
- Large amounts of energy are released when alkanols are burnt. Alkanols, particularly ethanol, are finding increasing use as fuels or fuel substitutes for transport and in industry. In the Caribbean, ethanol is mixed with gasoline to produce gasohol which is used as an automobile fuel called E10. E10 is gasoline blended woth 10% ethanol. The ethanol contains oxygen, which raises the octane level of gasoline to prevent engine knocking.

Figure 18.7
A mixture of gasoline and ethanol is used in Brazil as a fuel for cars. ▶

- Ethanol has germicidal properties. These properties come from its ability to attract water. Ethanol can therefore dehydrate micro-organisms, thus making them inactive.

- Ethanol has a low freezing point, and so can be used as an antifreeze. When added to water the ethanol/water mixture freezes at a temperature that is lower than 0 °C. Ethane-1,2-diol, an alkanol with two hydroxyl functional groups, is added to the water in radiators of cars during the winter months.

A&I
M&M

Experiment 18.3 Investigating the solubility of ethanol in water

Procedure

1 Pour some water into a measuring cylinder up to the 50 cm³ mark.

2 Use another measuring cylinder to transfer 50 cm³ of ethanol to the measuring cylinder containing the water.

3 Examine the mixture, describe it and note its final volume.

Questions

Q1 Are ethanol and water miscible liquids?

Q2 Suggest an explanation for the final volume of the mixture.

Anaerobic fermentation is the conversion of sugars to smaller molecules like carbon dioxide and ethanol. It takes place in the absence of oxygen.

Production of rum

Ethanol, the main ingredient in many alcoholic drinks such as beer, wine and rum, is obtained by the **anaerobic fermentation** of grapes, sugar, molasses and other carbohydrate materials. In the Caribbean, molasses, which is a by-product of cane sugar manufacture, is frequently used in fermentation.

▲ **Figure 18.8** Appleton Rum Tour.

▲ **Figure 18.9** Molasses produced at Appleton Estates Rum Tour.

The fermentation process can be summarised as follows:

sugar	+	enzymes	→	ethanol	+	carbon dioxide
starchy materials		supplied				
or molasses		by yeast				

for example:

$$C_6H_{12}O_6 \xrightarrow[\text{37 °C}]{\text{enzymes}} 2C_2H_5OH + 2CO_2$$

glucose, in
aqueous solution

ethanol

Yeasts are unicellular fungi and are the source of the enzymes that are necessary for fermentation. Yeast cells grow rapidly and bud freely if sucrose, water, specific nitrogen sources, vitamins and mineral salts are present.

Figure 18.10 summarises the stages in the manufacture of rum from molasses. The final distillation can be controlled to obtain ethanol for either industrial purposes (99.6% ethanol) or rum (40% ethanol).

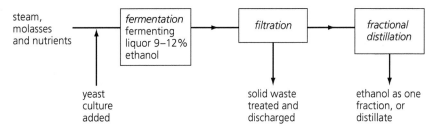

Figure 18.10
The stages in the production of rum ▶

The distillate is processed further. Some flavours are removed during the processing and the raw spirit is matured in oak casks before it acquires the characteristics (colour and taste) of rum.

Production of beer

Beer is an alcoholic beverage that is brewed mainly from malted barley. Barley is a cereal grain. Other ingredients involved in the brewing process include hops, yeast and water. Fruit, wheat and spices are sometimes also used. During the brewing process, the yeast turns the sugars in malt into alcohol, and the hops provide the bitter flavours in beer and the flowery aroma.

▲ **Figure 18.11** Barley and hops are two of the ingredients in brewing beer.

Figure 18.12
Red Stripe,
Jamaica ▶

We can summarise the process of brewing beer into seven major steps:.
Step 1: Malting
Step 2: Mashing
Step 3: Fermentation
Step 4: Storage
Step 5: Filtration
Step 6: Pasteurisation
Step 7: Packaging

The following flowchart shows the steps involved in beer brewing.

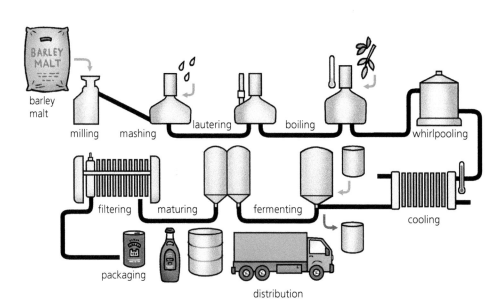

Figure 18.13
The beer brewing
process ▶

Experiment 18.4   Making wine at home (Group experiment)

Procedure

Follow the steps shown here to see how you could make wine at home.

Start with 0.5 kg crushed fruit, e.g. cherries.
1 Soak in 2 dm³ of water to extract flavour.
2 Set aside for 3–4 days.
3 Add 1 Campden tablet (which kills vinegar-producing bacteria).
4 Filter or strain into 3–4 dm³ container.

You now have the strained liquor.
5 Add 1.5 kg of sugar and stir to dissolve.
6 Prepare the wine yeast culture according to given instructions.
7 Add culture and attach the airlock (see Figure 18.14).

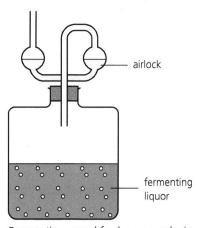

▲ **Figure 18.14**   Fermenting vessel for home-producing wine

You have now produced the 'must' – the sugary
solution that is fermented with yeast to form wine.
8 Add 5 g of citric acid or the juice of citrus fruit until the pH is about 4. Test with pH paper.
9 Set aside for one month.
10 Siphon off clear wine (dead yeast cells fall out).
11 Add one Campden tablet.
12 Add 15 cm³ of strong tea solution, to clear up haze.

This now leaves you with the racked wine.
13 Top up with sugar solution, set aside; rack occasionally (racking = siphoning clear wine from dead yeast cells).
14 Bottle your wine and cork it.

Questions

Q1 The airlock allows carbon dioxide to escape, but prevents air from entering the fermenting mixture. Explain how the airlock is able to achieve this.

Q2 Why is air kept out of the fermenting mixture?

Q3 What would happen if vinegar-producing bacteria entered the mixture?

18.3 The effects of ethanol on the body

Ethanol can have both short-term and long-term effects on the body. Many of the immediate effects are linked to the small size of the ethanol molecule, which allows it to pass through the lining of the stomach directly into the bloodstream. This is particularly the case if ethanol is consumed on an empty stomach. Body mass, size of liver, gender and type of alcoholic drink all influence how rapidly the effects of drinking ethanol become evident.

Some of the short-term effects of ethanol are:

- initial relaxation;
- euphoria;
- loss of inhibitions;
- impaired co-ordination;
- reflexes slow down;
- mental processes slow down;
- attitude changes.

Ethanol is a toxic drug. A concentration of 4 parts per 1 000 in the blood can lead to a deep anaesthetic effect or even death. When someone consistently consumes more ethanol than the body can break down, there is a significant rise in blood ethanol level, which leads, among other things, to:

- impaired circulation;
- accumulation of toxins;
- excessive loss of fluids via urination;
- problems of digestion, hence stomach problems;
- malfunction of organs such as the pancreas, liver and brain in the long term.

In addition, many common drugs, e.g. aspirin and antihistamines, contain ingredients that react unfavourably with ethanol. The simultaneous use of ethanol and such drugs can lead to complicated health problems.

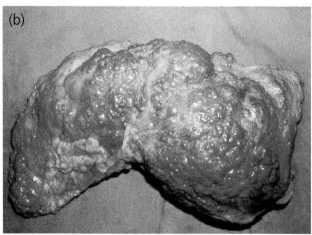

▲ **Figure 18.15** Alcohol is very damaging to the body: (a) a healthy and (b) a damaged liver.

Find out more

The social impact of alcohol consumption

Q1 What impact can high levels of alcohol consumption have on:
  a the family (or home);
  b society in general?

Summary

- Alkanols contain the –OH (hydroxyl) functional group, which forms hydrogen bonds with water molecules (the lower members are completely miscible with water).
- Reactions of alkanols:

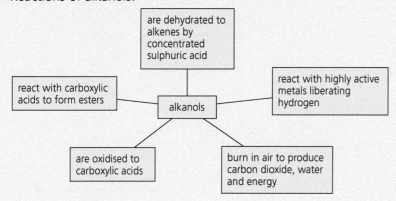

- Ethanol is produced industrially by fermentation of simple carbohydrates.
- Ethanol is used:
  - as a fuel. For example, gasohol, a mixture of gasoline and ethanol, is a common fuel for vehicles in some South American countries).
  - in drinks. Excessive consumption of alcohol can have a negative personal and societal impact.
  - as a solvent.

End-of-chapter questions

1 Which is the correct general formula for alkanols?
  A C_nH_{2n}
  B C_nH_{2n+2}
  C $C_nH_{2n+1}OH$
  D $C_nH_{2n+1}COOH$

2 Which reactions do alkanols undergo?
  I Addition
  II Hydration
  III Dehydration
  IV Combustion

  a I, II, III only
  b I, III and IV only
  c II, III and IV only
  d I, II, III and IV

3 Which of the following is the correct molecular formula for an alkanol with 10 carbon atoms?
  A $C_{10}H_{20}OH$
  B $C_{10}H_{22}OH$
  C $C_{10}H_{21}OH$
  D $C_{10}H_{21}COOH$

4 If 2 moles of propanol are burnt, what is the correct balanced equation for the combustion reaction?

A $2CH_3OH(l) + 3O_2(g) \rightarrow 2CO_2(g) + 4H_2O(g)$
+ heat

B $2C_3H_7OH(l) + 9O_2(g) \rightarrow 6CO_2(g) + 8H_2O(g)$
+ heat

C $2C_3H_7OH(l) + 4O_2(g) \rightarrow 3CO_2(g) + 5H_2O(g)$
+ heat

D $2C_2H_5OH(l) + 6O_2(g) \rightarrow 4CO_2(g) + 6H_2O(g)$
+ heat

5 How many of moles of oxygen will be required to burn 1 mole of ethanol?

A 1
B 2
C 3
D 4

6 Why are alkanols soluble in water?

A They have a hydroxyl functional group that makes them polar.
B They have a hydroxyl functional group that makes them non-polar.
C They have a large functional group that makes them soluble.
D They have an oxygen atom in their functional group that makes them attractive to water molecules.

7 Which of the following is an incorrect IUPAC name for an isomer of C_4H_9OH?

A Butan-1-ol
B Butan-2-ol
C Pentan-1-ol
D 2-methylpropan-2-ol

8 What is the IUPAC name for the compound produced when pentan-1-ol reacts with warm acidified $KMnO_4$?

A Pentanoic acid
B Pentan-2-ol
C Pentan-3-ol
D Pent-1-ene

9 Which statements about the reaction between ethanol and ethanoic acid are correct?

I It is a condensation reaction.
II They react to form an ester.
III It requires a catalyst.
IV It requires concentrated sulphuric acid.

A I, II and III only
B I, II and IV only
C II, III and IV only
D I, II, III and IV

10 Which of the following compounds undergo dehydration?

A Methanol
B Ethene
C Ethanoic acid
D Ethylethanoate

11 Look at the following reaction scheme.

H ←(conc H₂SO₄, 70 °C)— H—C—C—C—OH —(KMnO₄/H⁺, D + gas E)→ intermediate with carbonyl —(KMnO₄/H⁺)→ change blue litmus paper to red

A —Na→ D + gas E

B

C —→ salt F + gas G

a What is the general formula of compound A?
b Draw and give the IUPAC name for the compounds B, C, D, F and H.
c Describe two simple experiments to test for gases E and G.

d Compounds C and A can react to produce the sweet-smelling compound X.
  (i) Write a balanced equation for the formation of compound X. Include the reagents and conditions.
  (ii) To what homologous series does compound X belong?
  (iii) Draw and name the structural displayed formula for compound X.
  (iv) Circle the area in part (d) (iii) that is responsible for the sweet fragrance.
  (v) Describe one test that you could use to differentiate compound C from compound X.

12 The structure below is the active ingredient in aspirin, acetylsalicyclic acid.

a Circle the functional group that reacts with a carbonate to produce carbon dioxide.
b Draw the functional group that is responsible for the sweet fragrance in acetylsalicyclic acid.

13 The diagram below is an active ingredient in allergic medicines.

a Into which homologous series would you place this compound and why?
b Circle the functional group of the homologous series you have identified in part (a).
c Draw the structural displayed formula that results when the compound reacts with:
  (i) sodium metal;
  (ii) activated alumina heated to 450 °C;
  (iii) methanoic acid with a catalyst and concentrated H_2SO_4.
d Allegra can possess the functional group
   −COOH.
(i) Name the reaction that would allow Allegra to have the carboxylic acid functional group.
(ii) What reagent and condition is necessary for Allegra to possess the carboxylic acid functional group?

14 a Give the general formula of the alkanols.
  b What is the molecular formula of the alkanols with four carbon atoms?
  c Write three structural formulae corresponding to this molecular formula.

15 a Outline two different methods for obtaining ethanol on an industrial scale.
  b List three industrial uses of ethanol.
  c List two domestic uses of ethanol.

16 a Explain the principle on which the breathalyser works.
  b What impact can excessive intake of ethanol have on:
    (i) the body;
    (ii) society?

17 How does propanol react with:
  a sodium metal;
  b concentrated sulphuric acid;
  c acidified potassium dichromate(VI);
  d ethanoic acid and concentrated sulphuric acid?

18 a Name compounds A and B.

   b Give the reagent and conditions for this reaction.

   c What is the name given to this type of reaction?

   d A manufacturer starting with 1 500 kg of A obtained 960 kg of B. Calculate the percentage yield.

   e How does compound A react with a mixture of ethanoic acid and concentrated sulphuric acid? What is the role of the sulphuric acid?

   f How does compound B react with bromine dissolved in 1,1,1-trichloroethane?

   g How can compound B be reconverted to compound A?

$$H-\overset{\overset{\displaystyle H}{|}}{\underset{\underset{\displaystyle H}{|}}{C}}-\overset{\overset{\displaystyle H}{|}}{\underset{\underset{\displaystyle H}{|}}{C}}-\overset{\overset{\displaystyle H}{|}}{\underset{\underset{\displaystyle H}{|}}{C}}-\overset{\overset{\displaystyle H}{|}}{\underset{\underset{\displaystyle H}{|}}{C}}-OH \longrightarrow H-\overset{\overset{\displaystyle H}{|}}{\underset{\underset{\displaystyle H}{|}}{C}}-\overset{\overset{\displaystyle H}{|}}{\underset{\underset{\displaystyle H}{|}}{C}}-\overset{\overset{\displaystyle H}{|}}{\underset{\underset{\displaystyle H}{|}}{C}}=C\overset{\diagup H}{\diagdown_{H}} \quad + \quad H_2O$$

         A   B

19 a Give the names and formulae of the products formed when butan-1-ol is passed over heated alumina.

   b What is the function of the alumina in this reaction?

   c   **(i)** Write a balanced chemical equation for the reaction.

        (ii) Use the balanced equation to calculate the volume of the product formed at RTP, when 0.74 g of butan-1-ol are completely converted in this reaction.

20 When cherries are fermented under anaerobic conditions, wine is produced.

   a What do you understand by the term 'anaerobic'?

   b What gas would be formed during fermentation?

   c How would you test for the presence of this gas?

   d How could you tell when fermentation was complete?

   e How could you obtain a pure sample of ethanol from the fermentation product?

   f Suggest why it is dangerous to bottle wine before fermentation is complete.

Data analysis

20 Look at the table below.

Name	Boiling temperature (°C)	Solubility in H_2O (g/100 g H_2O)
methanol	64	miscible in all proportions
propanol	102	miscible in all proportions
butanol	118	7.9
pentanol	138	2.3
1,2-ethanediol ($C_2H_6O_2$)	197	miscible in all proportions
glycerol 290 (propane-1,2,3-triol)	290	miscible in all proportions

1,2-ethanediol ($C_2H_6O_2$) is a di-alcohol. Glycerol is a tri-alcohol.

   a Describe the trend in boiling points among the C_1 to C_5 alkanols inclusively.

   b Account for this trend in boiling points.

   c Identify and account for the trend in solubility of these alkanols in water.

19 Alkanoic acids and esters

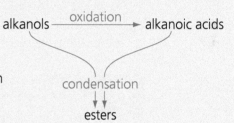
19.1 Introduction

Alkanoic acids are also known as carboxylic acids. Long-chain alkanoic acids, particularly those found in plants and animals, are known as fatty acids. Alkanoic acids and esters are constituents both of naturally occurring substances and of manufactured products with distinct flavours and odours.

Find out more

Alkanoic acids
Milk, milk products, such as cheese and butter, and some fruits contain alkanoic acids. Some examples are:

- lactic acid in milk (three carbon atoms, IUPAC name 2-hydroxypropanoic acid);
- tartaric acid in grapes (four carbon atoms, IUPAC name 2,3-dihydroxybutanedioic acid);
- butanoic acid (four carbon atoms), which is largely responsible for the rancid smell of butter.

▲ **Figure 19.1** Methanoic acid (formic acid)

Esters

Low boiling point, low relative molecular mass esters (formed by combination of alkanoic acids and alkanols) are mainly responsible for the flavours and fragrances of the flowers and fruits of many plants.

Alkanoic acids containing between 12 and 22 carbon atoms occur combined (as esters) in animal fats and plant oils.

19.2 Alkanoic acids

Alkanoic acids can be represented by their general formula $C_nH_{2n+1}COOH$, which can be shown more conveniently as R–COOH.

The –COOH group (the carboxyl group) is the functional group in organic acids. The carboxyl group gives this class of compounds:

- acidic properties;
- relatively high boiling points;
- solubility in water.

Understand it better

Understanding the properties of alkanoic acids
When the O—H bond in the –COOH functional group splits, it releases a hydrogen ion into solution, thus making this group of compounds acidic.

$$-C{\overset{\displaystyle O}{\underset{\displaystyle O-H}{}}}\text{(aq)} \quad \rightleftharpoons \quad -C{\overset{\displaystyle O}{\underset{\displaystyle O^2}{}}}\text{(aq)}+ \quad H^+\text{(aq)}$$

–COOH functional groups can hydrogen bond with each other, thereby increasing the intermolecular attractions and hence increasing the boiling points of alkanoic acids.

–COOH groups can also hydrogen bond with the solvent water and so alkanoic acids will dissolve in water.

Practice

1 The first member of the alkanoic acid homologous series has a value of $n = 0$. In this acid, the only carbon atom is in the –COOH group.

  a Write the molecular formula of each of the first four members of the series.

  b Write the full structural formula for the first and fourth member of the series.

Ethanoic acid

Ethanoic acid, CH_3COOH, has the following structural formula:

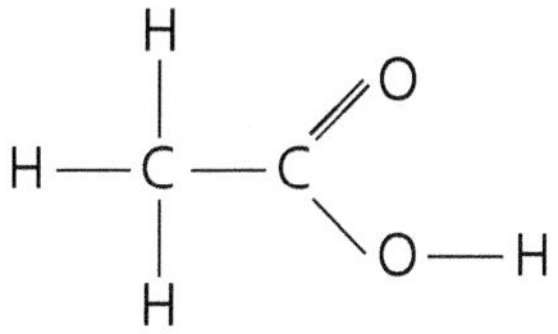

Ethanoic acid is a typical representative of the homologous series of alkanoic acids. It is found on the average kitchen shelf as vinegar, which is a 5% solution of ethanoic acid. Vinegar can be used as a preservative in pickling and also in other types of food preparations. Ethanoic acid is easily recognised by its 'sharp' smell and sour taste.

Preparation of ethanoic acid

A sample of ethanoic acid can be prepared by refluxing a mixture of ethanol and acidified potassium dichromate(VI) or acidified potassium manganate(VII). Then, the product mixture needs to be distilled. You need to collect the fraction that comes off with a boiling point near to 118 °C (see Section 5.3).

Physical properties of alkanoic acids

Alkanoic acids have higher boiling points than alcohols with similar mass. For example, methanoic acid with a mass of 36 g will have a higher boiling point than ethanol, which also has a mass of 36 g. The boiling point of methanoic acid is higher because carboxylic acid forms **dimers** in which hydrogen bonds are formed between the polar groups in the carboxylic group. The formation of the hydrogen bond results in a higher boiling point.

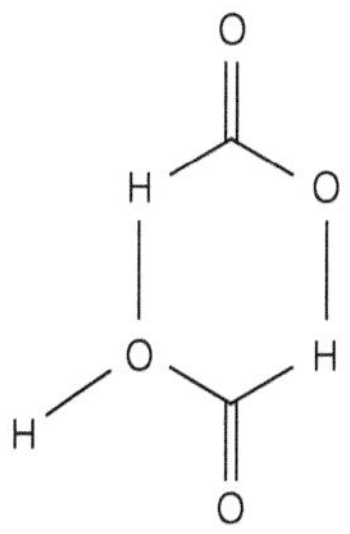

A **dimer** is a chemical structure formed from two identical compounds joined by bonds that can either be strong or weak, covalent or intermolecular.

▲ **Figure 19.2** Methanoic acid dimer

▼ **Table 19.1** Boiling points of alcohols and alkanoic acids with similar boiling points

Compounds	Molar mass (g/mol)	Boiling point (°C)
methanoic acid	36	100.8
ethanol	36	78.1
ethanoic acid	60	118
propanol	60	97.2
propanoic acid	74	141
butanol	74	117.7

Alkanoic acids are soluble in water because of the polar (−OH) group. This hydroxyl group, attached to the carbonyl group in the carboxyl functional group, forms hydrogen bonds with many water molecules.

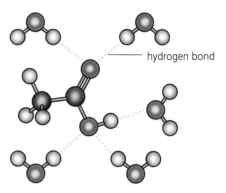

hydrogen bond

Figure 19.3
Alkanoic acids form hydrogen bonds with water molecules. This is what makes them soluble in water. ▶

As the length of the hydrocarbon chain increases, the water solubility of alkanoic acids decreases, i.e. the solubility decreases with increasing molar mass. Carboxylic acids with between one and four carbon atoms are completely miscible in water. Smaller-chain carboxylic acids are more water soluble because of the strong attractive force of the −OH group in the compound. The attractive force of the −OH reduces with increasing hydrocarbon chain length. The hydrocarbon chain attached to the carboxyl functional group is hydrophobic, so as the number of carbon atoms in the chain increases, the less soluble the carboxylic acid becomes. Carboxylic acids with a chain length of up to 10 carbon atoms are sparingly soluble. Carboxylic chains with more than 10 carbon atoms are insoluble.

▼ **Table 19.2**   Solubility and boiling points of alkanoic acids

IUPAC name	Boiling point (°C)	Solubility in water
methanoic acid	101	very soluble
ethanoic acid	118	very soluble
propanoic acid	141	very soluble
butanoic acid	164	very soluble
pentanoic acid	187	slightly soluble
hexanoic acid	205	slightly soluble
benzoic acid	250	slightly soluble

The reactions of alkanoic acids

Alkanoic acids are weak monobasic acids. They are incompletely dissociated. The alkanoic acids show the typical reactions of mineral acids, but they react less vigorously because they are weak acids, whereas the mineral acids are strong.

Combustion of alkanoic acids

Anhydrous alkanoic acids burn to give carbon dioxide and water. Alkanoic acid burns with a bright flame and gives off a vinegary scent:

$$CH_3COOH + 2O_2(g) \rightarrow 2CO_2(g) + 2H_2O(l)$$

Neutralisation

Alkanoic acids are weak acids and react like mineral acids:

$$CH_3COOH(aq) \rightleftharpoons CH_3COO^-(aq) + H^+(aq)$$
ethanoate ion

Reactions with metals

Metals high in the reactivity series react with organic acids to produce hydrogen gas and the salt of the organic acid:

$$2Na(s) + 2CH_3COOH(aq) \rightarrow 2CH_3COONa(aq) + H_2(g)$$
sodium ethanoate
(a salt of ethanoic acid)

The reaction between reactive metals and organic acids is less vigorous than the corresponding reactions with mineral acids.

$$Mg(s) + 2CH_3COOH(aq) \rightarrow (CH_3COO)_2Mg + H_2(g)$$
magnesium
ethanoate

The balanced equations indicate that ethanoic acid is a monobasic acid. Note that only one hydrogen atom, that of the –COOH group of the acid, is 'acidic'.

Reactions with metal oxides and hydroxides

The oxides and hydroxides of metals react with organic acids to form the salt of the acid and water. This is a neutralisation reaction:

$$MgO(s) + 2CH_3COOH(aq) \rightarrow (CH_3CO_2)_2Mg(aq) + H_2O(l)$$
oxide   salt   water

$$NaOH(aq) + CH_3COOH(aq) \rightarrow CH_3COONa(aq) + H_2O(l)$$
hydroxide

Reactions with carbonates and hydrogencarbonates

Carbonates and hydrogencarbonates react with organic acids to yield carbon dioxide, the salt of the acid and water:

$$2CH_3COOH(aq) + Na_2CO_3(aq) \rightarrow 2CH_3COONa(aq) + H_2O(l) + CO_2(g)$$
$$CH_3COOH(aq) + NaHCO_3(aq) \rightarrow CH_3COONa(aq) + H_2O(l) + CO_2(g)$$

Practice

2 Can the reaction with highly reactive metals be used to distinguish between alkanols and alkanoic acids? Justify your answer.

Condensation reactions take place by the elimination of a small molecule, such as water, ammonia or hydrogen chloride, between certain functional groups in organic compounds.

Reactions with alkanols

Esterification

Esterification is a reaction in which anhydrous alkanoic acids react with alkanols to form esters and water. Because water is being produced, esterification is considered to be a **condensation reaction**.

The mixture of the alkanoic acid, the alkanol and a few drops of concentrated sulphuric acid should be heated in a water bath for a few minutes. The mixture then needs to be poured into a concentrated solution of sodium chloride.

A water molecule is lost between the functional group of an alkanoic acid and the functional group of an alkanol producing an ester. The removal of water between the two molecules is made possible by the dehydrating property of the sulphuric acid in the mixture.

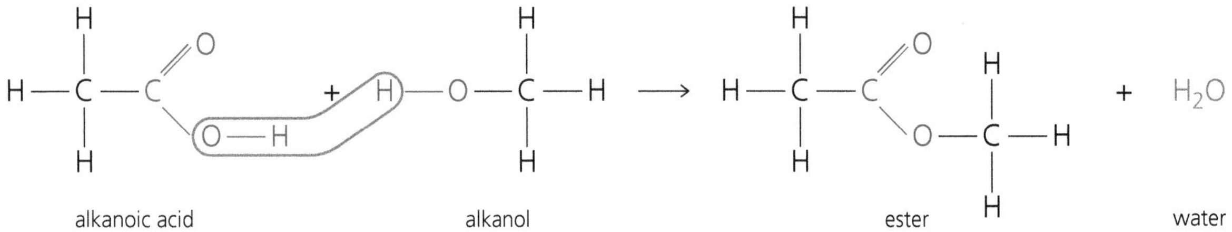

▲ **Figure 19.4** A condensation reaction

The $-C\overset{\displaystyle O}{\underset{\displaystyle O-}{}}$ group (the carboxylate group) retained in the ester is the functional group of the ester and the new C—O bond formed is the ester linkage.

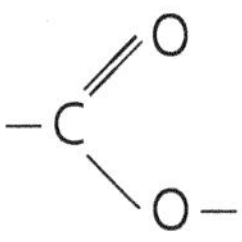

$$CH_3COOH + C_2H_5OH \rightleftharpoons CH_3COOC_2H_5 + H_2O$$

ethanoic acid   ethanol (ester)   ethyl ethanoate   water

The names of esters are derived from both the alcohols and the organic acids from which they are derived. Look at the following names and see if you can work out how the name of the ester is produced:

alkanol	+	organic acid	→	ester
methanol	+	ethanoic acid	→	methyl ethanoate
ethanol	+	ethanoic acid	→	ethyl ethanoate
propanol	+	methanoic acid	→	propyl methanoate

The '-oate' comes from the acid and the alkyl part comes from the alcohol.

Note, however, that in writing the molecular formula or the structural formula of the ester, the acid part is normally written first. Look at the structure of ethyl propanoate in Figure 19.6 and link the two parts of the name to the formula.

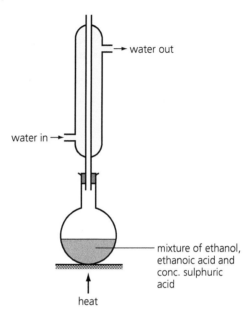

Figure 19.5
Ethyl propanoate ▶

Experiment 19.1 Preparing ethyl ethanoate (Teacher demonstration)

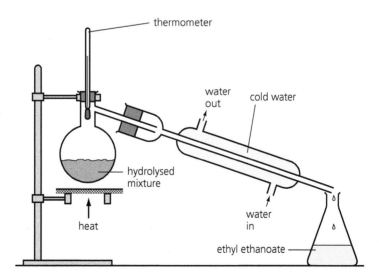

Figure 19.6
Refluxing the mixture ▶

Figure 19.7
Separating the ester by
distillation ▶

When ethanol is refluxed with excess ethanoic acid, in the presence of sulphuric acid as a catalyst, the ester ethyl ethanoate is formed. Ethyl ethanoate (boiling point 77 °C) can be obtained from the product mixture by distillation.

Practice

3 Write (i) the name and (ii) the structure of the esters formed when the following pairs of substances react:
  a Ethanoic acid and butanol
  b Methanol and propanoic acid
  c Butanoic acid and ethanol
  d Propanol and butanoic acid

19.3 Hydrolysis of esters

Hydrolysis of an ester involves the splitting of the ester linkage by the action of water to reform the alkanoic acid and the alkanol.

The ester linkage is formed by the elimination of water and can therefore be broken by the addition of water. This process is known as **hydrolysis**.

The hydrolysis of esters proceeds very slowly and can be catalysed by refluxing the ester either with dilute mineral acid (acid hydrolysis) or an excess of dilute alkali (alkaline hydrolysis).

Acid hydrolysis

Acid hydrolysis of ethyl ethanoate forms ethanol and ethanoic acid, as shown in this equation:

$$CH_3COOC_2H_5(aq) + H_2O(l) \rightleftharpoons CH_3COOH(aq) + C_2H_5OH(aq)$$

```
         O
         ||
CH₃— C — O — CH₂CH₃ + H₂O
     ethyl ethanoate          water
                              H⁺
         O
         ||
CH₃— C — OH + HO — CH₂CH₃
     ethanoic acid          ethanol
```

▲ **Figure 19.8** Acid hydrolysis

Alkaline hydrolysis

With alkaline hydrolysis, the alkanoic acid is formed first. The alkanoic acid is then neutralised by the excess alkali, to form the sodium salt of the acid. The other product is the alkanol:

$$CH_3COOC_2H_5(aq) + NaOH(aq) \rightarrow CH_3COO^-Na^+(aq) + C_2H_5OH(aq)$$

```
         O
         ||
CH₃— C — O — CH₂CH₃ + NaOH
     ethyl ethanoate          sodium hydroxide


CH₃CH₂— OH + Na⁺⁻O — CH₂CH₃
  ethanol        sodium   ethanoate
                  ion       ion
```

▲ **Figure 19.9** Alkaline hydrolysis

Alkaline hydrolysis of simple esters is sometimes referred to as **saponification**. However, the term is more appropriately applied to the process of soap manufacture.

The process of alkaline hydrolysis and the separation of the products of the reaction can be carried out in the laboratory by heating the mixture of ester and alkali under reflux and separating the products of the reaction by distillation. The apparatus is similar to that shown in Figures 19.7 and 19.8.

Experiment 19.2   Saponification

Apparatus
- 250 ml beaker
- evaporating dish
- oil
- 4 molar sodium hydroxide
- pH meter
- stirring rod
- tripod stand
- Bunsen burner
- wire gauze

Procedure

1 Boil some water in a large beaker.

2 Place an evaporating dish on the large beaker.

3 Place 5 ml of oil and 20 ml of 4 molar sodium hydroxide into an evaporating dish.

4 Heat the evaporating dish and its contents, stirring continuously.

5 Remove the evaporating dish from the heat and add sodium chloride crystals to precipitate the soap.

6 Allow the mixture to cool, and then filter the mixture and wash with distilled water. Measure the pH of the wash-water to test whether it is safe to use the soap.

7 You can now add chemicals such as glycerol (for transparency and moisturising), colourings, perfume and disinfectants.

Practice

4 a Write the equation for the alkaline hydrolysis of ethyl butanoate.
   b Which of the two producrs will collect as the distillate? Explain your choice.
   c How can butanoic acid be obtained from the sodium salt formed in the hydrolysis?
   d Propanol and butanoic acid

19.4   Uses of alkanoic acids and esters

Alkanoic acids

The use of ethanoic acid as a preservative has already been mentioned. Ethanoic acid is also used in the manufacture of plastics, in cellulose ethanoate, textiles and in a wide range of pharmaceuticals.

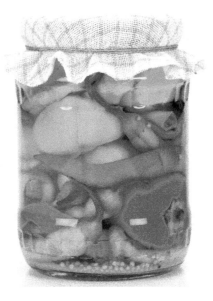

Figure 19.10
Acetic acid can be used to preserve pepper sauce. ▶

Esters

Many esters occur naturally. Some of the low boiling point, short-chain esters, for example butyl ethanoate and pentyl ethanoate, are pleasant smelling. They are responsible for the fragrances and flavours of flowers and fruits. Artificial essences or flavourings are usually mixtures of esters dissolved in ethanol.

Esters with long carbon chains are commonly found in vegetable oils and animal fats. They are widely used in food preparations and in making soap. Fats and oils are important constituents of our diets. They are concentrated energy sources, they make food more palatable, and contain vitamins A, D and E.

Experiment 19.3   Finding the concentration of ethanoic acid in vinegar

The problem statement

Three different brands of vinegar are sold at a local supermarket. Plan and design an experiment to show which of the three different brands of vinegar has the highest concentration of ethanoic acid.

The hypothesis

Fixed volumes of the different brands should require different volumes of a standard solution of an alkali for neutralisation.

Apparatus and materials

- volumetric flask
- conical flask
- measuring cylinder
- funnel
- burette
- pipette
- pipette filler
- 0.1 mol dm^{-3} sodium hydroxide solution
- samples of vinegar
- phenolphthalein indicator

Procedure

1 Dilute the vinegar sample to obtain a solution that is approximately 0.1 mol dm^{-3}.

2 Pipette 25.0 cm^3 of the diluted vinegar into a conical flask and add 2 drops of phenolphthalein indicator.

3 Fill the burette with the sodium hydroxide solution.

4 Deliver the alkali from the burette into the conical flask containing the vinegar/indicator mixture until a pale pink permanent colour is obtained.

5 Record the volume of standard alkali needed for the neutralisation of the 25 cm^3 of the diluted vinegar.

6 Repeat the titration as many times as is necessary to obtain consistent results.

Results

Results of titrations are usually tabulated, as shown in Table 19.3.

▼ Table 19.3

Burette readings (cm^3)	Trial	Accurate		
		1	2	3
final burette reading	18.50	36.50	18.25	36.55
initial burette reading	0.00	18.30	0.00	18.25
titre	18.50	18.20	18.25	18.30

average volume of NaOH used $= \dfrac{18.20 + 18.25 + 18.30}{3} = 18.28$ cm^3

Summary: 25.0 cm^3 of the diluted ethanoic acid (vinegar) required 18.28 cm^3 of dilute sodium hydroxide for neutralisation.

Conclusion

The sample of vinegar that required the largest volume of standard alkali for neutralisation in the experiment will be the sample with the highest concentration of ethanoic acid.

Discussion

In titration experiments, it is important that the concentration of the standard is not changed at any point during the experiment. One way of ensuring this is to rinse the equipment in which the standard is to be placed first with distilled water, then with the standard solution.

In this experiment, the burette should be rinsed with the alkali. Since we also want to maintain the concentration of the vinegar, the pipette should be rinsed with the diluted vinegar solution prior to dispensing the acid into the conical flask. Note that the conical flask should be rinsed only with distilled water, since exactly 25.0 cm³ of the weak acid is being analysed in the experiment. Standard solutions should show certain characteristics, some of them being:

- high molar mass;
- stability in air.

Sodium hydroxide has deliquescent properties and will therefore absorb moisture from the atmosphere. Furthermore, if left to stand for too long, the alkali can react with the carbon dioxide in the atmosphere. These factors can change the actual concentration of the alkali. Sodium hydroxide is therefore not an ideal standard for use in this type of analysis.

Despite these problems with using sodium hydroxide as a standard, the procedure is useful to compare the concentrations of the vinegar samples.

Practice

4 a Write the equation for the alkaline hydrolysis of ethyl butanoate.
b Which of the two products will collect as the distillate? Explain your choice.
c How can butanoic acid be obtained from the sodium salt formed in the hydrolysis?

19.5 Detergents

Detergents are organic cleaning agents that remove grease and dirt from porous surfaces. Porous surfaces are surfaces such as fabrics, clothes or non-treated wood. We classify detergents into two groups: soaps and syndets.

- Soaps: Soaps are soapy detergents made by saponification, which is the alkali hydrolysis of esters. If the soap is a salt of sodium, we say that it is a hard soap. If the soap is a salt of potassium, we call it a soft soap. Example of soaps are sodium stearate and sodium palmitate. Stearic acid is octodecanoic acid and palmitic acid is hexadeanoic acid.

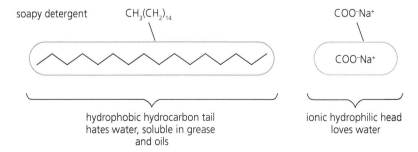

Figure 19.11
Soapy detergent ▶

- Syndets: Syndets, which are SYNthetic DETergents, are soapless detergents. Syndets are salts of sodium or potassium and sulphonated petrochemicals. Examples of soapless detergents are straight-chain alkyl benzenesulphonates such as sodium dodecyl benzenesulphonate. These are the most widely used detergents in liquid and powdered soapless laundry detergents and also in dishwashing liquids.

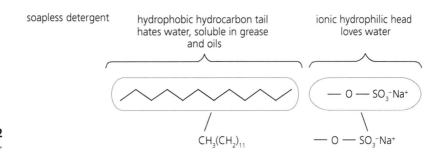

Figure 19.12
Soapless detergent. ▶

19.6 Structure of a detergent molecule

We can think of soaps and syndets as having a polar or ionic water-loving (hydrophilic) head and a non-polar hydrocarbon chain, which is grease-loving (hydrophobic).

The tail represents the hydrocarbon part of fatty acid molecules such as palmitic acid ($C_{15}H_{31}-$), stearic acid ($C_{17}H_{35}-$) and oleic acid ($C_{17}H_{31}-$). The tail of a non-biodegradable soap has several branches that inhibit microbe attack. They serve as a link between the grease/oil and the water

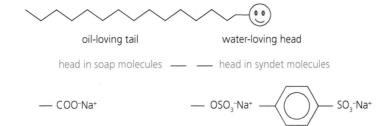

Figure 19.13
Both the tail and the head serve as a link between the grease/oil and the water. ▶

19.7 The action of detergents

The major problem that we try to deal with when cleaning objects is to remove grease, oils and fats that stick to surfaces. Water, by itself, is not an effective cleaner because it:

- is not a 'grease-loving' molecule – water molecules are polar whereas grease molecules are non-polar and do not readily dissolve in water (see Sections 4.5 and 4.9);

- has poor wetting properties, arising from its high **surface tension**.

Surface tension is that property of a liquid that makes it behave as if its surface is enclosed in an elastic skin. Surface tension arises from forces between the molecules of the liquid.

You can see surface tension in action when water forms beads on surfaces such as glass and fabric rather than spreading out and wetting the fabric.

Certain chemicals, such as soaps and soapless detergents, are able to improve the cleaning properties of water. Such chemicals are called **surfactants**.

Surfactants (surface active agents):

- lower the surface tension of water;

- loosen dirt;

- disperse and hold the dirt particles in suspension in water (as an emulsion) until they can be washed away.

The cleaning action of detergents

The dirt on the surface of fabrics or clothes is held there by oil and grease. When the surface is placed into soapy water, the non-polar tails of the detergent molecules attach themselves to the grease while the polar heads remain dissolved in the water. With constant agitation, the grease and the dirt are dislodged from the surface, as shown in Figure 19.14.

Detergent molecules then attack the grease from all sides. Since the heads are still radiating outward with the same charge, they cannot settle or reattach themselves to the surface, and instead they remain dispersed in the water, which is then thrown out. A follow-up rinse will then remove the remaining detergent-grease structures.

(a) 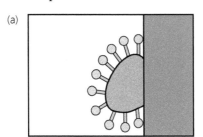   (b)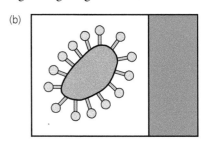

▲ **Figure 19.14**   How detergents work. (a) Detergent molecules begin to surround the grease on the fabric. The grease-loving tails attach themselves to the grease. The water-loving heads point away from the grease. (b) The piece of grease has been lifted off. It is completely surrounded by detergent molecules and floats away as part of the detergent/grease emulsion.

- Examples of soaps are sodium stearate and sodium palmitate (stearic acid is octadecanoic acid and palmitic acid is hexadecanoic acid).

- Examples of soapless detergents are straight-chain alkyl benzenesulphonates such as sodium dodecyl benzenesulphonate. These are the most widely used detergents in liquid and powdered soapless laundry detergents, and also in dishwashing liquids.

Understand it better

You can get a good idea of the emulsifying action of surfactants by carrying out the following simple activity.
1 Take two identical clear glasses or test-tubes.
2 To each glass, add two tablespoons of water followed by a tablespoon of cooking oil. Shake both glasses vigorously.
3 Let the glasses stand and observe what happens.
4 Now add a squirt of dishwashing liquid to one glass. Shake both glasses vigorously for a few seconds.
5 Let the glasses stand and observe what happens now.
Account for what happens in steps 3 and 5.

▼ **Table 19.4** Comparison of soapy versus soapless detergents

Soapy detergents	Soapless detergents
are all biodegradable	may be non-biodegradable
are made from plant oils and animal fats, which are renewable	made from petro-chemicals, which are non-renewable
form scum (insoluble calcium or magnesium salt) with hard water	lather easily in any water
do not cause pollution	pollute nearby streams due to phosphate content, leading to eutrophication (overgrowth of algae, which starves fish of oxygen and shade underwater plants)
are used mainly for laundering clothes; overuse leads to the loss of colour in fabrics	used in industry to emulsify oil slicks
soaps are better cleaners in soft water than hard water	
soaps cannot be used in strongly acidic solution	

19.8 Chemicals used in laundering

An **active ingredient** is the part of a product that actually does what the product is designed to do.

Surfactants are the major **active ingredients** in detergents. Other ingredients, such as water softeners and emulsifiers, help the surfactants to be more effective.

However, we want our laundry not only to be clean but to look clean and bright, and to feel good. Many other chemicals are used in laundering to achieve these results. Some chemicals, such as bleaches, remove stains and others, such as fabric softeners, starch and optical brighteners, help the fabrics feel and/or look better. Most of these products are available as separate products or they may be used as additives in a single detergent.

Table 19.5 summarises the ingredients that may be found in a typical laundry detergent.

▼ **Table 19.5** Some of the ingredients in laundry detergents

Type of ingredient	Function(s)	Examples and remarks
surfactants	• lower surface tension • disperse dirt and grease • stabilise suspensions of dirt in water	soapless detergents, such as linear alkyl benzenesulphonates, or soaps such as sodium stearate
anti-redeposition agents	• help to prevent loosened dirt from resettling on the cleaned fabrics	sodium carboxyl methyl cellulose (SCMC)
water softeners	• decrease hardness in water by removing calcium, magnesium and iron ions	complex phosphates, sodium silicate, sodium carbonate

fabric softeners	• make fabrics softer and fluffier • decrease static electricity	quaternary ammonium compounds (these also have surfactant properties)
optical brighteners/ fluorescers	• makes clothes appear whiter and brighter in daylight	these are not cleaning agents
bleaches	• help to remove stains • whiten and brighten fabrics	chlorine bleach, e.g. sodium hypochlorite, oxygen bleach, such as sodium perborate

Summary

- Organic acids (alkanoic acids) have the general formula $C_nH_{2n+1}COOH$.
- The first four members of the series are:
  - methanoic acid, $HCOOH$;
  - ethanoic acid, CH_3COOH;
  - propanoic acid, C_2H_5COOH;
  - butanoic acid, C_3H_7COOH.
- Alkanoic acids are weak monobasic acids.
- Ethanoic acid:
  - reacts with active metals, liberating hydrogen;
  - is neutralised by metal oxides and hydroxides;
  - reacts with carbonates, liberating carbon dioxide;
  - reacts with alkanols to form esters in a condensation reaction.
- Esters are hydrolysed by alkalis to produce an alkanol and the salt of the organic acid.
- Acid hydrolysis of the ester produces the alkanol and the carboxylic acid.
- Vinegar is 5% ethanoic acid and is used in pickling, seasoning and preservation of foods.
- Ethanoic acid is used in the manufacture of textiles, plastics and pharmaceuticals.
- Detergents are cleaning agents. They may be soapy or soapless.
- Soapy detergents:
  - are made from hydrolysis of fats;
  - are biodegradable;
  - cause scum formation with hard water;
  - are excellent for laundering if water is soft;
  - lead to loss of fabric colour following continued use in laundry.
- Soapless detergents:
  - are made from petroleum;
  - many are non-biodegradable;
  - do not cause scum with hard water;
  - their effectiveness and cleaning power are not affected by hard water.

- Soapy and soapless detergents act as surfactants. They lower surface tension and disperse grease and dirt.
- Other common ingredients in a typical soapless detergent are:
  - dirt-suspending agents to keep removed dirt in suspension until it can be washed away;
  - phosphates, which soften water, remove dirt and help the removed dirt to stay in suspension;
  - anhydrous sodium sulphate and sodium silicate, which provide bulk and keep the product dry;
  - fluorescers to hide discolorations in garments;
  - bleaching agents, e.g. sodium perborates.

End-of-chapter questions

1 Which of the following acetic acids is the most soluble?
   A Propanoic acid
   B Butanoic acid
   C Pentanoic acid
   D Hexanoic acid

2 Carboxylic acids have considerably higher boiling points than alcohols of similar molecular mass because they:
   A Are more acidic
   B Have a greater oxygen content
   C Form stable hydrogen-bonded dimers
   D Are hydrophobic

3 What is the IUPAC name for the salt produced when propanoic acid reacts with sodium hydroxide?
   A Sodium propanoic acid
   B Sodium propanoate
   C Sodium propanoxide
   D Sodium ethylmethanoate

4 What is the IUPAC name for the compound below?

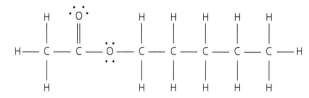

A Methyl pentanoate
B Ethyl pentanoate
C Pentyl ethanoate
D Butyl ethanoate

5 Which of the following are the correct products for the acid hydrolysis of methylethanoate?
   A Methanoic acid and ethanol
   B Methanol and ethanoic acid
   C Sodium methanoate and ethanol
   D Sodium ethanoate and methanol

6 What is the common name for methanoic acid?
   A Formic acid
   B Acetic acid
   C Pimelic acid
   D Lactic acid

7 Which of the following is a soap?
   A $C_{25}H_{51}COONa$
   B $C_{18}H_{37}COOH$
   C $C_3H_8O_3$
   D $C_4H_{10}O_4$

8 Which of the following is a known environmental problem associated with the use of detergents?
   A Global warming
   B Eutrophication
   C Soil erosion
   D Flooding

9 Which of the following occurs when a greasy plate is scrubbed in soapy water?

   A The hydrophobic head of the soap bonds with the grease.

   B The hydrophobic tail of the soap bonds with the grease.

   C The hydrophilic head bonds with the water molecules.

   D The hydrophilic tail of the soap bonds with the water molecules.

10 Why is it better to use soap in soft water rather than in hard water?

   A It washes the clothes cleaner and does not form scum.

   B It causes eutrophication in nearby rivers.

   C It is used mainly for laundering clothes.

   D It is made from plant oils and animal fats, which are renewable.

---

11 Draw an ethanoic acid dimer and indicate the hydrogen bonds present.

12 Describe two disadvantages of soapless detergents.

13 Which alkali must be used to ensure that a detergent is soapless?

14 If the following chemical is inside a corked bottle, what would happen when you removed the cork? Explain your answer.

15 For each of the following, give the correct IUPAC name.

   a $C_7H_{15}COOH$

   b $C_2H_5COOCH_3$

   c

   d

16 a What is the functional group in organic acids?

   b Write the formula of the organic acid that contains four carbon atoms.

   c Name this organic acid.

17 With reference to propanoic acid, explain what is meant by the term 'a weak acid'.

18 a Describe the laboratory preparation of ethanoic acid from ethanol.

   b Draw a simple line diagram of the apparatus that can be used to carry out this preparation. Label your diagram.

   c State one domestic use of ethanoic acid.

   d State two industrial uses of ethanoic acid.

19 a Write the general formula for the homologous series to which ethanoic acid belongs.

   b Draw the fully displayed structure for ethanoic acid.

   c Ethanoic acid is the second member of this homologous series. Give the molecular formula and name of the first member of the series.

20 Functional groups are responsible for characteristic reactions of organic compounds.
    a What is the functional group in alkanoic acids?
    b Explain how the presence of the functional group is responsible for:
        (i) the relatively high boiling point of ethanoic acid;
        (ii) the solubility of ethanoic acid in water.

21 Write an equation for each of the following reactions:
    a ethanoic acid with magnesium;
    b propanoic acid with sodium hydrogencarbonate;
    c methanoic acid with sodium hydroxide;
    d propanoic acid with methanol.

22 a Describe, giving experimental details, how ethanoic acid reacts with ethanol.
    b What type of chemical reaction is this?
    c To which group of organic compounds does the product of the reaction belong?
    d Name the compound formed in the reaction.

23 Name the organic product(s) formed when:
    a pentanoic acid reacts with butan-1-ol;
    b propanoic acid reacts with potassium.

24 a What do you understand by the term 'alkaline hydrolysis' of an ester?
    b What products are formed when propyl ethanoate is hydrolysed by sodium hydroxide?
    c Describe how the hydrolysis reaction is carried out experimentally.

25 Starting with ethene, describe how you would obtain a sample of ethanoic acid. Your description should include the conditions for each step of the reaction and balanced equations for the reactions described.

26 Starting with ethanol as the only organic compound, describe how you would obtain a sample of ethyl ethanoate.

27 Propanoic acid and ethanol are warmed together in the presence of concentrated sulphuric acid. $C_2H_5COOC_2H_5$ and water are formed, as shown in the following equation:

$$C_2H_5COOH + C_2H_5OH \rightleftharpoons C_2H_5COOC_2H_5 + H_2O$$

    a What is the chemical name of the compound with formula $CH_3COOC_2H_5$?
    b What is meant by the sign $\rightleftharpoons$ in the equation?
    c How would you tell that $C_2H_5COOC_2H_5$ was produced?
    d What is the function of the concentrated sulphuric acid in this reaction?

28 Some precautions are listed on a can of oven cleaner:
    • For use only on porcelain, enamel, iron, stainless steel and glass surfaces.
    • Do not use on exterior oven surfaces, aluminium and chrome.
    • Keep off all electrical connections, such as heating elements, thermostats and bulb receptacles.
    a What is the active ingredient in oven cleaners?
    b Why should this product not be used on aluminium surfaces? Give supporting chemical equations.
    c Having regard to the nature of the active ingredient of oven cleaners, what precautions should you take in handling this product?
    d Design an experiment to determine the concentration of the active ingredient in a can of oven cleaner.

In this chapter, you will study the following:

- how to draw diagrams to show the structure of monomers;
- how monomers are linked to form polymers;
- how to draw diagrams to show the structure of polymers;
- how to differentiate between monomers and polymers by means of chemical tests;
- the properties and uses of polymers.

This chapter covers
Objectives 4.1–4.3 of Section B of the CSEC Chemistry Syllabus.

Over the years, materials such as metals and their alloys, glass, timber and cement have been used for a variety of purposes world-wide. Their use was wide ranging and included utensils, electrical fixtures and furniture in the home, building construction, the manufacture of cars and various other items. There were limitations to the use of some of these materials; for example, glass breaks easily, metals soften at high temperatures and corrode on exposure to water and air.

The invention of synthetic polymers introduced new materials with more desirable properties such as:

- lower densities;
- an ability to withstand high impact without bending or breaking;
- a lack of reactivity to chemicals;
- thermal and electrical insulators.

Today, plastics are used globally.

Unfortunately, this huge increase in the use of polymers has had a negative impact on our environment. Plastics (polymers) are very difficult to dispose of because:

- they are not readily degradable;
- many are flammable and release toxic fumes, which add to air pollution.

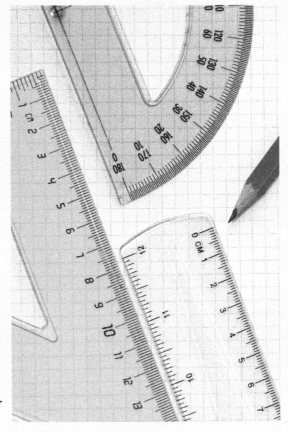

▲ **Figure 20.1** Perspex

20.1 What are polymers?

The word **'polymer'** means 'many units'. Polymers consist of long, chain-like molecules that are formed when several thousands of small molecules, known as 'monomers', combine chemically.

Monomers are small molecules that can be linked to form long-chain molecules with high molar masses known as polymers.

Naturally occurring **polymers** such as cotton, starch and rubber were familiar materials for years before synthetic polymers such as polyethene and Perspex appeared on the market.

Linking monomers

Monomers are considered to be the 'building blocks' of polymers. The nature of the monomer determines how they bond together and the type of polymer that is formed. The number of monomers used determines the molar mass of the polymer formed. There are two distinct types of monomers, as shown in Figures 20.2 and 20.3.

Unsaturated monomers

Unsaturated monomers are unsaturated hydrocarbons, which include alkenes and alkynes. Examples of unsaturated monomers include ethene and some of its derivatives, such as chloroethene and phenylethene.

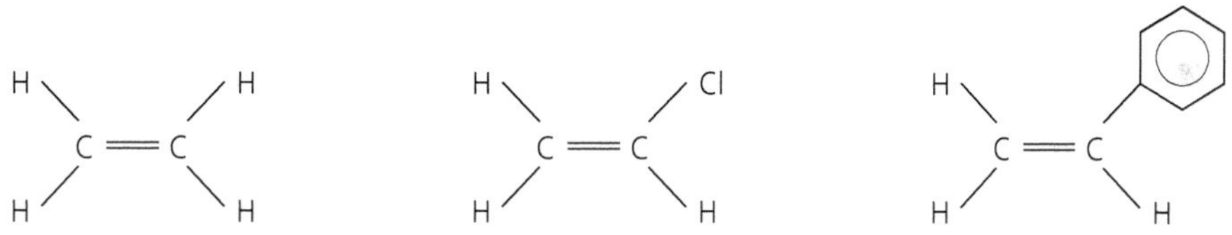

ethene chloroethene phenylethene

Figure 20.2 The structural formulae of some unsaturated monomers ▶

Since unsaturated monomers contain double or triple bonds, they will bond by addition reactions to form **addition polymers**. The process by which the polymers are formed is called addition **polymerisation**.

Bifunctional monomers

Bifunctional monomers are molecules that contain two functional groups. The functional groups in a bifunctional monomer can be the same or different. Examples of these monomers are ethane-1,2-diol, 1, 6-diaminohexane and amino acids.

Many bifunctional monomers contain two identical functional groups. These monomers rarely bond with each other, so we usually use two different bifunctional monomers, and they bond together in a condensation reaction. The polymer formed is a condensation polymer and the process by which it is formed is **condensation polymerisation**.

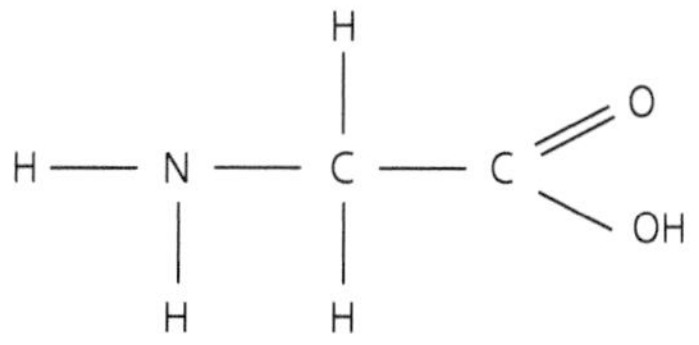

Figure 20.3 Glycine amino acid ▶

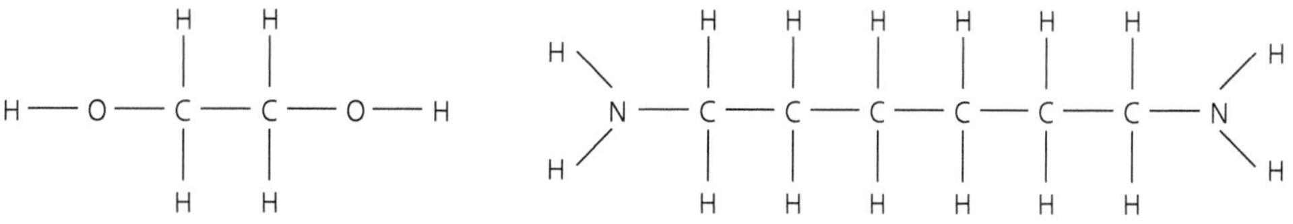

ethane-1,2-diol hexane-1,6-diamine

Figure 20.4 The structural formulae of some monomers containing two functional groups ▶

Practice

1   State and explain whether each of the molecules listed here may be used as monomers in a polymerisation process.
   a C_2H_5OH
   b C_3H_6
   c CH_3CHCH_2
   d $FCHCH_2$
   e $H_2N(CH_2)_4COOH$
   Hint: You may find it helpful to draw the structure of these molecules.

Classifying polymers

We can group polymers based on three sets of criteria.

1   Are they synthesised in living organisms (plants and animals) or are they produced industrially? Polymers are therefore either **naturally occurring** or **synthetic**.

Figure 20.5
Tom Cringle's Cotton Tree. Cotton is a naturally occurring polymer. ▶

Tom Cringle's Cotton Tree was named after a character in a popular novel of the 19th century called *Tom Cringle's Log*. This ceiba or silk cotton tree was located in front of the Ferry Police Station in St Catherine, Jamaica. It was said to measure some 18–20 ft in diameter. Like the cotton tree in Half Way Tree, (named by the English for the tree's location half way between Kingston Harbour and Spanish Town), Tom Cringle's Cotton Tree was used as a marker between the three plantations once owned by the wealthy Spaniards Liguaney, de Yalis and Lizama. It was used as a directional aid on several maps and as the 100th milestone marker for Kingston. It also provided a resting place for slaves and English soldiers on long marches. Tom Cringle's Cotton Tree collapsed on Monday, January 18, 1971.

2   What type of chemical reaction is used to form polymers? Polymers are either **addition polymers**, if formed by an addition process, or **condensation polymers**, if formed by condensation process.

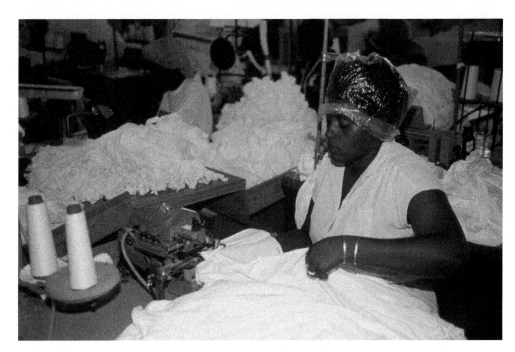

Figure 20.6
Factory worker stitching fabric polymer at a factory in Montego Bay, Jamaica ▶

3 Are one or two types of monomers used? Polymers are either homo-polymers or co-polymers.

Figures 20.7 and 20.8 show the first two sets of classifications, with examples.

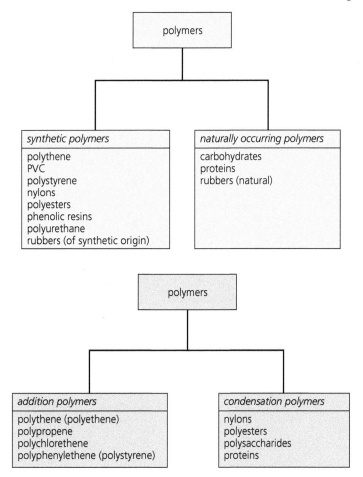

Figure 20.7
Polymers are often put into two groups: synthetic and naturally occurring. ▶

Figure 20.8
Addition and condensation polymers ▶

▼ **Table 20.1** Types of monomers

Molecules	Type of monomer	Type of polymer	Repeat unit	Examples
ethene	unsaturated	addition		polythene containers
vinyl chloride	unsaturated	addition		PVC pipes
styrene	unsaturated	addition		polystyrene containers
propene	unsaturated	addition		polypropene carpets
diol + dioc acid	bifunctional	condensation (ester linkages)		polyester fabric; examples include dacron, terylene and crimplene used for wrinkle-proof clothes and GRP for rust-proof structures

continued

▼ **Table 20.1** *continued*

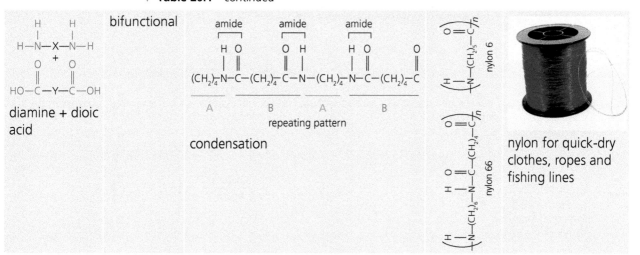

nylon for quick-dry clothes, ropes and fishing lines

20.2 Forming addition polymers

Addition polymerisation is generally achieved by subjecting the unsaturated monomers to heat and pressure in the presence of a suitable catalyst. The sequence in the formation of polyethene is shown below.

A
double bond is intact

B
double bond is broken

• represents an unshared electron.

Species B (which is highly reactive) then attacks another carbon–carbon double bond, leading to the formation of species C.

B   +   A   →   C

Species C is itself reactive and will, in turn, attack another ethene molecule, eventually leading to the formation of the polymer.

C   +   A   →   D

part of polymer chain

The process of addition polymerisation can also be represented by using a general representation of the monomer as shown:

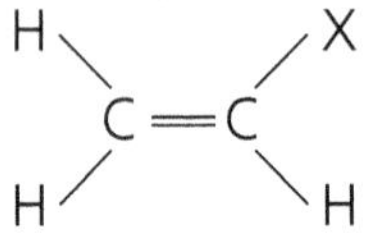

$$\begin{matrix} H & & X \\ & C=C & \\ H & & H \end{matrix}$$

X = H for ethene, X = Cl for chloroethene (vinyl chloride) and X = a benzene ring for phenylethene (styrene)

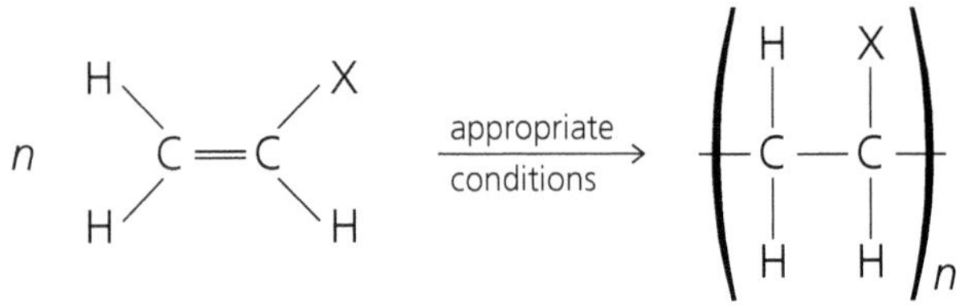

Figure 20.9
Illustrating addition
polymerisation ▶

$$n \quad \begin{matrix} H & & X \\ & C=C & \\ H & & H \end{matrix} \quad \xrightarrow[\text{conditions}]{\text{appropriate}} \quad \left(\begin{matrix} H & X \\ | & | \\ C - C \\ | & | \\ H & H \end{matrix} \right)_n$$

Since there is no loss of material during addition polymerisation, the empirical formula of the polymer is the same as that of the monomer.

Practice

2 a Using chloroethene as the starting material, show how four molecules can bond to form part of a polychloroethene polymer chain.

  b Will polychloroethene be a saturated or unsaturated molecule? Explain your answer.

Industrial preparation of polyethene and other addition polymers

Polyethene, a well-known polymer, exists in two forms:
- Low-density polyethene (LDPE)
- High-density polyethene (HDPE)

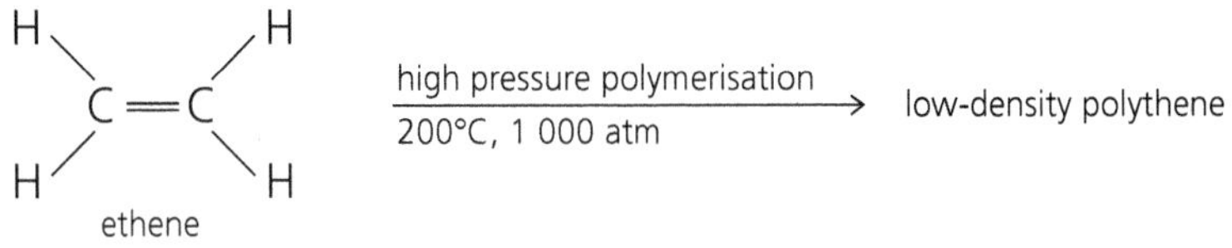

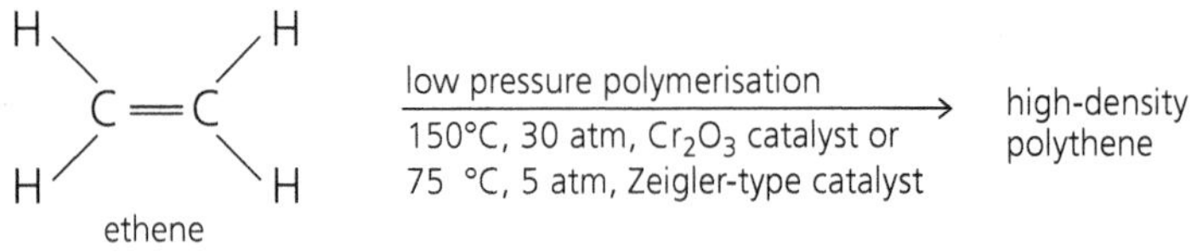

$$\begin{matrix} H & & H \\ & C=C & \\ H & & H \end{matrix}$$
ethene

$\xrightarrow[\text{200°C, 1 000 atm}]{\text{high pressure polymerisation}}$ low-density polythene

$$\begin{matrix} H & & H \\ & C=C & \\ H & & H \end{matrix}$$
ethene

$\xrightarrow[\substack{\text{150°C, 30 atm, Cr}_2\text{O}_3 \text{ catalyst or} \\ \text{75 °C, 5 atm, Zeigler-type catalyst}}]{\text{low pressure polymerisation}}$ high-density polythene

Figure 20.10
These objects are made of polyethene, commonly referred to simply as polythene. Low-density polyethene is used for washing-up bowls, squeezy bottles and polyethene bags. High-density polyethene is used to make rigid items such as crates and packaging material. ▶

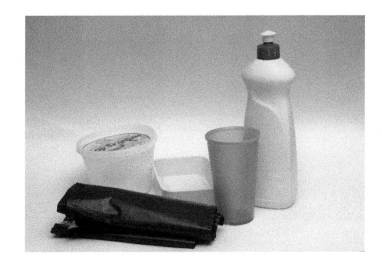

Low-density polyethene and high-density polyethene are structurally different materials. This is because they have different properties such as:

- the length of the polymer chains;
- the spacing between the chains; and
- the extent to which chains are interlinked.

The conditions of polymerisation can have an effect on the properties of the polymer.

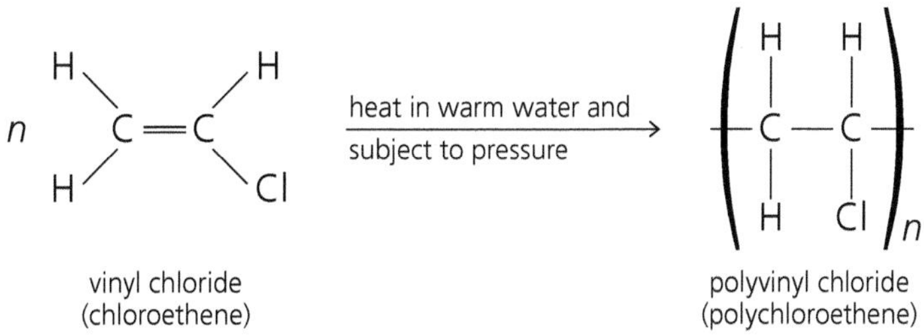

Figure 20.11
The formation of PVC by addition polymerisation ▶

n C=C (vinyl chloride (chloroethene)) → (heat in warm water and subject to pressure) → polyvinyl chloride (polychloroethene)

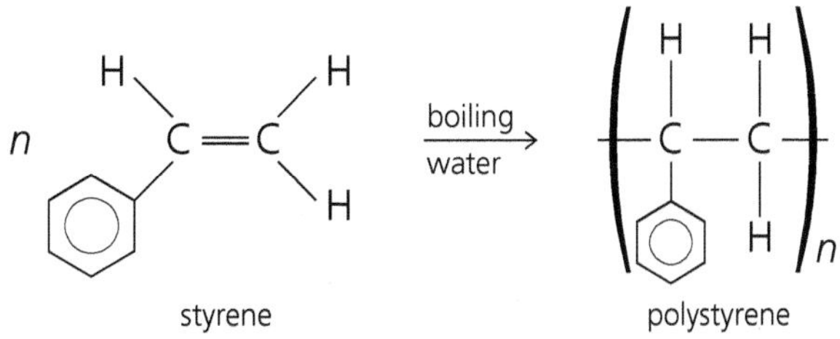

Figure 20.12
The formation of polystyrene by addition polymerisation ▶

n C=C (styrene) → (boiling water) → polystyrene

Naming addition polymers

In general, addition polymers are named after their monomers by inserting the prefix 'poly-' before the name of the monomer. Polyethene, for example, is formed from ethene and polystyrene is formed from the monomer styrene, and so on.

Differences between monomers and polymers

Although addition monomers have the same empirical formula as the polymers that they form, their properties are distinctly different. Table 20.2 compares the properties of ethene and polyethene.

▼ **Table 20.2** Properties of ethene (a monomer) and polyethene (a polymer).

Properties	Ethene	Polyethene
empirical formula	CH_2	CH_2
type of molecule	small and unsaturated	large and saturated
reactivity	highly reactive: double bonds are present	generally unreactive as only single bonds present
type of reaction shown	addition	can be cracked, otherwise unreactive
physical state	gas	solid

20.3 Condensation polymerisation

Condensation polymerisation is essentially a method of joining monomers by eliminating small molecules such as water. This is in contrast to addition polymerisation, as this only produces the polymer. Condensation polymers do not have the same composition as the monomers from which they are made.

Polyesters

Polyesters are polymers that contain many ester linkages. They are formed in condensation reactions between di-acids and di-alcohols (usually called diols). Remember that a di-acid has two organic acid (–COOH) groupings and that a di-alcohol contains two alcohol (–OH) groupings.

Here are examples of a di-acid (benzene-1,4-dicarboxylic acid, commonly called terephthalic acid) and a di-alcohol (ethane-1,2-diol).

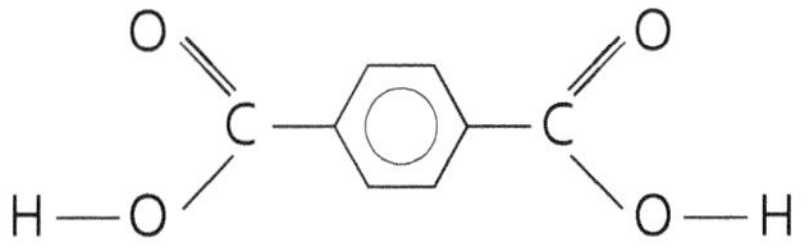

▲ **Figure 20.13** A di-acid 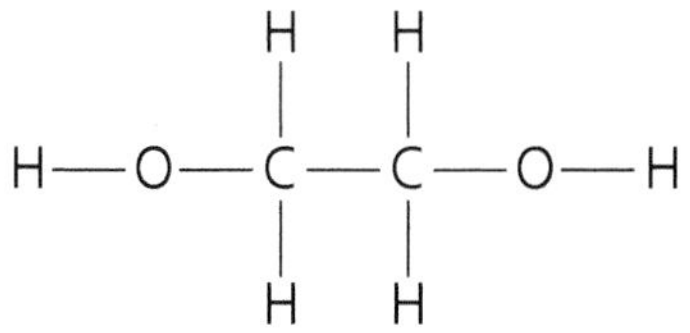   ▲ **Figure 20.14** A di-alcohol

In the sequence below, a di-acid (E) and a di-alcohol (F) combine to form a further compound (G), a simple ester that is capable of condensing on both ends.

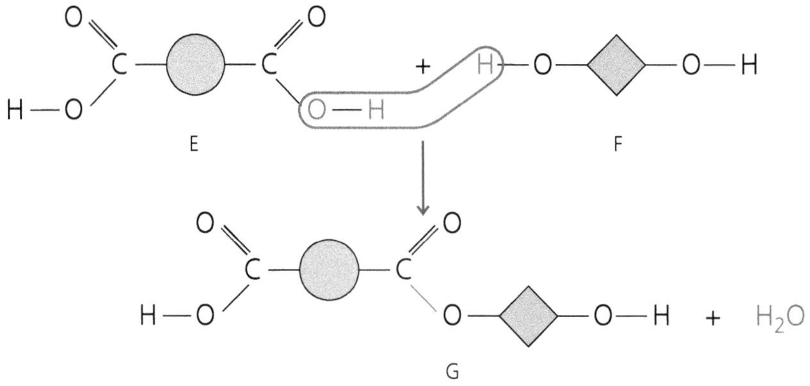

Compound G has the following groups:

- An ester grouping
- An uncombined alcohol grouping
- An uncombined acid grouping

Compound G then undergoes further condensation reactions at both ends of the molecule to form compound H.

Even more condensation reactions will eventually lead to a polyester of molecular mass of 10 000–20 000.

a polyester

Terylene is an example of a long-chain polyester made from benzene-1,4-dicarboxylic acid and ethane-1,2-diol.

terylene

Polyamides

Polyamides are polymers that contain many amide (peptide) groups. The carbon to nitrogen bond within this group is the amide (peptide) linkage.

◄ The amide group

Nylon is a typical polyamide. It can be prepared as shown in the following sequence.

Consider the reaction between the di-acid $HOOC(CH_2)_4COOH$, which we shall represent as:

and the diamine $H_2N(CH_2)_6NH_2$ which we shall represent as:

Both these starting materials can be made from phenol, which is obtainable from the naphtha fraction of oil.

When the di-acid (J) reacts with the diamine (K), then species L is formed:

Species L has an amide linkage, but it also contains two other functional groups. These two functional groups are capable of further condensation to produce the polymer which is represented by one repeat unit:

Experiment 20.1   Preparing a sample of nylon – the nylon rope trick (Teacher demonstration)

Nylon can be prepared by reacting 1,6-diaminohexane with hexanedioyl chloride (a more reactive version of a di-acid).

Figure 20.15
Part of the nylon (polyamide) chain ▶

Procedure

1 Dissolve 1,6-diaminohexane in water.

2 Add this to a solution of hexanedioyl chloride in a chlorinated organic solvent.

Observations

• Since water and the organic solvent are immiscible, two distinct layers are formed.

• A weak but continuous white strand can be pulled from the boundary between the two liquids.

Figure 20.16
This photograph shows the nylon rope being pulled out from between 1,6-diaminohexane and hexanedioyl chloride. ▶

Question

Q1 Why is nylon formed at the interface between the two liquids?

Naming condensation polymers

Condensation polymers are named after the functional group that is repeated along its chain.

Polyesters contain many ester linkages (–COO–):

Polyamides contain many peptide linkages (–CONH–):

Practice

3 The monomers used in condensation polymerisation may be represented by simple diagrams showing the functional groups on each monomer. Use the following diagrams to illustrate how the monomers are linked in this type of polymerisation:

a

a di-amine   a di-carboxylic acid

b

a di-alcohol   a di-carboxylic acid

4 The repeat unit of a polymer shows part of the polymer formed when two monomers condense. Draw a section of the two polymers formed in question 3, showing two repeat units.

5 How will you name the polymers formed in question 3?

Testing for amino acids

• Reagent: Ninhydrin solution.
• Procedure: Add 1 cm³ of ninhydrin to food or solution of amino acid. Boil for 1–2 minutes, then allow to cool.
• Results: A blue colour indicates that amino acids (or proteins which contain a free amino acid group) are present.

Testing for proteins

• Reagent: NaOH(aq) and copper sulphate(aq) (known as the biuret reagent).
• Procedure: Add biuret reagent to food material in water and heat.
• Results: A violet colour indicates the presence of proteins.

Another test for protein, but not as precise as the test given, is to heat the protein foods with a base.

• Mix the food sample with soda lime (a mixture of sodium hydroxide and calcium oxide).
• Place the mixture in a hard glass test-tube.
• Heat the mixture, placing a moist strip of red litmus at the mouth of the test-tube.

If the food contains proteins, ammonia gas will be given off, turning the red litmus paper blue.

20.4 Polysaccharides

Polysaccharides are polymers that belong to a group of naturally occurring compounds containing carbon, hydrogen and oxygen, based on the formula $C_x(H_2O)_y$.

The simplest carbohydrates are the monosaccharides, which have a general formula of $C_nH2_nO_n$. Some examples of these simple sugars are glucose, fructose and galactose, which you may have studied in your Biology course. These all contain six carbon atoms and have the same molecular formula, $C_6H_{12}O_6$, but they have different structures and are therefore isomers (see Section 17.3).

Figure 20.17 shows one way that the structure of a glucose molecule can be represented.

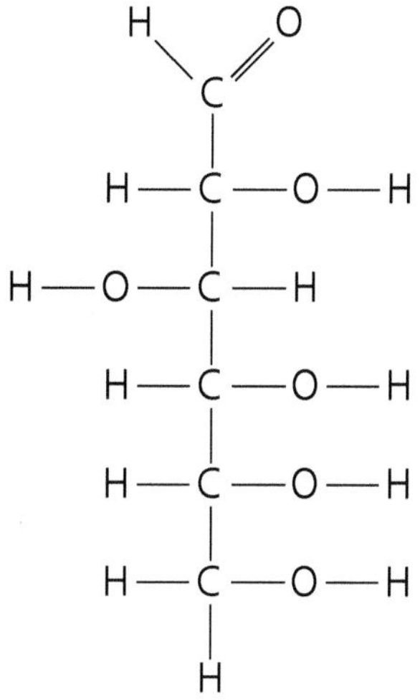

Figure 20.17
One way of representing glucose ▶

Glucose molecules are the building blocks for the carbohydrate polymers. The condensation of glucose molecules can occur in a stepwise manner forming disaccharides, trisaccharides and eventually more complex carbohydrates, the polysaccharides. Some monosaccharides, and the disaccharides and polysaccharides formed from them are shown in Table 20.3.

▼ **Table 20.3** Examples of monosaccharides and the disaccharides and polysaccharides derived from them

Type of carbohydrates	Examples and their composition	Source
monosaccharides photosynthesis	glucose	primary sugar formed in
	galactose	a sugar in milk
	fructose	a sugar in honey
disaccharides	sucrose: glucose + fructose	table sugar
	lactose: galactose + glucose	found in milk
	maltose: glucose + glucose	produced when starch is digested

▼ **Table 20.3** *continued*

polysaccharides	cellulose: long, rigid straight chains of glucose molecules	the main structural material in plants
	glycogen: branched chains of thousands of glucose molecules	carbohydrate stores in animals – insoluble,
	starch: two types • amylose – unbranched chains of glucose molecules • amylopectin – highly branched chains of glucose molecules	food reserve of plants

Polysaccharide formation

Disaccharides and polysaccharides contain an ether linkage, that is, the C–O–C linkage. This linkage is commonly called the 'glycosidic linkage' in Biology. Disaccharides have two reactive (end) –OH groups. These reactive groups may undergo further condensation reactions to yield a trisaccharide and, eventually, the polysaccharides.

The following structure is used to represent a glucose molecule:

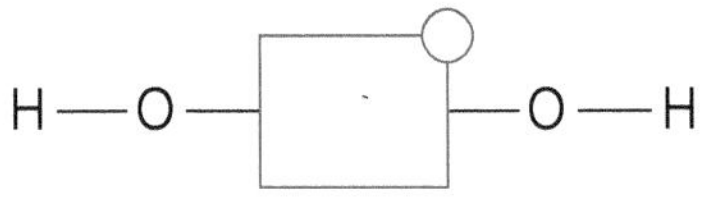

H—O—[ ]—O—H

where the block 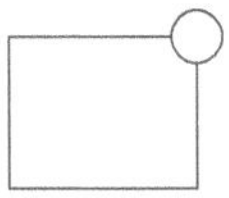 [ ] represents the rest of the molecule.

Two monosaccharides can condense to yield a disaccharide. For example:

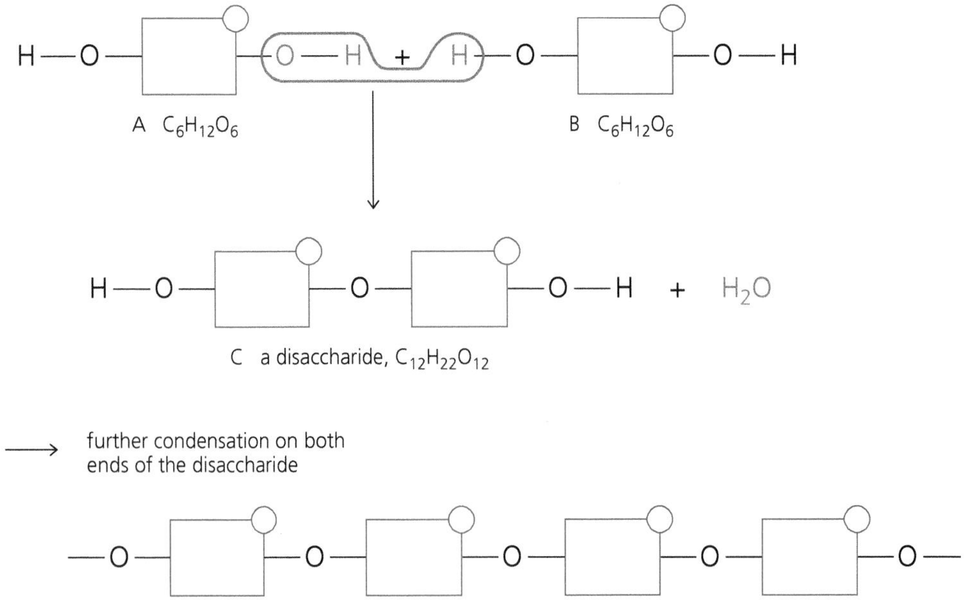

H—O—[ ]—O—H + H—O—[ ]—O—H

A $C_6H_{12}O_6$ B $C_6H_{12}O_6$

↓

H—O—[ ]—O—[ ]—O—H + H_2O

C a disaccharide, $C_{12}H_{22}O_{12}$

⟶ further condensation on both ends of the disaccharide

—O—[ ]—O—[ ]—O—[ ]—O—[ ]—O—

Figure 20.18
The polymerisation of glucose molecules ▶

Hydrolysis of carbohydrates

The polymerisation of glucose molecules to form a polysaccharide, such as starch, is a condensation reaction. Hydrolysis is the reverse of the condensation process. Starch molecules, for example, are broken down partially to maltose or completely to glucose.

Find out more

From polymers to plastics

Polymeric materials undergo many modifications (such as the addition of stabilisers, fillers, pigments and dyes), which reduce the cost and enhance the appearance of the finished product.

Plastics are classified as thermoplastics or thermosets, depending on their behaviour when heated. Table 20.4 highlights the characteristics of these two types of plastics.

▼ **Table 20.4**   Characteristics of thermoplastics and thermosets

	Thermoplastics	Thermosets
Structure	long, thin and flexible polymer chains	cross-links between polymer chains gives a more rigid polymer
Effects of heating	soften when warmed without decomposition	decompose and release poisonous fumes when heated
Moulding properties	can be remoulded	cannot be remoulded
Examples	polyethene, nylon, polychloroethene (PVC)	Bakelite, urea-formaldehyde (a plastic resin)

Some examples of plastics and their uses are given in Table 20.5.

▼ **Table 20.5**

Plastic	Major uses
Polyethene	Low-density polyethene is used to make plastic bags, containers for cleaning agents, mixing bowls and other household items. High-density polyethene is used to make more rigid packaging materials such as crates.
Polypropene	Does not crack easily. Used for making car accessories, e.g. battery housing, making toys, domestic ware, shoe heels, crates and carpets.
PVC (polychloroethene)	Artificial leather, records, water pipes, floor tiles, waterproof articles, sheeting and insulating material for cables and wires.

Plastic	Major uses
Perspex	Safety glass (e.g. windshield glass), lighting fittings, display symbols, traffic signs, contact lenses.
Polystyrene	A wide variety of polystyrene plastics are manufactured. Used in making toys, packaging materials, bottle caps, measuring jugs, light fittings.
Nylon and other polyamides	Used in making rope, textiles, carpets, bushes and bearings, trolley wheels.
Terylene and other polyesters	Clothing and ropes. When reinforced with fibre glass, it has a high-impact strength, so used for roofing sheets, motor car bodies, hulls for boats.

Find out more

Carbohydrates and energy

The monosaccharide glucose is the primary product of photosynthesis.

Leaves, which can be thought of as 'biochemical factories', contain the important substance chlorophyll. This green substance can absorb energy from sunlight and use this energy to convert carbon dioxide and water to simple sugars. In this way, the sun's energy is stored as chemical energy.

$$6CO_2(g) + 6H_2O(l) \xrightarrow[\text{chlorophyll}]{\text{light}} C_6H_{12}O_6(aq) + 6O_2(g)$$

Plants utilise the sugar produced in a number of ways:

1 Sugar provides energy for growth. This energy is released during respiration, which is in some sense a reverse of the photosynthesis reaction.

2 Sugar is converted to starch and stored. Starch is therefore an energy reserve.

3 Sugar is converted to cellulose.

Animals get energy by feeding on plants. Complex sugars are hydrolysed during digestion and the simple sugars produced are used to release energy in respiration. Humans store carbohydrates in the liver as glycogen. When needed, this is broken down to disaccharides and into glucose, which can be used up in respiration to produce carbon dioxide, energy and water.

We can hydrolyse disaccharides and polysaccharides in the laboratory by acid, as shown in Experiment 20.2. In the human body, hydrolysis is brought about by digestive enzymes.

Experiment 20.2   Enzyme hydrolysis of starch

Procedure

1 Add 1 cm³ of amylase solution to 5 cm³ of starch solution. (Amylase is an enzyme found in saliva.)

2 Spot one drop of this solution onto a spotting tile containing the iodine reagent.

3 Repeat the test in step 2 at 1-minute intervals until the blue-black starch/iodine colour no longer develops.

4 Test one portion of the final solution with Benedict reagent.

Results expected

The solution after starch is hydrolysed gives a visible change with Benedict reagent.

Deductions

Some reducing sugar was formed. If a portion of the final solution is treated with acid, further hydrolysis occurs and the resulting solution gives a more definite positive test with Benedict reagent.

Summary

- The size of a polymer is dependent on the number of monomers linked in the polymerisation process.
- Polymers may be classified as addition or as condensation polymers.
- Addition polymers are made, generally, from one type of monomer.
- Unsaturated molecules such as ethene, propene, chloroethene and styrene are examples of monomers which are used to make addition polymers.
- Condensation polymers are generally made from two different monomers.
- Monomers used in condensation polymerisation must contain at least two functional groups.
- Nylon (a polyamide) and Terylene (a polyester) are examples of condensation polymers.
- Some plastics (thermoplastics) can be repeatedly reshaped by heating, whereas others (thermosetting plastics) cannot be reshaped after being formed.
- Plastics are widely used.

End-of-chapter questions

1 What is meant by the terms (a) monomer and (b) polymer?

2 a What feature or grouping in chloroethene makes it suitable for making the addition polymer polychloroethene, also known as PVC?

 b Use four chloroethene molecules to illustrate the formation of part of the polychloroethene polymer chain.

 c State two ways in which the polymer is different from the monomer.

3 What is meant by the terms (a) 'thermoplastics' and (b) 'thermosetting plastics'?

4 Which plastic would be best for the following purposes?
 a Coating electrical wires and cables
 b Coating the inside of a frying pan
 c The hulls of small boats
 d Pipes for drainage
 e Ropes
 f The handles of tools such as screwdrivers
 g Fabrics for clothing

5 The table below gives the properties of two plastics A and B. Which of A and B is a thermoplastic? Justify your choice.

| A | very stiff | chars, but does not melt | swells and cracks |
| B | flexible | melts (softens) | burns, but no smoke given out |

6

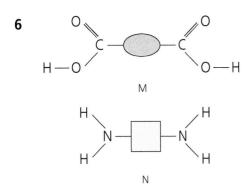

M

N

a Name the functional group(s) present:
 (i) in compound M;
 (ii) in compound N.

b Name the linkage(s) formed when compound M reacts with compound N.

c Use a suitable block diagram to show part of the structure of the polymer formed when compound M reacts with compound N. Show at least two repeat units.

d Give the name of a synthetic material that is formed in this way.

e Give the name of one type of naturally occurring material that has the same linkages as the synthetic material in part (d).

f What type of polymerisation takes place when the material in part (d) is formed from compounds M and N?

g Explain why the reaction between ethanoic acid (CH_3COOH) and ethylamine ($C_2H_5NH_2$) does not lead to the formation of a polymer.

7 Read the extract below and answer the questions that follow.

Dangerous plastics

FIREFIGHTERS today controlled a blaze at an abandoned sewage plant that spewed clouds of smoke containing hydrogen chloride for 21 hours …

The smoke forced more than 10 000 people to flee their homes temporarily …

The blaze, possibly caused by a worker's torch, began at 2.30 p.m. Monday and was confined to the plant, where the Styrofoam-filled roof collapsed and burned. The fire released hydrogen chloride, a chemical that causes nausea and headaches, irritates the eye and throat, and poses dangers for people with heart or lung problems …

The officials said the fire consumed polyvinyl chloride (PVC) in the doomed roof, which produced hydrogen chloride.

New York Times, July 16th 1985

a State the three conditions that must be satisfied for a fire to start or be sustained.

b In the case of the fire reported above, fire-fighters used bulldozers to pile dirt on the fire. Which of the three conditions identified in part

(a) did this action remove?

c What symptoms develop as a result of extended exposure to hydrogen chloride fumes?

d Polyvinyl chloride is a polymeric material formed from the vinyl chloride monomer by an addition polymerisation process.

(i) What is meant by the terms 'monomer' and 'addition polymerisation'?

(ii) Write the formula of the vinyl chloride monomer.

e It is unwise to dispose of plastics by burning them. Many plastics produce toxic fumes when they burn. Is burial a sensible way of disposing of plastics? Support your answer.

21 Metals and their compounds

In this chapter, you will study the following:

- the physical and chemical properties of metals;
- the reactivity of metals with oxygen, water, and dilute non-oxidising acids;
- displacement reactions with metals and solutions of salts;
- the ease of decomposition of metal hydroxides, nitrates and carbonates;
- the reactions of metal oxides, hydroxides and carbonates with dilute non-oxidising acids;
- the metal reactivity series;
- how to deduce the order of reactivity from data or experimental results.

This chapter covers
Objectives 1.1, 1.2, 2.1 and 2.2 of Section C of the CSEC Chemistry Syllabus.

In Chapter 3, you saw that most of the elements in the Periodic Table are metals. We compared the properties of metals with those of non-metals. In Chapter 4, you saw how the properties of metals relate to the type of bonding in their structure. In Chapter 13, you saw that there is an order of preferential discharge of metal cations from an electrolyte. We called this list the reactivity series. In Chapter 11, you saw that acids have characteristic reactions with metals, metal oxides and hydroxides. You also learnt that many bases are oxides and hydroxides of metals. Then, you saw that soluble salts can be formed by reacting an acid with either an alkali (a soluble base) using titration or with a metal or an insoluble base in excess.

21.1 Properties of metals

Physical properties

Generally, metals have the following physical properties. They:
- are solids at room temperature;
- have high melting and boiling points;
- have high densities;
- are hard (high tensile strength);
- are lustrous;
- are malleable and ductile;
- are good conductors of heat;
- are good conductors of electricity, both in the solid and liquid state.

▼ **Table 21.1**   Melting points and densities of some metals

Metal	Melting point (°C)	Density (g dm^{-3})
sodium	98	0.97
magnesium	650	1.74
aluminium	660	2.70
calcium	839	1.54
iron	1 535	7.87
copper	1 083	8.96
zinc	420	7.14
silver	962	10.50
tin	232	7.26
gold	1 064	19.30
mercury	−39	13.53
lead	327	11.34

Mercury is the only metal that is a liquid at room temperature (its melting point is −39 °C). Sodium is a solid that has a low melting point. Tin, lead and zinc have fairly low melting points. Magnesium, aluminium and calcium have melting points that are not as high as the more typical metals such as copper and iron.

The density of sodium is less than that of water (0.97 g dm^{-3}), and so it floats in water. Calcium, magnesium and aluminium are not very dense. Zinc, copper and iron have the high densities that are typical of metals.

Group I metals, which includes sodium, are soft enough to be cut with a knife. Tin, lead, calcium, magnesium and zinc are not as hard as iron.

Metals are lustrous, i.e. they have a shiny surface. They are malleable and can be hammered into thin sheets. They are ductile, which means they can be drawn out into thin wires.

All metals can conduct heat because they have mobile electrons, which vibrate faster as they absorb more heat energy. Of all the known metals, silver and copper are the best conductors of heat. Metals are also very good conductors of electricity, both in the solid and liquid state. In the solid state, the electrons in metals are already mobile. When a difference in electrical potential is applied, these electrons move. When the metal is heated until it melts, the electrons are still mobile. Silver and copper are also the best conductors of electricity.

Figure 21.1
Sodium floating on water. Sodium reacts with water to form sodium hydroxide and hydrogen. The reaction is highly exothermic, and the heat produced causes the hydrogen to ignite. ▶

Chemical properties

Generally, a metal will:

- react with oxygen to form metal oxides;
- react with water to form hydroxides and give off hydrogen gas;
- react with dilute hydrochloric and sulphuric acid to form salts and hydrogen gas;
- show reducing properties;
- form ionic compounds.

21.2 The reactions of metals with oxygen, water and non-oxidising acids

Chemical reactivity of the metals

▲ **Figure 21.2** Sodium reacting with oxygen in a glass jar. Sodium gives off a bright yellow light when it burns.

In your study of Groups I and II (see Section 6.1) and Period 3 (see Section 6.3), you saw that metals typically display a certain set of chemical properties:

- They react with oxygen to form basic oxides.
- They react with water to form a base and to liberate hydrogen gas.
- They react with dilute acids to form a salt and to liberate hydrogen gas.
- They show reducing properties.
- They form ionic compounds.

The **relative chemical reactivity** of a metal is an indication of the ease with which metal atoms ionise. The greater the ease of ionisation of the metal, the more rapidly it reacts with chemical reagents.

Although many metals show these properties, there are differences in their **reactivity** and some metals do not give the typical reactions listed. The reactivity of the metals is linked to their ease of ionisation.

Metallic character decreases from Group I to Group III, but increases as a group is descended.

- Compare potassium and calcium: Potassium is in Group I and calcium is in Group II, and they are both in Period 4 (so they are adjacent to each other in the Periodic Table). Potassium reacts more vigorously with oxygen, water and dilute acids than does calcium.
- Compare calcium and magnesium: Both of these metals are in Group II, but magnesium is above calcium. Calcium is more reactive than magnesium.

The greater the ease of ionisation of the metal, the more reactive it is. The more reactive metals have larger atomic radii and fewer valence electrons than the less reactive ones.

In this chapter, we will compare and contrast the reactivity of the following metals: zinc, iron, sodium, calcium, magnesium, aluminium, lead and copper.

Practice

1 Find the elements zinc, iron, sodium, calcium, magnesium, aluminium, lead and copper in the Periodic Table. Based on their positions in the Periodic Table:

   a can you classify them into transition metals and non-transition metals;

   b can you place them in their groups?

2 Can you arrange these metals in order of increasing reactivity?

Comparing the reactivity of metals

Comparing the reactivity of metals from varied positions in the Periodic Table presents more of a challenge than for metals in the same period or group. If you are given data on the ease of ionisation of these metals (see Section 6.1), you will be able to make a more accurate statement about their reactivity. Chemists can also determine the reactivity of metals by conducting experiments in the laboratory in which the metals are exposed to a variety of reagents, such as:

- oxygen;
- water;
- dilute non-oxidising acids;
- metal salts in aqueous solution.

We will take the experimental approach in this chapter.

Guidelines for conducting experiments

As you conduct these experiments, a number of questions must be answered. Table 21.2 shows these and how you can determine their answers.

▼ **Table 21.2** What to look for in reactivity experiments.

Question	What to look for
Does the metal react or not?	Signs of reaction such as: • effervescence (a gas is given off); • formation of a precipitate; • a colour change; • disappearance of the metal.
How fast is the reaction when compared with the other metals under similar conditions?	How quickly do the changes listed above occur? Is there a delay in the reaction?
If there is no reaction under one set of conditions; does changing the conditions cause a reaction?	Does the reaction occur at room temperature, or is heat required?

Precautions to be taken

Many factors affect the rate of a reaction (see Chapter 14). It is important to control these variables carefully so that your comparisons can be fair. It would be unwise, for example, to use one metal in its powdered form and another metal as granules since differences in surface area can affect reaction rates. The quantities of metal used should also be controlled.

ORR

M&M

Experiment 21.1   To determine the reactivity of some metals

⚠ Oxygen gas facilitates combustion. Keep away from flames. Reactions with alkali metals can be violent. They should be carried out as teacher demonstrations. Use tiny pieces of the metal.

Procedure

Before you begin read the precautionary note above and look at Table 21.2.

A Reaction of the metals with oxygen

Oxygen gas can be prepared and collected prior to carrying out this experiment.
1 Quickly drop a small piece of metal into a gas jar of oxygen.
2 Take the necessary precautions to prevent loss of the gas.
3 Observe what happens over a period of time.
4 If there is no reaction at room temperature, heat the metal before adding it to the oxygen gas.
5 Record your observations.

B Reaction of the metals with water
1 Add a small piece of each metal to water in a test-tube.
2 Test any gas evolved with a lit splint.
3 If there is no sign of a reaction, heat the water before adding the metal.
4 If there is still no sign of reaction, set up the apparatus as shown in Figure 21.2 and pass steam over the heated metal.
5 Record your observations.

C Reaction of the metals with dilute non-oxidising acids
1 Pour dilute hydrochloric or dilute sulphuric acid into a test-tube.
2 Add a small piece of metal to the acid.
3 Test any gas evolved with a lit splint.
4 Record your observations.

Questions

Q1 Which reaction was most vigorous in each case?

Q2 Which reaction was least vigorous in each case?

Q3 In which case(s) did no reaction take place?

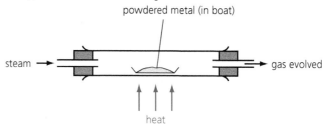

Figure 21.3
Apparatus for the reaction of a metal with steam ▶

Figure 21.4
Magnesium reacting with dilute acid. This is a very vigorous reaction, producing a large number of bubbles of hydrogen gas. ▶

21.3 Displacement reactions of metals with solutions of salts

We will now compare the reactivity of metals by looking at how they react with solutions of salts that contain different metal cations. Will the metal go into solution as a metal cation and cause the metal cation of the solution to become a metal?

ORR **Experiment 21.2** Displacement reactions, used to determine the relative reactivity of zinc, magnesium and copper

Procedure

1 Add a small strip of magnesium ribbon to a dilute solution of copper sulphate in a beaker.

2 Observe the contents of the tube over 10–15 minutes.

3 Add a small granule of zinc to a dilute solution of copper sulphate in a test-tube. Observe the contents of the tube over 10–15 minutes.

4 Add a strip of magnesium to an aqueous solution of zinc chloride.

5 Add a small granule of zinc to an aqueous solution of magnesium chloride.

6 Design a table to record your results.

Questions

Q1 Write ionic equations for any reactions that took place.

Q2 How would you expect copper to react with (a) zinc sulphate and (b) magnesium sulphate?

Displacement reactions

Patterns in reactivity of metals can be established by reacting a chosen metal with the solutions of salts of other metals.

You may have seen the changes that occur in a blue solution of copper(II) sulphate after a piece of magnesium or zinc metal has been dropped into it, as in Experiment 21.2. The result is that the blue solution fades, and eventually becomes colourless, as the metal that was added is used up.

In this reaction, the more reactive magnesium displaces the less reactive copper from a solution containing copper(II) ions. The full equation for the reaction is:

$$Mg(s) + CuSO_4(aq) \rightarrow MgSO_4(aq) + Cu(s)$$

An ionic equation shows the essential changes:

$$Mg(s) + Cu^{2+}(aq) \rightarrow Mg^{2+}(aq) + Cu(s)$$

No reaction occurs if we add copper to a solution of magnesium sulphate.

▲ **Figure 21.5** Zinc displaces copper from a solution of copper sulphate.

Explaining reactivity patterns

All of the chemical reactions of the metals described above are examples of redox reactions, in which the metal atoms are oxidised to cations. For this to happen, the metals must release electrons, that is, they must behave as reducing agents (see Section 12.7).

$$M \rightarrow M^{n+} + ne^-$$

Metal atoms lose electrons and are oxidised. The electrons that are released from the metals reduce another species.

- Oxygen molecules are reduced to oxide ions. This happens when metals react with oxygen:

$$O_2(g) + 2e^- \rightarrow O^{2-}(s)$$

  Oxygen molecules gain electrons and are reduced.

- Hydrogen cations are reduced to hydrogen molecules (hydrogen gas). This happens when metals react with water or dilute acids. Only metals which are more powerfully reducing than hydrogen will show this reaction:

$$2H^+(aq) + 2e^- \rightarrow H_2(g)$$

- Metal ions in a salt solution are reduced to metal atoms. This happens when a more reactive metal (M) is added to a salt solution of a less reactive metal (X):

$$M(s) + X^{2+}(aq) \rightarrow X(s) + M^{2+}(aq)$$

Note that the metal displacing the metal ion in solution has a greater ease of ionisation than the metal displaced. The greater the ease of ionisation of a metal, the more powerful the reducing action of the metal and the greater its reactivity.

The results

A Reactivity of the metals with oxygen

All metals react with oxygen to form ionic compounds known as metallic oxides.

- Some metals, such as sodium and calcium, react spontaneously with oxygen in the air.
- Other metals, such as magnesium and aluminium, react more slowly and become covered with an oxide layer.
- Zinc and iron react even more slowly in dry air.
- Copper articles will tarnish only over many years.

Note that the more spontaneous the reaction, the more reactive the metal is, provided the factors that affect reaction rates are controlled.

Equations for the reactions of a few metals with oxygen are shown below:

$$4Al(s) + 3O_2(g) \rightarrow 2Al_2O_3(s)$$
aluminium oxide

$$3Fe(s) + 2O_2(g) \rightarrow Fe_3O_4(s)$$
iron(II,III) oxide

$$2Cu(s) + O_2(g) \rightarrow 2CuO(s)$$
copper(II) oxide

Corrosion is the reaction of metals with substances in their environment, producing oxides and, in some cases, sulphides, carbonates, hydroxides and sulphates.

Note that of the three oxides of iron, FeO (iron(II) oxide), Fe_2O_3 (iron(III) oxide) and Fe_3O_4 (iron(II,III) oxide), the latter is the main product when the heated metal reacts with oxygen.

It is these reactions of metals with oxygen that contribute to the **corrosion** of metals exposed to the environment.

The section on 'Rusting and its prevention' at the end of Chapter 13 is relevant here.

B Reactivity of the metals with water

- Metals such as potassium, sodium and calcium react with cold water to form metalhydroxides.
- Iron and zinc react with steam, forming oxides.
- Copper and silver give no reaction with either water or steam.

Equations for some reactions are shown below:

$$2K(s) + 2H_2O(l) \rightarrow 2KOH(aq) + H_2(g)$$
potassium hydroxide

$$2Na(s) + 2H_2O(l) \rightarrow 2NaOH(aq) + H_2(g)$$
sodium hydroxide

$$Ca(s) + 2H_2O(l) \rightarrow Ca(OH)_2(aq) + H_2(g)$$
calcium hydroxide

$$3Fe(s) + 4H_2O(g) \rightarrow Fe_3O_4(s) + 4H_2(g)$$
iron(II,III) oxide

$$Zn(s) + H_2O(g) \rightarrow ZnO(s) + H_2(g)$$
zinc oxide

C   Reactivity of the metals with dilute non-oxidising acids

The acids used here are dilute hydrochloric acid or dilute sulphuric acid. The products of this reaction are a salt and hydrogen gas. Remember that nitric acid, even when dilute, has oxidising properties.

- Metals such as magnesium, zinc and iron can all react safely and with decreasing vigour with dilute acids.
- A metal such as lead shows little reactivity, partly because the salt formed is insoluble.
- A metal such as copper gives no reaction.

Equations for some reactions are shown below:

$$Mg(s) + 2HCl(aq) \rightarrow MgCl_2(aq) + H_2(g)$$

The equation for zinc is similar.

$$Fe(s) + H_2SO_4(aq) \rightarrow FeSO_4(aq) + H_2(g)$$
$$Pb(s) + H_2SO_4(aq) \rightarrow PbSO_4(s) + H_2(g)$$

A summary

Table 21.3 summarises how the selected metals react with oxygen, water and dilute non-oxidising acids.

▼ **Table 21.3**   The reactions of selected metals with oxygen, water and dilute non-oxidising acids

Metal	Reaction with oxygen	Reaction with water	Reaction with dilute non-oxidising acids*
sodium	smoulders and eventually catches fire to form sodium oxide	reacts with cold water to give hydrogen gas and the metallic hydroxide	reacts violently to form salt and hydrogen
calcium	less spontaneous reaction than for sodium; calcium oxide is formed	calcium reacts like sodium, but less rapidly, to form the hydroxide and hydrogen gas	gives quite a vigorous reaction, forming salt and hydrogen gas
magnesium	when lit, burns with a brilliant flame producing a white ash	reacts slowly with hot water; reaction is significantly increased with steam	reacts quickly to form salt and hydrogen gas
aluminium	rapidly tarnishes in the cold; a thin coating of aluminium oxide forms	if pure, reacts with water, liberating hydrogen and a lot of heat	liberates hydrogen, and a salt is formed
zinc	zinc oxide forms – slowly in the cold, but more rapidly when heated	displaces hydrogen from steam	liberates hydrogen, and a salt is formed
iron	forms an oxide if heated; rusts in the cold if moisture is present	reacts reversibly with steam to produce hydrogen	liberates hydrogen, and a salt is formed
copper	forms black copper oxide if strongly heated	no reaction	no reaction

*Dilute sulphuric acid gives essentially the same reactions as dilute hydrochloric acid. The exception is the reaction of calcium, where the reaction with dilute sulphuric acid is slower because the sparingly soluble salt calcium sulphate precipitates on the metal, lessening contact between metal and acid.

Practice

3 Can you explain why there is sometimes a delay in the visible reaction when a metal such as aluminium or magnesium reacts with a dilute acid?

4 Use the solubility rules for salts to explain why calcium reacts more vigorously with hydrochloric acid than with sulphuric acid.

5 What determines whether a metal forms the hydroxide rather than the oxide when it reacts with water?

6 Use your knowledge of the reactivity series to predict the results of the reactions in Table 21.4 between metals and various aqueous solutions.

    a Complete the table using '1' where a reaction occurs and '2' where no reaction occurs.

    b Write ionic equations for all the reactions you predicted.

▼ **Table 21.4**

Metal	Solution					
	$MgSO_4$	$ZnSO_4$	$FeSO_4$	$Al_2(SO_4)_3$	$Pb(NO_3)_2$	$CuSO_4$
magnesium						
zinc						
iron						
aluminium						
lead						
copper						

21.4 The ease of decompostion of metal nitrates, carbonates and hydroxides

We now investigate how certain metal compounds react to being heated.

The relative stability of the metal nitrates, carbonates and hydroxides can be determined by heating them.

- Thermally stable compounds are not easily decomposed by heat.

- The higher the temperature needed to decompose a compound, the more thermally stable that compound is.

Observations and interpretation

In the experiments that follow, it is important to record whether the decomposition takes place at lower temperatures (soon after heating begins) or whether much higher temperatures are needed (as indicated by prolonged heating before the gas is evolved). Compounds that are more stable, are decomposed at higher temperatures than less stable ones.

The information in Table 21.5 will serve as a guide to help you interpret your observations.

▼ **Table 21.5** Action of heat on nitrates, carbonates and hydroxides

Action of heat	Observations	Deduction
nitrate	droplets of liquid	salt is losing water of crystallisation
	brown gas	nitrogen dioxide gas is evolved, so the compound is decomposing
	glowing splint is rekindled	oxygen gas is evolved, so compound is decomposing
carbonate	droplets of liquid	salt is losing water of crystallisation
	colourless gas gives white precipitate with lime water	carbon dioxide gas is evolved, so compound is decomposing
hydroxide	droplets of liquid	compound is decomposing

Heavy metallic nitrates (those containing divalent or trivalent metal ions) may give additional observations as they decompose (see below).

 Find out more

What changes do heavy metal nitrates undergo on heating?
Heavy metallic nitrates, containing divalent or trivalent metal ions, make a crackling sound (decrepitation) as gases burst out of the solids during decomposition. It is not unusual for these nitrates to change to a liquid soon after the heating begins, a reaction that can be mistaken for melting. The crystals actually dissolve in their water of crystallisation. The resulting solution must be heated to remove the water and form the solid anhydrous salt. Further heating then decomposes the salt to produce the brown gas and oxygen.

M&M
ORR
Experiment 21.3   To determine the relative stability of metallic nitrates, carbonates and hydroxides

- Avoid over-heating by moving the test-tube in the flame.
- Slant the test-tube and occasionally heat near the mouth of the test-tube to evaporate liquid and prevent it running back into the hot test-tube.
- The coloured gas NO_2 is poisonous and should not be allowed to escape into the laboratory. USE A FUME CUPBOARD.
- Remember to direct the mouth of the test-tube away from your classmates and yourself.

Procedure

A Action of heat on metallic nitrates

1 Set up the apparatus as shown in Figure 21.6.
2 Heat samples of different metallic nitrates.
3 Test liquid droplets on the upper wall of the test-tube with blue cobalt chloride paper.
4 Test for oxygen gas.
5 Observe if another coloured gas is released.

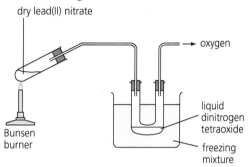

Figure 21.6
Testing the action of
heat on nitrates ▶

B Action of heat on metallic carbonates

Use the apparatus shown in Figure 21.7.

1 Heat different samples of metallic carbonates.
2 Test any liquid on the upper cooler part of the test-tube with blue cobalt chloride paper.
3 Test for carbon dioxide gas.
4 Note any colour changes in the solid.

⚠ When testing with lime water, remove the test-tube containing lime water before removing the hot test-tube from the flame.

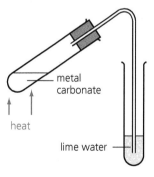

Figure 21.7
Testing the action of
heat on carbonates ▶

C Action of heat on metallic hydroxides

Use the apparatus shown in Figure 21.8.

1 Heat different samples of metallic hydroxides.
2 Look for liquid droplets on the upper cooler part of the test-tube.
3 Test droplets of liquid with blue cobalt chloride paper.
4 Look for possible colour changes in the solid.

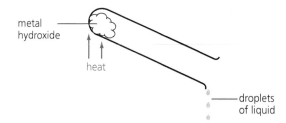

metal
hydroxide

heat

droplets
of liquid

Figure 21.8
Testing the action of
heat on hydroxides ▶

The most stable compounds, such as sodium carbonate and sodium hydroxide, will show no reaction when heated. Less stable compounds will decompose.

More reactive metals form very stable compounds. The higher the metal in the reactivity series, the more resistant is its compound to decomposition by heat.

▲ **Figure 21.9**   Nitrates decompose when heated to form a brown gas

Equations that show the action of heat on metallic nitrates, carbonates and hydroxides are shown in Table 21.6.

▼ **Table 21.6**   The action of heat on metal nitrates, carbonates and hydroxides.

	Potassium, sodium	**Calcium, magnesium, zinc, iron(II), lead(II), copper(II)**
Nitrate	$2MNO_3 \rightarrow 2MNO_2 + O_2$ nitrate nitrite oxygen	$2M(NO_3)_2 \rightarrow 2MO + 4NO_2 + O_2$ nitrate oxide oxygen
Carbonate	no reaction; Group I carbonates are stable to heat	$MCO_3 \rightarrow MO + CO_2$ carbonate oxide carbon dioxide
Hydroxide	no reaction; Group I hydroxides are stable to heat	products are oxide + water $M(OH)_2 \rightarrow MO + H_2O$ hydroxide oxide water

In these equations, 'M' represents any metal.

21.5 The reaction of metal oxides, hydroxides and carbonates with acids and alkalis

We will now investigate how certain metal compounds react with dilute acids or alkalis.

M&M
ORR

Experiment 21.4 Testing the oxides, hydroxides and carbonates of selected metals with acids and alkalis

Obtain samples of the oxides, hydroxides and carbonates of sodium, calcium, magnesium, aluminium, iron(II), zinc and copper(II).

Procedure

1 Shake small portions of oxides, hydroxides and carbonates with water and test the resulting solution with universal indicator solution. Record all your observations in a table.

2 React a portion of each sample with dilute hydrochloric acid, warming if necessary. Record your observations in a table. Carry out tests to identify any gas evolved.

3 Repeat step 2 using dilute sulphuric acid in place of the hydrochloric acid.

4 Shake a portion of each sample of the oxides and hydroxides with aqueous sodium hydroxide, warming if necessary. Record all your observations.

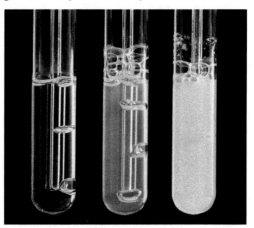

Figure 21.10
Testing for carbon dioxide using lime water. A white precipitate forms. ▶

Questions

Q1 If iron(II) hydroxide were not available in your school's store room, explain how you would have attempted to obtain a sample suitable for testing.

Q2 Which of the oxides or hydroxides dissolved in or reacted with the acids but not the sodium hydroxide solution? How are these oxides and hydroxides classified?

Q3 Which of the oxides dissolved in or reacted with both the acids and aqueous sodium hydroxide? How are these oxides classified?

Q4 Which of the carbonates, if any, dissolved in water?

Q5 a What was the only gas evolved in this experiment?

   b Which anion releases this gas on treatment with acid?

   c Did all metallic compounds containing this anion give the gas?

The set of reactions in Experiment 21.4 do not show the variation in reactivity that was shown when metallic compounds are heated.

The oxides and hydroxides of the metals are all bases and will therefore react with dilute hydrochloric acid and dilute sulphuric acid to form a salt and water.

Group I metals:

$$M_2O(s) + 2HCl(aq) \rightarrow 2MCl(aq) + H_2O(l)$$

Group II metals and other metals with a 2+ oxidation state:

$$MO(s) + 2HCl(aq) \rightarrow MCl_2(aq) + H_2O(l)$$

Group III metals and other metals with a 3+ oxidation state:

$$M_2O_3(s) + 6HCl(aq) \rightarrow 2MCl_3(aq) + 3H_2O(l)$$

The ionic equation for the reaction is:

$$O^{2-}(s) + 2H^+(aq) \rightarrow H_2O(l)$$

These reactions are described as neutralisation reactions (see Section 12.5).

The carbonates of all the metals react with dilute acids to form a salt, water and carbon dioxide gas.

For example:

$$Na_2CO_3(s) + H_2SO_4(aq) \rightarrow Na_2SO_4(aq) + H_2O(l) + CO_2(g)$$

or:

$$Na_2CO_3(s) + 2H_2SO_4(aq) \rightarrow 2NaHSO_4(aq) + H_2O(l) + CO_2(g)$$

Sulphuric acid is dibasic and can form an acid salt or a normal salt, dependent on the ratio of acid to metallic carbonate.

The oxides and hydroxides of zinc, aluminium and lead are amphoteric and will also react with alkalis.

$$Zn(OH)_2(s) + 2NaOH(aq) \rightarrow Na_2Zn(OH)_4(aq)$$
sodium zincate

In this reaction, a solid insoluble reactant is chemically changed to a soluble product. You should observe the solid hydroxide dissolving in the reaction.

The equation can be re-written, replacing the formula of sodium zincate by the dehydrated form of the salt ($Na_2ZnO_2 + 2H_2O$). Similar equations can be written for the reaction of lead(II) hydroxide with sodium hydroxide.

The ionic equation is:

$$M(OH)_2(s) + 2OH^-(aq) \rightarrow [M(OH)_4]^{2-}(aq)$$

Aluminium hydroxide reacts in a similar manner but the aluminate ion has the formula of either $[AlO_2]^-$ or $[Al(OH)_4]^-$.

Practice

7 a State what you will observe in each of the following reactions in part (b).

  b Write one or more balanced ionic equations for the reactions taking place:

    (i) Copper(II) carbonate is heated in a dry hard glass test-tube.

    (ii) A strip of magnesium metal is added to a solution of iron(II) sulphate solution.

    (iii) Lead(II) nitrate is heated in a hard glass test-tube, gently and then more strongly.

    (iv) Aqueous sodium hydroxide is added to zinc hydroxide with stirring.

21.6 The reactivity series of metals

Chemists have used experiments similar to those described above to rank metals in order of their reducing power. The most powerful reducing metal is placed at the top of the list and the least powerful is placed at the bottom of the list. Dealing with a list containing all the known metals is not practical, so we will use a shortened form of the metal reactivity series. However, if you are supplied with practical information on any metal you should be able to place it into its correct position in the series. The series that we will use is shown in Figure 21.11.

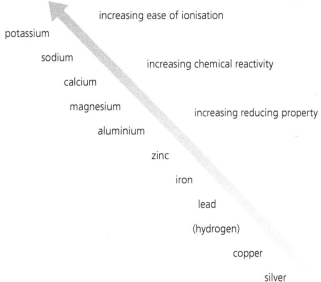

Figure 21.11
The metal reactivity series ▶

Note that hydrogen is included in this list, mainly as a reference point. This helps our predictions of how metals are expected to react with reagents, such as acids and water, which contain hydrogen.

General conclusions about the reactivity series of metals

The position of a metal in the reactivity series of metals gives general clues as to its chemistry.

It indicates the following:

- How easily the metal atom loses electrons to form cations – metals higher in the series ionise more easily.

- Its reducing power – metals high in the series are more powerful reducing agents.

- Its reactivity – the higher up in the series the metal is, the more reactive it is. For example, sodium is more reactive than iron, which is more reactive than copper.

It is used to predict a metal's behaviour in the displacement reactions:

- Metals high in the series displace those lower down from aqueous solutions of their salts.
- Metals above hydrogen in the series will displace hydrogen from aqueous solutions of acids and from water or steam.
- Whether the metal's compounds (such as the hydroxides, nitrates and carbonates), will be decomposed by heat

It is also used to select an appropriate method by which the metal can be extracted from its compounds.

21.7 How to deduce the order of reactivity of metals based on data

- The more vigorous a reaction is with oxygen, water, or a non-oxidising acid, the higher up the reactivity series is the metal.
- Metals higher up the reactivity series will displace metals lower down the reactivity series from solutions of their salts.
- Nitrates, carbonates and hydroxides of metals higher up the reactivity tend to be stable to heat.

Worked example 1

Deduce the order of reactivity of the three metals A, B and C, given the results of the tests shown in Table 21.7.

▼ **Table 21.7**

	Metal A	Metal B	Metal C
Reaction with water	no reaction	extremely rapid effervescence with cold water	very slow reaction in cold water
Reaction with solution of chloride of A	not attempted	reacts with water to form hydrogen gas	a brown solid forms
Reaction with solution of chloride of B	nothing occurs	not attempted	very slow reaction in cold water
Reaction with solution of chloride of C	nothing occurs	reacts with water to form hydrogen gas	not attempted
Heat on nitrate	forms a brown gas that turns blue litmus red	forms a colourless gas that relights a glowing splint	forms a brown gas which turns blue litmus red

Solution

Metal B is less reactive with water. It is so reactive that it reacts with the water of any solution. The nitrate of B does not decompose completely to form brown nitrogen dioxide gas. Metal B must be high in the reactivity series.

Metal C is the least reactive with water. It seems to displace metal A from a solution of one of its salts. It does not seem to be able to displace metal B from a solution of one of its salts. The nitrate of C decomposes to form brown nitrogen dioxide gas. So, metal C is less reactive than metal B and lower than B in the reactivity series.

Metal A is unreactive in water. It cannot displace metal B or metal C from a solution of one of its salts. The nitrate of A decomposes to form brown nitrogen dioxide gas. So, metal A is less reactive than metal B and metal C, and lower than B and C in the reactivity series.

The metals, in increasing order of reactivity, are A, C and B.

Summary

- The physical properties of metals include that they are solid at room temperature, they have a high melting and boiling point, a high density, they are hard (a high tensile strength), lustrous, malleable, ductile, and are good conductors of heat and electricity both in the solid and liquid state.
- Metals show a range of reactivities with oxygen, water and non-oxidising acids.
- The arrangement of metals in the reactivity series reflects how easily the metal is ionised.
- The relative positions of metals in the series can be determined experimentally.
- The more reactive the metal, the more vigorous is its reaction with oxygen, water and dilute acids.
- A more reactive metal displaces a less reactive metal from a solution of a salt of the less reactive metal.
- The more reactive the metal, the more thermally stable are its compounds, and the higher the temperature needed to decompose them.
- The order of increasing reactivity for the metals in this study is Cu, Pb, Fe, Zn, Al, Mg, Ca, Na.
- All of the metals form basic oxides and hydroxides, which react with acids to form salt and water.
- ZnO, PbO and Al_2O_3 show acidic properties in addition to their basic properties, and are therefore amphoteric.
- Amphoteric oxides react with alkalis to form salt and water.
- Compounds such as carbonates, nitrates, oxides and hydroxides formed by the more reactive metals are relatively more stable to heat than the compounds of the less reactive metals.

Table 21.8 summarises some key reactions of metals.

▼ **Table 21.8** A summary of some key reactions of metals

	Reactions of metals:			
	... with water to form hydrogen	... with steam to form hydrogen	... with dilute hydrochloric acid to form hydrogen	... with solutions of compounds of other metals
Potassium, sodium	violent reaction	violent reaction	violent reaction	displace hydrogen from water
Calcium	steady reaction			water
Magnesium	very slow reaction	the metal burns	steady reaction	displace a metal which is lower in the series
Aluminium, zinc	no reaction			
Iron		reversible reaction		
Lead		no reaction		
(Hydrogen), copper, silver			no reaction	

End-of-chapter questions

1 Which of these properties do metals have?
I   High boiling point
II   Conducts electricity only when molten
III   Poor conductor of heat
IV   Does not react with water
A I only
B I and II only
C I, II and III
D I, II, III and IV

2 Which of the following statements is (are) not true?
I   All metals react with oxygen.
II   All metals react with water.
III   All metals react with non-oxidising acids.
IV   A metal can displace another metal that is lower in the reactivity series from a solution of one of its salts.
A IV only
B III and IV only
C II and III only
D I, II, III only

3 Which metal is highest in the reactivity series?
A Aluminium
B Calcium
C Copper
D Magnesium

4 Which of the following statements is (are) not true?
I   All nitrates decompose to form a brown gas.
II   All carbonates decompose to form carbon dioxide gas.
III   All hydroxides decompose to form water vapour.
IV   All carbonates react with non-oxidising acids.
A I only
B I and II only
C I, II and III only
D I, II, III and IV

5 Which of the following statements is (are) not true?
I   All metals can displace another metal from a solution of one of its salts.
II   Metals high in the reactivity series are more easily ionised.
III   A more vigorous reaction of a metal with a non-oxidising acid means that the metal is higher in the reactivity series.
IV   All metals react with oxygen to form oxides that can react with non-oxidising acids.
A IV only
B III only
C II, III and IV
D I only

6 What are the distinguishing features of metals?

7 Compare the reactivity of aluminium, zinc and copper with:
 a oxygen, water or steam;
 b dilute hydrochloric acid and aqueous sodium hydroxide.

8 Describe, with the aid of equations, the action of heat on:
 a sodium nitrate;
 b calcium nitrate.

9 a What would you observe if a sample of copper(II) nitrate is heated until there was no further change?
 b Write a balanced chemical equation for the reaction.
 c How will the solid residue left in the tube react with:
  (i) carbon, if heated;
  (ii) dilute sulphuric acid?

10 Copper(II) carbonate is decomposed by heat whereas sodium carbonate is not.
 a Explain the differences in the behaviour of these compounds to heat.
 b Write a balanced chemical equation, including state symbols, for the thermal decomposition of copper(II) carbonate.
 c How will you test to identify the gaseous product of the reaction?

11 Calcium hydroxide is strongly heated in air until there is no further change.
 a Write a balanced chemical equation for the reaction.
 b Calculate:
  (i) the mass of residue obtained if 7.4 g of calcium hydroxide were used in the experiment;
  (ii) the mass of gaseous product formed.

12 Part of the reactivity series of metals is shown below:

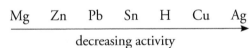

Mg   Zn   Pb   Sn   H   Cu   Ag

decreasing activity

 a Which metals will displace zinc metal from a solution of a zinc salt?
 b Would you expect silver metal to displace copper metal from a solution of a copper salt? Give reasons for your answer.
 c Which of these metals is expected to have an oxide that is the least stable to heat?
 d Which of the following reactions would you expect to release more energy per mole of product formed?
  (i) $Mg^0(s) + Cu^{2+}(aq) \rightarrow Mg^{2+}(aq) + Cu(s)$
  (ii) $Mg^0(s) + Zn^{2+}(aq) \rightarrow Mg^{2+}(aq) + Zn(s)$
  Give reasons for your choice.

13 A metal M is positioned between aluminium and zinc in the reactivity series. What reaction, if any, will you expect M or its compound to undergo in each of the following cases?
 a The molten metal oxide is electrolysed.
 b The metal oxide is heated with carbon.
 c The metal hydroxide is added to sulphuric acid.
 d The metal hydroxide is added to sodium hydroxide solution.
 e The metal is added to a solution of magnesium chloride.
 f The metal is added to a solution of silver nitrate.

22 Metals: Extraction and uses

In this chapter, you will study the following:

- the principles underlying the extraction of metals from their ores;
- how to use the reactivity series to select an appropriate method for extracting a metal;
- the methods used to extract iron and aluminium from their ores;
- why metal alloys are often used in place of the metals;
- how the uses of metals and their alloys relate to their properties.

This chapter covers
Objectives 2.3, 3.1 and 3.2 of Section C of the CSEC Chemistry Syllabus.

Metals are not usually found as unreacted metal. The exceptions to this are silver and gold. Silver and gold are very low in the metal reactivity series. Metals are normally found combined with other elements as compounds. Metals have differing reactivities, as described in the metal reactivity series.

22.1 Principles underlying the extraction of metals

Ores are naturally occurring metallic compounds from which metals can be extracted. They are usually contaminated by other materials and are therefore impure.

Most metals occur in the Earth's crust as impure ionic compounds, from which they can be extracted. The compounds are known as **ores**. Many ores are oxides, carbonates or sulphides of the metals.

Since ores are ionic compounds, they contain the cations (positive ions) of the metals. Extracting the metal from its ore therefore involves the conversion of the metal cations to metal atoms by reduction. The process involves gain of electrons by the cations (see Section 12.7).

$$M^{2+}(s) + 2e^2 \rightarrow M(s)$$

Stages involved in the extraction of metals

The first stage in the extraction of a metal involves mining the ore. The ore must then be purified before the reduction is carried out. Even so, the metal produced may still contain impurities from the reduction process and must be purified or refined. Figure 22.1 outlines the stages involved in the extraction process.

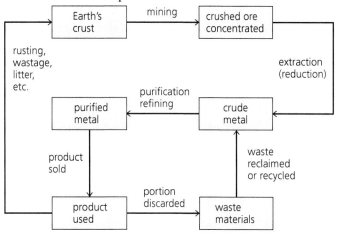

Figure 22.1
How metals are obtained ▶

Reduction of the ore to the metal

The method used to reduce the ore to the metal depends on:

- the nature of the ore;
- the reactivity of the metal.

The main methods of reduction used in the process of extracting a metal from its ore are electrolytic reduction and chemical reduction.

Figure 22.2
'Red mud' waste produced in the purification of bauxite to form pure alumina. This is used for extracting aluminium. ▶

Electrolytic reduction

Electrolytic reduction is the most powerful method of reduction (see Section 22.3). It is used where the ore is a compound of a very reactive metal. The molten purified ore is used as the electrolyte.

Examples are obtaining sodium from molten $NaCl$ and aluminium from purified Al_2O_3.

Chemical reduction with carbon, carbon monoxide or hydrogen

Carbon, carbon monoxide and hydrogen are all reducing agents. Carbon monoxide requires much higher temperatures than carbon and hydrogen in order to act as an effective reducing agent. Carbon monoxide is a poisonous gas, so it is produced in the furnace from carbon in a two-stage reaction:

$$C(s) + O_2(g) \rightarrow CO_2(g)$$
$$CO_2(g) + C(s) \rightarrow 2CO(g)$$

Hydrogen is less powerfully reducing than either carbon or carbon monoxide. It can remove oxygen from the oxides of less reactive metals at the relatively low temperatures supplied by a Bunsen burner in the laboratory.

Chemical reduction is a suitable method to extract metals below aluminium in the reactivity series of metals. If the ore is a metallic carbonate that can be decomposed to its oxide at high temperatures, reduction of the oxide formed with carbon monoxide is again recommended.

Special methods

Other methods have been used to extract metals from their ores. The extraction of copper falls into this category.

Find out more

One ore from which copper is extracted is iron pyrites ($CuFeS_2$). The extraction can be represented by two equations:

$$2CuFeS_2(s) + 4O_2(g) \longrightarrow Cu_2S(s) + 3SO_2(g) + 2FeO(s)$$
copper(I) sulphide

The iron(II) oxide is removed as a slag by heating with silica (see Section 22.4). Copper(I) sulphide is heated in a limited supply of air:

$$Cu_2S(s) + O_2(g) \rightarrow 2Cu(s) + SO_2(g)$$

The copper formed is impure and can be refined by electrolysis (see Section 13.4).

Practice

1 The reactivity of cobalt is roughly similar to that of iron. What method of reduction would you use to obtain cobalt from cobalt oxide?

2 An element is obtained commercially from its fused (molten) chloride by electrolysis. Assign an approximate place to it in the reactivity series.

ORR
A&I

Experiment 22.1   Looking at the ease of reduction of metal oxides with carbon and hydrogen

 Materials

- magnesium oxide
- aluminium oxide
- iron(III) oxide
- iron(II) oxide
- lead(II) oxide
- copper(II) oxide

Procedure

Test the oxides in turn, using the apparatus shown in Figures 22.3 and 22.4.

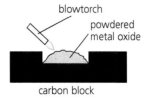

▲ **Figure 22.3**   Reacting a metal oxide with carbon.

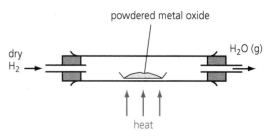

▲ **Figure 22.4**   Reacting a metal oxide with hydrogen.

Results and observations

1 Determine which oxides are reduced by:
   a carbon;
   b hydrogen.

2 Record your results in a suitable table.

3 Why should a metal formed in Figure 22.4 be cooled in a stream of hydrogen gas?

22.2 Using the metal reactivity series to select an appropriate method for extracting a metal

Metals that are very reactive are difficult to obtain by reduction from compounds. The metal cations prefer to remain as cations rather than be reduced by accepting electrons to form the metal. In such cases, a lot of energy needs to be used to force the metal cation to accept electrons to form the metal. Therefore, we have to use an electrolytic method. All metals from aluminium and up in the metal reactivity series can only be obtained using an electrolytic method.

All metals below aluminium in the metal reactivity series are less reactive than aluminium. Their cations can be made to accept electrons and become reduced to form the metal more easily than aluminium. It is not necessary to use an electrolytic method, which uses a lot of energy. We can obtain most of these metals from their compounds by chemical reduction with carbon, carbon monoxide or hydrogen.

Summary of extraction methods

Table 22.1 gives a general guide to the selection of an appropriate extraction procedure.

▼ **Table 22.1** Some of the most common methods of metal extraction

Metal	Most common ore	Method of extraction	Comments
potassium, sodium	chlorides	electrolysis of the molten chloride or oxide	the metals are high in the reactivity series
calcium, magnesium	chlorides, carbonates		
aluminium	oxide		
zinc, iron, tin	oxides, sulphides	chemical reduction of the oxide with carbon, carbon monoxide or hydrogen	these metals are moderately high in the reactivity series
lead, copper	sulphides	special method	these metals are low in the reactivity series

22.3 Extraction and production of aluminium

Bauxite plays an important role in the economies of some countries and is the only important ore of aluminium. Aluminium – one of the more abundant elements in the Earth's crust – is never found in the free state. It occurs chemically combined with other elements, mainly in the form of aluminosilicates. Bauxite (major ingredient $Al_2O_3.xH2O$) is the chief ore of aluminium.

The crude (mined) bauxite is either:

- heated to 3 000 °C to produce calcined bauxite; or
- converted to pure alumina (Al_2O_3), the anhydrous compound.

Aluminium is obtained from alumina by electrolysis. Pure alumina, which melts at 2 050 °C, is dissolved in molten cryolite (sodium aluminium fluoride), Na_3AlF_6. The addition of the cryolite lowers the melting temperature to 950 °C. (Remember the presence of an impurity lowers the melting point of a substance.) The presence of the cryolite also gives the melt better conducting properties and, in addition, it does not mix with the aluminium metal formed in the electrolysis.

Figure 22.5
Bauxite conveyor, Jamaica. An overhead conveyer belt on wheels that runs on wire ropes 3.4 km long, carrying loads of 1 200 tonnes of bauxite per hour from the mine to the processing plant. it generates its own energy. ▷

The electrolysis of alumina

- The electrolyte is a molten mixture of aluminium oxide and cryolite.
- Power requirements are 5 V and 100 000 A.
- The electrodes are graphite for the cathode and graphite or titanium alloys for the anodes. The apparatus used is illustrated in Figure 22.6.

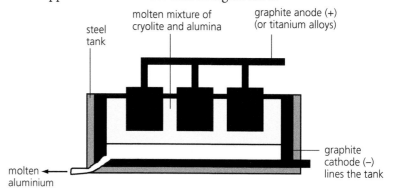

Figure 22.6
The manufacture of aluminium from purified bauxite ▷

The cathode reaction is:

$$Al^{3+}(l) + 3e^- \rightarrow Al(l)$$

This equation shows the reduction of aluminium cations to the metal. Note that three moles of electrons are needed to discharge one mole of aluminium.

The anode reaction is:

$$2O^{2-}(l) - 4e^- \rightarrow O_2(g)$$

Oxygen gas is released at the anode.

Since 3 moles of electrons are needed to discharge 1 mole of aluminium, it should be appreciated that aluminium production consumes a huge amount of electrical energy. Jamaica and Guyana are two Caricom countries in which large deposits of bauxite are located. The ores are purified in these countries but are sent to Canada for the extraction of the metal. This is because the extraction of aluminium requires an enormous amount of electricity and so generally takes place in parts of the world where electricity is produced cheaply by hydroelectric power stations.

Figure 22.7
The raw aluminium-containing ore is mined and lightly processed at the Bauxite Company, Linden, Guyana, and then exported for aluminium extraction in Canada. ▶

Practice

3 a Use the ionic equation for the cathode reaction to calculate the quantity of electricity needed to extract one tonne (1 000 kg) of aluminium.

  b What mass of purified aluminium oxide must you start with to produce one tonne of aluminium? Assume that the purification process is 90% efficient.

4 Graphite electrodes are ACTIVE in the presence of oxygen gas, but the more expensive titanium anodes do not react with oxygen.

  a Suggest a reason or reasons why graphite is still used in aluminium smelters.

  b Can this process be used as a commercial source of oxygen when graphite anodes are used? Explain your answer.

22.4 Extraction of iron

The production of iron

More iron is produced than any other metal and most of the iron produced is converted to steel. The raw materials for the iron industry are:

- iron ore, usually an oxide of iron;
- an energy source;
- a reducing agent, usually coke or natural gas;
- a flux, usually limestone, to form a slag with silicates and other impurities.

Iron has traditionally been produced in blast furnaces. Modern blast furnaces, which can run continuously for as long as 10 years, are fully automated.

Iron is extracted from its ore in the following stages.

- Dried heated iron ore, limestone and coke are fed into the top of the furnace.

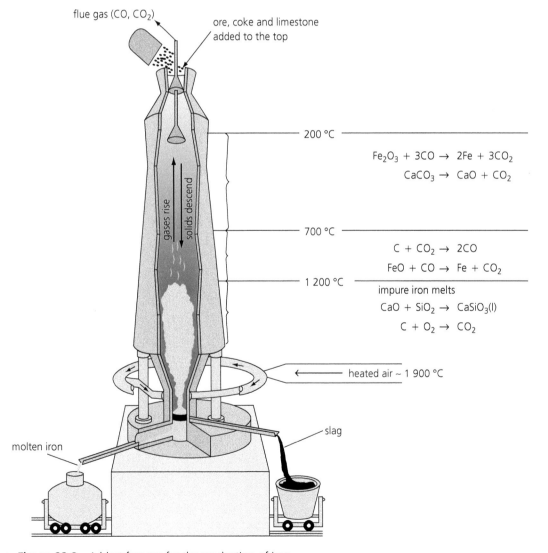

flue gas (CO, CO$_2$)

ore, coke and limestone added to the top

gases rise

solids descend

200 °C

$$Fe_2O_3 + 3CO \rightarrow 2Fe + 3CO_2$$
$$CaCO_3 \rightarrow CaO + CO_2$$

700 °C

$$C + CO_2 \rightarrow 2CO$$
$$FeO + CO \rightarrow Fe + CO_2$$

1 200 °C

impure iron melts

$$CaO + SiO_2 \rightarrow CaSiO_3(l)$$
$$C + O_2 \rightarrow CO_2$$

heated air ~ 1 900 °C

slag

molten iron

▲ **Figure 22.8** A blast furnace for the production of iron.

- Hot air is blown into the furnace near the bottom. The coke burns in the blast of hot air, producing carbon dioxide and generating a great deal of heat:

$$C(s) + O_2(g) \rightarrow CO_2(g); \Delta H = -ve$$

- The carbon dioxide formed is reduced to carbon monoxide by the hot coke:

$$CO_2(g) + C(s) \rightarrow 2CO(g)$$

4 The carbon monoxide reduces the hot iron ore to molten iron:

$$Fe_2O_3(s) + 3CO(g) \rightarrow 2Fe(l) + 3CO_2(g)$$

or

$$Fe_3O_4(s) + 4CO(g) \rightarrow 3Fe(l) + 4CO_2(g)$$

The molten iron runs to the bottom of the furnace.

The iron produced in a blast furnace would have very large amounts of impurities if limestone were not used. The limestone, at the temperature of the furnace, breaks down:

$$CaCO_3(s) \rightarrow CaO(s) + CO_2(g)$$

The calcium oxide formed combines with silicon dioxide (sand), the main impurity in the iron ore, to form a molten slag:

$$CaO(s) + SiO_2(s) \rightarrow CaSiO_3(l)$$

The slag does not mix with the molten iron but floats on it. The molten iron and slag are run off separately.

The iron formed in a blast furnace is an impure form of the metal. Because the molten metal is allowed to solidify in shallow trays known as casts, the iron is called 'cast iron' or 'pig' iron. Cast iron is quite brittle and contains 2–4% carbon and other impurities (such as sulphur, phosphorus and silicon). Most of the cast iron is converted into steel.

The blast furnace reaction is not the only one used in the manufacture of cast iron. More modern industrial processes use a mixture of carbon monoxide and hydrogen to reduce the iron ore to the metal. Two chemical reduction processes are in operation:

$$Fe_2O_3(s) + 3CO(g) \rightarrow 2Fe(s) + 3CO_2(g)$$

as happens in the blast furnace, and:

$$Fe_2O_3(s) + 3H_2(g) \rightarrow 2Fe(s) + 3H_2O(l)$$

The reducing gases, carbon monoxide and hydrogen, are obtained by steam reforming, a process in which a hydrocarbon such as methane is reacted with steam in the presence of a heated nickel catalyst.

▲ **Figure 22.9** Iron and steel is produced at the Arcelormittel Company at Pt Lisas, Trinidad.

Iron produced in this process is again used for the production of steel.

$$CH_4(g) + H_2O(g) \rightarrow CO(g) + 3H_2(g)$$

Production of steel

The raw materials for the steel industry are:

- a flux, usually limestone;
- materials for alloying;
- cast iron.

Limestone is decomposed to calcium oxide (CaO, quick lime), which is then used to form a slag with silicates and other impurities in the cast iron. The other materials added include carbon and various transition metals such as nickel, manganese, chromium, cobalt, tungsten, vanadium and molybdenum.

Impure molten iron is mixed with powdered calcium oxide (lime) in a large container. Oxygen, at high pressure, is blown onto the surface of the mixture. This blast of oxygen stirs up the mixture and oxidises the impurities in the iron. The oxidised impurities can escape as gases or form a slag, which is then poured off. The molten metal left is pure iron, to which is added carbon or other metals to form the various steels.

22.5 Uses of some metals and their alloys

Uses of metals

The uses of a particular metal are determined by the unique physical properties of that metal. Table 22.2 links the uses of the metals aluminium, iron, copper and lead to their properties.

▼ **Table 22.2** Uses of aluminium, iron, copper and lead, linked to their properties

Metal	Distinctive properties	Uses
aluminium	forms a thin, compact protective oxide coating; resistant to corrosion; forms alloys easily; highly reflective when pure; non-toxic; non-magnetic; non-combustible	aircraft manufacture; cooking utensils; window frames; engines; high tension cables; foil for packaging food
iron	very strong but rusts easily; forms a wide range of alloys	all largely as steels reinforcement for concrete; bodies of cars; tools; castings for cars, machinery, pipes; girders in bridges; cooking utensils
copper	a very good conductor of heat and electricity; soft, malleable and easy to bend; low in reactivity series, hence its resistance to corrosion	roof sheeting; electrical cables; cooking utensils; cylinder heads; car radiators; making coins

continued

▼ **Table 22.2** *continued*

lead	high density; malleable but not ductile; low melting point	shielding to absorb X-ray and other nuclear radiations; making car batteries; sheathing electrical cables; roofing and pipes; making solder

▲ **Figure 22.10** Aeroplane bodies are made of aluminium alloy, because they must be lightweight. What other alloys are used in objects in this photograph?

Alloys and their uses

An **alloy** is a mixture of a metal with one or more other elements. Often the other element is also a metal.

Today, metals are rarely used in their pure form. Instead, most metals are used as **alloys**.

Alloys are made by melting the individual metals, mixing them, and allowing the molten mixture to solidify. Alloys can be thought of as solid solutions. Further treatment of the alloy by rolling or heating can cause considerable hardening of some alloys. An alloy may also be treated to reduce or lessen a property that was shown by one of its metallic components, but is unsuitable for the particular use for which the alloy is intended.

Many alloys have properties that are quite different to the metals from which they are made. Here are some examples:

- Copper and nickel have high electrical conductivities, but they form alloys whose conductivities are sufficiently low to be used in wires with high electrical resistance.
- Bronze, an alloy of copper and tin, is used for decorative and ornamental purposes since it is harder than pure copper.
- Brass, an alloy of copper and zinc, is used for making coins and water fittings.

Steel

Steel is an alloy of iron and the non-metal carbon (0.5–1.5%) and some other transition metals, as shown in Table 22.4.

There is an iron and steel plant at Point Lisas in Trinidad and Tobago.

▼ **Table 22.4**   Alloy steels are widely used and illustrate the wide range of properties achieved by alloying.

Name and alloying metal	Properties	Typical uses
manganese steel	tough	railway points; safes
tungsten steel	high melting; tough	tools
chromium steel	hard; high tensile strength	ball bearings
stainless steel (high percentage of Ni, Cr and Mo)	non-rusting	car accessories; cutlery
cobalt steel	easily magnetised	magnets
titanium steel	withstands high temperatures	turbine blades; aircraft
vanadium steel	very hard	used for cutting other steels

Aluminium alloys

Aluminium alloys are widely used in aircraft manufacture because aluminium is light and aluminium alloys are stronger than pure aluminium.

▼ **Table 22.5**   The common alloys of aluminium

Name	Alloying metal	Properties	Uses
Magnalium	aluminium + 10–30% magnesium	castings of high strength and low density	aircraft parts; car wheels; sheathing uranium in nuclear power plants
Duralumin	contains 4% copper	high strength, low density	kitchen utensils; aircraft

Solder

Solder is a common alloy of lead and is used to join metals together. Soft solder, an alloy of about equal parts of tin and lead, has a melting point of 200–300 °C. This type of solder is used in making electrical connections.

Summary

- Most metals are obtained by extraction from their ores.
- Highly reactive metals (like calcium, magnesium and aluminium) are extracted from their ores by electrolysis.
- Less reactive metals (like iron and lead) are extracted from their ores by chemical reduction.
- The only important ore of aluminium is bauxite.
- Aluminium is extracted by electrolysis of a mixture of molten alumina and cryolite.
- Extraction of aluminium requires large amounts of electricity.
- Iron is extracted by the reduction of its oxide with carbon monoxide at high temperatures in a blast furnace.
- The cast iron formed is impure and can be converted to wrought iron and steel.
- The unique physical properties of a metal determine the uses to which a metal can be put.
- Alloys are made to enhance and in some instances to downplay some properties of the contained metal. Alloys extend the range of uses of a given metal.
- Steels are commercial alloys made by reducing the quantity of carbon dissolved in cast iron or wrought iron and mixing the iron with small amounts of other elements.

End-of-chapter questions

1 Which of the following statement(s) is (are) true?
   I   The extraction of a metal from its ore requires a chemical reduction process.
   II The metal ore, obtained by mining, first has to be purified.
   III All metal cations easily accept electrons and become metal atoms.
   IV The reducing process always involves the use of carbon.
   A I only
   B I and II only
   C I, II and III
   D I, II, III and IV

2 Which of the following statement(s) is (are) true?
   I   Obtaining a very reactive metal from its ore requires a powerful reduction process.
   II Electrolysis is a reduction process in which the ore has to be heated until it becomes molten.
   III To obtain a metal that is low in the reactivity series from its ore requires a reduction process that involves the use of carbon, carbon monoxide or hydrogen.
   IV Heating the ore and the reducing agent is done in a blast furnace.
   A I only
   B I and II only
   C I, II and III
   D I, II, III and IV

3 In the production of aluminium from its ore, which of the following statements is not true?
   A The melting point of the alumina is reduced by adding an impurity.
   B The electrolytic process requires a huge amount of electrical energy.
   C The anode is usually made of graphite.
   D The aluminium rises to the surface of the molten electrolyte.

4 In the production of iron from its ore, which of the following statements is true?
  A Limestone is used as a reducing agent.
  B Coke is an impurity that must be removed before the ore is heated.
  C The molten iron floats on the mixture being heated.
  D The iron produced by the blast furnace is impure.

5 Which of the following pairs of statements correctly relates a property to the use of a metal or alloy?
  A Aluminium is used to make door frames. Aluminium is lightweight.
  B Steel is used in the construction of buildings. Steel has high tensile strength.
  C A tin/lead alloy is used to make solder. Tin is a shiny grey metal.
  D Lead is used to make the electrodes in car batteries. Lead is very heavy.

6 75% of the Earth's crust is accounted for by the non-metallic elements silicon and oxygen. Metals account for the other 25%. The most abundant of the metals are aluminium (8%), iron (5%), calcium (3.6%) and magnesium (2.1%). Draw a pie chart to show the abundance of the metals in the Earth's crust.

7 Discuss the principles used in the extraction of metals from their ores.

8 Describe the extraction of iron from its ore with reference to:
  a the materials needed for the process;
  b the generation of the effective reducing agent;
  c the conversion of the ore to the metal;
  d the removal of the impurity from the ore.
  Write relevant equations for the reactions described.

9 a What is meant by the term 'alloy'?
  b Give the approximate composition of four different alloys.
  c Why are alloys preferred to pure metals for some purposes? Illustrate your answer with specific examples.

10 The iron produced in the blast furnace is brittle, mainly because it contains a dissolved non-metal. The cast iron can be commercially treated to produce wrought iron or steel, thus providing materials with improved tensile strength and better resistance to corrosion.
  a Identify the element primarily responsible for the brittleness of cast iron.
  b Explain how the element gets into the cast iron.
  c What type of chemical reaction reduces the quantity of the element when cast iron is treated commercially? Write a balanced chemical equation for the reaction.
  d (i) What type of elements are added to reduce the tendency of steel to corrode?
    (ii) How do these elements achieve their function?

11 Suggest the possible reactions that can occur when a mixture of copper(II) carbonate and charcoal (carbon) is heated. Give balanced equations for all the reactions you discuss.

12 Use a suitable diagram to illustrate the general steps in the conversion of a crude metallic ore to a refined purified metal.

13 Extraction processes can sometimes pollute the environment. Discuss the ways in which this can happen when:
  a bauxite is converted to purified aluminium oxide (alumina);
  b aluminium oxide is converted to aluminium;
  c iron is produced from iron ore by the blast furnace process.

14 Which of the following oxides would you expect to be most difficult to reduce: copper(II) oxide, zinc oxide or magnesium oxide? Justify your answer.

15 Compare aluminium, iron, copper and lead in terms of electrical conductivity, hardness, relative density, strength and the uses to which they are put.

23 Metals in the environment

In this chapter, you will study the following:

- the conditions necessary for corrosion;
- corrosion in iron and aluminium;
- the metals important to living systems;
- the harmful effects of metals on living systems and the environment.

This chapter covers
Objectives 4.1, 4.2 and 4.3 of Section C of the CSEC Chemistry Syllabus

We find metals and their alloys everywhere in our world. We use them for constructing buildings, bridges and vehicles of all types. We also use them on land, in the sea, in the air and in space. We can make metals into many different types of objects that we can use in industry or in our homes. While many metals are very useful, some metals are very toxic to living things. The differences in the reactivity of metals help us to arrange them in a reactivity series. One metal that reacts with the environment and deteriorates is iron. In Chapter 13, you learnt that we can prevent iron from rusting in two ways that involve electrolysis: use of a sacrificial anode and galvanising. Aluminium is a metal that is resistant to corrosion because it forms a protective oxide layer. Anodising enhances this protection by causing a thicker oxide layer to form on an aluminium object.

23.1 Conditions necessary for corrosion of metals

Corrosion is a process in which metals react with their environment, usually with oxygen.

- In copper and brass objects, corrosion results in a variety of colours, which we refer to as a patina. The most distinctive patina is the blue-green colour seen in the Statue of Liberty in New York, USA.

- Aluminium forms an oxide coating that protects a metal from further corrosion.

- Corrosion of iron is commonly called rusting. Rusting weakens the metal and eventually leads to its destruction.

Corrosion in iron

Rust is hydrated iron(III) oxide ($Fe_2O_3.xH_2O$). Rust does not stick to the iron that is underneath it, but flakes off, leaving a fresh surface exposed for further reaction. This process continues, weakening the metal until it is completely destroyed. Rusting of iron is

▲ **Figure 23.1** The green patina on the copper of the Stature of Liberty.

431

significant because iron and steel (an alloy of iron) are widely used in constructing buildings, roofs, vehicles, bridges and all other types of equipment. When the iron is weakened by rust, it can lead to accidents and possible loss of life. Replacing rusted iron parts, for example, replacing a rusted roof, can be very expensive.

Rusting is an oxidation reaction. Iron loses electrons according to the equation:

$$Fe(s) - 2e^- \rightarrow Fe^{2+}(aq)$$

This reaction occurs in the parts of the iron that are not exposed to air. In the parts of the iron where there is a good supply of oxygen and water, the following reaction occurs:

$$O_2(g) + 2H_2O(l) + 4e^- \rightarrow 4OH^-(aq)$$

The Fe^{2+} ions react with the OH^- ions to form iron(II) hydroxide.

$$Fe^{2+}(aq) + 2OH^-(aq) \rightarrow Fe(OH)_2(s)$$

Oxygen from the air oxidises the iron(II) hydroxide to hydrated iron(III) hydroxide.

$$2Fe(OH)_2(s) + O_2(g) \rightarrow Fe_2O_3.xH_2O(s)$$

Figure 23.2
Collapsed bridge over Shark River, Trinidad, 1999. What do you think caused this bridge to collapse? ▶

In Chapter 13, you conducted an experiment to work out the conditions for rusting.

Corrosion in aluminium

A thin layer of aluminium oxide forms on aluminium objects. This layer adheres strongly to the metal, unlike the rust that forms on the surface of iron. This layer makes aluminium very resistant to corrosion.

However, aluminium can undergo galvanic corrosion. If another metal that is lower in the metal reactivity series is attached to a piece of aluminium, and there is an electrolyte present (for example, water), this forms an electrolytic cell. The aluminium becomes the anode and undergoes corrosion. Because it is the more electropositive metal, it loses electrons.

$$Al(s) - 3e^- \rightarrow Al^{3+}(aq)$$

The less electropositive metal becomes the cathode and is protected from corrosion. This is what happens when a sacrificial anode is used to prevent the rusting of iron (see Chapter 13).

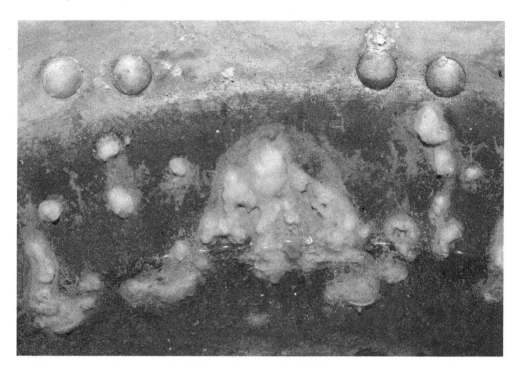

Figure 23.3
The corrosion of
aluminium ▶

23.2 Metals important to living systems

There are many metal ions that are important to living things. Some of these metal ions are found attached to carbon compounds by coordinate covalent bonds. These compounds are sometimes referred to as organometallic compounds.

- **Sodium** controls water balance in body fluids, assists in the transmission of nerve impulses and assists in muscle contraction.

- **Potassium** is found in high concentrations in all living cells and it is necessary for the cells to function properly. The transport of nerve impulses and the contraction of muscle cells depend on the movement of sodium and potassium ions.

- **Calcium** is an important part of the structure of plant cell walls and cell membranes. In animals, calcium is essential in building strong bones and teeth. It is also involved in muscle contraction and prevents muscle cramps.

- **Magnesium** is important to living systems for many different reasons. Mg^{2+} ions are found in all living cells. These ions are important in the function of DNA, which controls all the activities in cells, especially protein manufacture and reproduction. A magnesium ion plays an essential role in the role of ATP. Energy is stored in ATP molecules and released when needed for chemical reactions to occur in cells. Many enzymes also work only when they have magnesium ions bonded to them.

A magnesium ion is part of a chlorophyll molecule. Chlorophyll is the green pigment found in plant leaves. This compound absorbs energy from sunlight, allowing plants to manufacture food in the process of photosynthesis. Green plants are the start of all food chains, and so they are essential in keeping animals alive.

Figure 23.4
Chlorophyll-a with Mg²⁺
ion ▶

- **Iron** is found in haemoglobin, which gives blood cells their red colour. Red blood cells carry oxygen from the lungs (or gills) to the tissues, where the oxygen is used to release energy from food. This energy is needed for the many chemical reactions that sustain life. A lack of haemoglobin causes a disease called anaemia. Haemoglobin is made up of four protein molecules, each with a haem group. Each haem molecule is a coordination complex containing an Fe^{2+} ion held by four (coordinate) dative covalent bonds. An oxygen molecule can attach reversibly to the Fe^{2+} ion. The amount of iron found in an adult human body is almost 4 g.

R = globin protein

oxyhemoglobin

Figure 23.5
Haem molecule with
Fe^{2+} ion and O_2
molecule bound ▶

Trace metals

Trace metals are metals that are needed in very small quantities. They are often referred to as micronutrients.

Trace metals needed by plants

- **Magnesium** helps in cell division and in the formation of chlorophyll. Signs of magnesium deficiency include the yellowing of young leaves.

- **Manganese** activates enzymes and assists chlorophyll synthesis and nitrate use. Signs of manganese deficiency include dark green bands on the main veins of the leaves.

- **Zinc** is essential for the formation of chlorophyll and carbohydrates. It is found in plant hormones and enzymes that control growth. Signs of zinc deficiency include stunted growth and leaves on shoot tips that look like roses.

- **Boron** is important for the formation of cell walls and cell division, in germination of pollen grains, and in seed formation. Signs of boron deficiency include older leaves turning yellow and fruit may be badly formed.

- **Copper** activates enzymes, and plays a part in photosynthesis and reproduction. Signs of copper deficiency include leaf tips turning yellow and leaves drooping.

- **Molybdenum** has an important role in plants being able to use nitrates and phosphates. It helps in the formation of root nodules in legumes. Signs of molybdenum deficiency include stunted growth.

- **Cobalt** is necessary for fixation of nitrogen in legumes. Signs of cobalt deficiency include older leaves turning yellow.

Figure 23.6
These leaves show chlorosis (yellowing of the leaves) as a result of iron deficiency. ▶

Trace metals needed by humans

- **Cobalt** is found in vitamin B12 (cobalamin), which is the only vitamin with a metal ion. This vitamin is connected to the regeneration of folic acid (vitamin B9), which the body cannot manufacture and so has to be obtained from the diet. Cobalamin or folic acid is needed in the synthesis of DNA and in cell division, including the manufacture of new red blood cells. When vitamin B12 is absent, there is lack of red blood cells, which results in a condition called anaemia. Folic acid deficiency interferes with the proper development of the nervous system of unborn babies, and so folic acid is an essential part of a pregnant mother's diet. Vitamin B12 deficiency, even when folic acid is present, results in damage to the brain and nervous system.

- **Zinc** is important for hair growth, wound healing and cell division.

- **Copper** is the key element in an enzyme that is essential for respiration: cytochrome C oxidase.

- **Manganese** is found in enzymes that are involved in many different reactions, for example, the formation of bone connective tissue and genetic proteins.

- **Molybdenum** is found in the active sites of some enzymes. Deficiency can result in toxic levels of sulphites and damage to nerves. It can also result in an increased risk of cancer of the oesophagus.

23.3 Harmful effects of metals and their compounds on living systems and the environment

Many transition metals are needed in very small amounts by the human body, but they can become toxic at higher levels. Mercury, arsenic and lead are some of the metals that are not known to be useful to be useful to the human body, and are poisonous to us.

Lead

Lead is released into the environment from these sources:

- The exhaust fumes of motor vehicles that use leaded gasoline. Leaded gasoline contains the additive tetraethyl lead, $Pb(C_2H_5)_4$, which is used to reduce engine 'knock'. 1,2-dibromoethane is also added to ensure that the lead does not build up in the engine and is instead released in the exhaust fumes as the volatile compound, lead bromide.
- Lead-based paint in older buildings, which flakes off and form dust. Yellow paint used to be made from lead chromate and white paint used to be made with lead carbonate.
- Discarded motor vehicle batteries. The electrodes of batteries are made of lead.
- Lead pipes that carry water in very old buildings.
- Lead weights used on fishing nets.
- Lead shot from ammunition used in hunting.

Lead can enter the body:

- through the lungs by inhaling motor vehicle exhaust or dust from paint;
- through the skin, when in contact with leaded gasoline, with soil contaminated with motor vehicle exhaust particles (near roadways), broken lead batteries, or even at work in lead factories;
- through the gut from contaminated water or food, because lead can enter ground water and get into rivers and then the sea, and animals used for food may have ingested lead on vegetation.

Lead shot can be ingested by birds, together with the gravel that they eat to help their digestion. Weakened, lead-poisoned animals become easier prey. The amount of lead in larger carnivores increases, because they eat many smaller animals. Lead poisoning has been found in eagles in the USA. This is because they hunt and feed on carcases left after being shot with bullets containing lead.

Signs of lead poisoning begin to manifest when the lead levels are sufficiently high. Children, who have smaller bodies than adults, are more affected by lead poisoning. Lead poisoning mainly effects the nervous system.

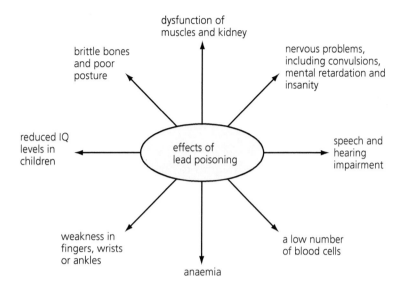

Figure 23.7
Effects of lead
poisoning ▶

Mercury

Mercury can enter the environment in the following ways:

- Emissions from volcanoes
- Combustion of fossil fuel
- Release from mercury amalgam used in the extraction of gold
- Waste water containing mercury released from the manufacture of caustic soda and chlorine by the flowing mercury cathode cell
- Broken fluorescent bulbs releasing the mercury vapour they contain
- Broken thermometers releasing mercury liquid

Figure 23.8
Mercury is the only
metal that is liquid at
room temperature. ▶

What to do when a mercury thermometer breaks

- Ask everyone else to leave the room.
- Do not let anyone walk through the spilled mercury.
- Clothing or shoes that have come into direct contact with the metal should be safely discarded in a garbage bag at a toxic waste disposal facility.

- Open all windows and doors that lead outside.
- Close doors to other rooms.
- Put on rubber gloves.
- Do not let children help clean up the spill.
- Pick up the broken pieces of glass with care, fold them in a paper towel and place and place them in a sealable bag.
- Use a squeegee or piece of cardboard to gather up the beads of mercury.
- Use an eyedropper to carefully suck up the mercury and place it on a damp paper towel. Fold the paper towel and place it in a sealable bag.
- Use a flashlight at a low angle, close to the floor in a darkened room to locate beads of mercury as they reflect the torch light. (Mercury can roll a long way.)
- Use sticky tape/duct tape to collect very small beads of mercury (and very small pieces of glass) that may be on the floor or in cracks. Place the sticky tape in a sealable bag.
- Sulphur powder may be sprinkled on the area to bind to any very small beads of mercury that were missed and also to prevent the formation of mercury vapour.
- Put all the material from the clean-up, including the gloves, in a garbage bag and take it to a toxic waste disposal facility.
- Keep the area well ventilated for at least 24 hours after successfully cleaning up the spill.

Never:
- use a broom – it will break up the mercury into smaller droplets and cause them to spread further;
- use a vacuum cleaner – it will produce more mercury vapour;
- wash clothes or other items that have come into direct contact with mercury – it will contaminate the washing machine and release mercury into the environment.

Mercury can enter the body in the following ways:
- Through the lungs by inhaling the mercury vapour that is released when fluorescent bulbs break
- Through the skin when in contact with mercury compounds, and to a lesser extent, elemental mercury
- Through the gut from contaminated water or food as fish and shellfish readily take in methyl mercury, and it accumulates going up the food chain

Mercury poisoning results in 'pins and needles' in hands and feet; loss of coordination of movement; muscle weakness; impaired speech, hearing and vision, tremors; irritability; mood swings; memory loss; reduced mental function; kidney failure; respiratory failure and death.

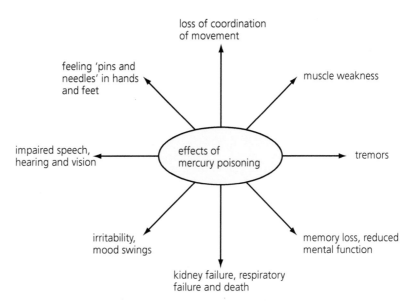

Figure 23.9
Effects of mercury poisoning ▶

loss of coordination of movement

feeling 'pins and needles' in hands and feet

muscle weakness

impaired speech, hearing and vision

effects of mercury poisoning

tremors

irritability, mood swings

memory loss, reduced mental function

kidney failure, respiratory failure and death

A foetus in the womb exposed to mercury can experience severe effects. These include mental retardation, lack of coordination, inability to speak, blindness, seizures and crippling deformities.

Dental amalgam, used in filling teeth, is made with mercury. Research has not conclusively shown that dental amalgam is harmful.

A compound of mercury, thiomersal, is used to preserve vaccines. It is widely believed to trigger medical problems in children receiving the vaccine.

Gold mining in Guyana

Gold mining in Guyana uses mercury to form an amalgam to extract gold particles from the soil. The water that washes the sediment from the heavier gold amalgam washes some mercury into the rivers. Mercury vapour also escapes into the atmosphere when the amalgam is heated to release the gold. High levels of mercury have been found in the bodies of the miners who work with mercury and also in the jewellers who work with the raw gold. High levels of mercury are also found in soil, in rivers and in the bodies of people who live downstream from gold mines. Even after a mine has closed, there is still a high level of mercury in the soil and the fish found in rivers downstream of disused gold mines.

Figure 23.10
Using mercury to extract gold in Guyana ▶

Minamata, Japan

In 1923, in a town called Minamata in Japan, a chemical factory began producing acetaldehyde using mercury sulphate as a catalyst. The waste water from the factory was discharged directly into the sea. When the fish haul of the fishermen began to decrease, the company offered them compensation. Fish were dying, floating and washing ashore. Cats started to walk strangely (which the locals called 'dancing cat disease'), have seizures and die.

In 1956, the first child was born that showed signs of nervous system damage, including difficulty in walking, in speaking and having convulsions. As more affected children were born and people started dying, university researchers reported that sea food contaminated with a heavy metal was the suspected cause.

The company conducted research, the results of which showed that their waste contained dangerously high levels of many heavy metals, but they withheld this information from outside researchers. Instead, they redirected the waste water line and funded research to find some other cause for the disease. Protests by fishermen and victims of the disease led to the company offering compensation. In 1959, the company set up a wastewater treatment plant to satisfy the protestors and the government, but the treatment process did not solve the problem of the level of mercury in the waste water. Children continued to be born with nervous system damage, because the pollution continued for many years, until 1968 when the company stopped using mercury in the production process. That year, the government finally declared that the company was responsible and, in 1976, the courts found the former president and former plant manager guilty of negligence. However, years of legal battles for better compensation followed. Fishing had stopped and nets were spread to stop contaminated fishes from migrating out. Also, the sea floor was dredged. By 1997, the level of mercury had fallen to a level where fishing could resume.

Cadmium

Cadmium is released into the environment through the:
- production of zinc, lead and copper from their ores;
- improper disposal of nickel/cadmium rechargeable batteries;
- recycling of iron coated with cadmium as protection against corrosion;
- burning of fossil fuels, and household and industrial waste.

Cadmium enters the body in the following ways:
- Through contaminated food and water
- By inhaling contaminated air in or near workplaces that deal with cadmium

Cadmium is extremely toxic to all forms of life. It contaminates soil and enters plants. Fish and other animals become contaminated. Cadmium poisoning causes 'itai-itai' disease, which was first described in Japan in 1912, when cadmium was released into rivers when mining for lead and zinc. The symptoms include severe pain in the joints and spine, weak deformed or brittle bones, kidney failure and death.

Arsenic

Arsenic can enter the environment in the following ways:

- In groundwater from shallow wells
- From mining and smelting copper and zinc
- In old wood preserved with arsenic
- In certain agricultural insecticides and feed additives

Arsenic can enter the body by:

- drinking contaminated well water;
- eating contaminated food, for example, rice and leafy vegetables;
- breathing in smelter dust.

Organic compounds of arsenic are not very poisonous. Inorganic arsenic compounds are very powerful poisons that interfere with the process of cell respiration. In the past, inorganic arsenic compounds have been used to poison people, but they also have some medical uses, for example, treating one form of leukemia. They are also used as a poison to preserve wood, in agricultural insecticides and as a feed additive. Arsenic is used in some types of semiconductors in electronic devices.

Problem of solid waste disposal

Most of the solid waste produced in homes and factories is sent to a landfill. A sanitary landfill is made by first removing topsoil to create a pit. Then the area is lined with clay to prevent liquid contaminants before leaching into the surrounding soil and groundwater. Each day, the waste collection vehicles dump their load, which is spread out and compacted by bulldozers to reduce its volume. At the end of the day, the waste is covered with a layer of soil so that it is less accessible to pests and vermin, for example, rats, roaches and flies. When the pit is full, it is covered with a layer of soil and may be converted into a recreational area.

Figure 23.11
A landfill site ▶

A huge amount of waste is sent to landfill sites. The large volume of waste causes landfills to fill up quickly. Using up more land for new landfill sites reduces the amount of land available for agriculture and housing in small island states. A lot of the material found in this solid waste can be recycled, for example, paper and cardboard, plastic, glass, and metal. It is important for small countries to consider recycling solid waste. We have become accustomed to throwing away many things that we could use for other purposes, for example, using an empty paint can or bucket as a plant pot. We have also become used to buying in small individual packages of food, which increases the amount of packaging that we throw away. We must begin to think in terms of reducing the amount of waste we create. One way of doing this is to buy goods in larger quantities and separate the goods into small reusable containers, for example, buying a gallon of cooking oil and pouring it into a small bottle for everyday use.

We can also separate metals from the waste and recycle them. There are a lot of iron cans that contained food and beverages. Iron is the easiest metal to separate from waste materials, because it is magnetic. We can even separate iron out after the waste has been incinerated. Recycling iron products saves energy, because less new iron has to be extracted from its ore, which is a very energy intensive process. It takes about 75% less energy to make iron from recycled iron products than it does from its ore.

Aluminium beverage cans are also easily recycled. This also saves energy. Making aluminium from recycled aluminium products takes about 5% of the energy needed to make aluminium from its ore.

Figure 23.12
Recycling aluminium
cans ▶

Other metals, such as copper and zinc, can also be recycled from waste. The presence of many types of alloys today is making recycling metals more difficult.

Electronic waste

With the rapid advances in electronic technology, there is a corresponding increase in electronic waste. Some electronic waste is difficult to recycle, for example, computer screens. Recycling electronic waste exposes workers to different heavy metals such as cadmium. Much electronic waste is exported to developing countries, for example, to Ghana.

Figure 23.13
Electronic waste ▶

Summary

- Corrosion is a process in which metals react with substances in their environment, usually oxygen.
- Iron rusts in the presence of both water and oxygen.
- Rust is hydrated iron oxide, $Fe_2O_3.xH_2O$.
- Aluminium forms an oxide coating that protects the metal from further corrosion.
- Aluminium can undergo galvanic corrosion if it is attached to another metal that is lower in the metal reactivity series.
- Metals important to living things include sodium, potassium, calcium, magnesium and iron.
- Trace metals required by plants include iron, manganese, zinc, boron, copper, molybdenum and cobalt.
- Trace metals required by humans include cobalt, zinc, copper, manganese and molybdenum.
- Metals that are toxic to living systems include lead, mercury, cadmium and arsenic.
- Most metals can be recycled, which would reduce the amount of solid waste that needs to be disposed of.

End-of-chapter questions

1 Why do iron window frames on a house close to the sea rust more quickly than ones further inland?
   A Sea spray corrodes all metals.
   B The sea air is more moist than air further inland.
   C Sea spray contains salt dissolved in water, both of which increase the rate of rusting.
   D The cooler nights near the sea increase the rate of rusting.

2 Young people who avoid eating all green, leafy vegetables are at risk of suffering from a deficiency in:
   A Iron
   B Magnesium
   C Potassium
   D Calcium

3 Persons who do not eat any meat may suffer from a lack of which element that is found in the blood of all animals?

A Iron

B Magnesium

C Potassium

D Calcium

4 Metals that can cause harmful effects on humans can be found in:

I   Car batteries

II Mercury thermometers

III Paints

IV Plastics

A I only

B I and II only

C I, II and III

D I, II, III and IV

5 In 1993, a child living in an unplanned settlement in East Trinidad was having stomach aches, had a decreased appetite and was losing weight. He was admitted to a hospital when he began having seizures. The dirt road in the area where he lived was filled with rubble that included discarded car batteries. What could he have been poisoned by?

A Mercury

B Lead

C Cadmium

D Arsenic

6 Mr. Joseph is building a home near the sea. The builder has suggested that he use aluminium door frames and aluminium louvres for his windows instead of iron ones.

a What is likely to happen to the iron ones if he uses them?

b What is the chemical composition of the substance that will form on the iron?

c Why would the same thing not happen to the aluminium door frames and louvres?

7 Mary Ramcharan recently began having her menstrual cycles (which is when blood passes out from the vagina once a month). The doctor prescribed iron tablets for her to take daily.

a What is the name of the substance that gives blood its red colour?

b What is the function of this substance in the body?

8 More than 2 000 years ago, Chinese farmers discovered that crop yields improved when they treated the soil with 'calcined' bones. (The term 'calcined' means heated strongly, leaving an ash.)

a Suggest a metal that you expect to be in bones that is a plant nutrient.

b Suggest a chemical compound that might be in the bones, and which contains that element.

c Write an equation to show the effect of heat on that compound.

9 Most residents of the Caribbean islands enjoy eating fish. The Environment Protection Agency of a certain country is advising pregnant women and young children to limit the amount of large predatory fish that they eat such as shark, swordfish, king mackerel and tuna, because of the possibility of heavy metal poisoning.

a Suggest the name of a heavy metal that may be found in high levels in these fish.

b Suggest why only these fish are mentioned and not all the fish that live in the same sea.

10 In your country, the landfill site is filling up. The government is proposing to open a new landfill site by clearing a forested area.

a Suggest two environmentally friendly alternatives for solid waste treatment.

b Suggest one other way individual householders can reduce the amount of waste that they send to landfill sites.

Multiple-choice questions for Chapters 16–23

1 All of the following structures show the same molecule, except one. Which structure is different?

A

H—C—C—C—C—H (with H atoms above and below each carbon)

B

(branched structure)

C $CH_3CH_2CH_2CH_3$

D

$H_3C—C—CH_3$ (with CH_3 above and H below central C)

2 Which one of the following hydrocarbons will decolorise acidified potassium manganate(VII) solution?

A CH_4
B C_3H_6
C C_5H_{12}
D C_4H_{10}

3 Which pair of terms correctly describes the reaction between ethene (C_2H_4) and water?

I   Hydrogenation
II Hydration
III Addition
IV Substitution

A I and III
B II and IV
C II and III
D I and IV

The information below relates to questions 4 and 5.

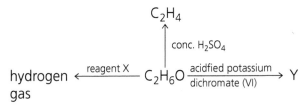

4 Reagent X is most likely to be:
A Water
B An acid
C A metal
D A base

5 Which property is compound Y likely to have?
A Turns blue litmus red
B Relights a glowing splint
C Decolorises potassium manganate(VII)
D Is insoluble in water

6 A substance with a pleasant smell that undergoes hydrolysis to form an alcohol as one of the products is likely to be:
A Ethanoic acid
B Butyl ethanoate
C Ethanol
D Sodium ethanoate

7 What is the correct name for the substance with the formula following?

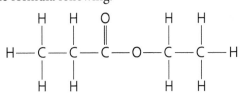

A Propyl ethanoate
B Ethyl propanoate
C Pentanoic acid
D Butanoic acid

8 The formula of ethanoic acid ($C_2H_4O_2$) shows four hydrogen atoms, but it is a monobasic acid. This means that it:
A Is a weak acid
B Has only one –OH group
C Reacts with only one base
D Forms only one salt with sodium

9 The monomer ethene and its polymer polythene have the same:
- **A** Physical properties
- **B** Functional group
- **C** Empirical formula
- **D** Structural formula

10 Which pair of polymers contains peptide linkages?
- **A** Polythene and proteins
- **B** Polypropene and starch
- **C** Nylon and proteins
- **D** Terylene and cellulose

11 Which of the following is most likely to form polymers by addition?

A

$$Cl-\overset{\overset{\displaystyle H}{|}}{\underset{\underset{\displaystyle H}{|}}{C}}-\overset{\overset{\displaystyle H}{|}}{\underset{\underset{\displaystyle H}{}}{C}}=\overset{\displaystyle H}{\underset{\displaystyle H}{C}}$$

B

$$H-O-\overset{\overset{\displaystyle H}{|}}{\underset{\underset{\displaystyle H}{|}}{C}}-\overset{\overset{\displaystyle H}{|}}{\underset{\underset{\displaystyle H}{|}}{C}}-\overset{\overset{\displaystyle H}{|}}{\underset{\underset{\displaystyle H}{|}}{C}}-O-H$$

C

$$\overset{\displaystyle H}{\underset{\displaystyle H}{N}}-\overset{\overset{\displaystyle H}{|}}{\underset{\underset{\displaystyle H}{|}}{C}}-\overset{\overset{\displaystyle H}{|}}{\underset{\underset{\displaystyle H}{|}}{C}}-\overset{\overset{\displaystyle O}{\|}}{C}\overset{O-H}{}$$

D

$$N-\overset{\overset{\displaystyle H}{|}}{\underset{\underset{\displaystyle H}{|}}{C}}-\overset{\overset{\displaystyle H}{|}}{\underset{\underset{\displaystyle H}{|}}{C}}-\overset{\overset{\displaystyle H}{|}}{\underset{\underset{\displaystyle H}{|}}{C}}-\overset{\displaystyle H}{\underset{\displaystyle H}{N}}$$

12 An alkene has the following structural formula:

$$H-\overset{\overset{\displaystyle H}{|}}{\underset{\underset{\displaystyle H}{|}}{C}}-\overset{\displaystyle H}{\underset{}{C}}=\overset{\displaystyle H}{\underset{}{C}}-\overset{\overset{\displaystyle H}{|}}{\underset{\underset{\displaystyle H}{|}}{C}}-H$$

How many isomers does the compound have?
- **A** none
- **B** 1
- **C** 2
- **D** 3

13 Which of the following is true for **ALL** members of a homologous series?
- I They have the same molecular formula.
- II They have identical physical properties.
- III They have the same functional group.
- IV They have similar chemical properties.
- **A** I only
- **B** I and III only
- **C** II and III only
- **D** III and IV only

14 Which of the following is true of the anaerobic fermentation of sugars?
- I Oxygen must be excluded.
- II An enzyme is needed.
- III An alcohol is formed.
- IV Each molecule of a simple sugar loses one molecule of carbon dioxide.
- **A** I only
- **B** I and II only
- **C** I, II and III only
- **D** I, II, III and IV

15 Consider an alcohol and an alkene containing the same number of carbon atoms. Which one of the following statements about these two compounds will be true?
- **A** The alcohol will have a lower molar mass than the alkene.
- **B** The alkene will be less volatile than the alcohol.
- **C** The alcohol will be more soluble in water than the alkene.
- **D** The alcohol will have a lower boiling point than the alkene.

16 Which equation best describes a cracking reaction?
- **A** $C_8H_{18} \rightarrow 2C_4H_8 + H_2$
- **B** $C_6H_{12} + H_2 \rightarrow C_6H_{14}$
- **C** $C_8H_{18} \rightarrow C_8H_{16} + H_2$
- **D** $C_4H_8 + C_2H_4 \rightarrow C_6H_{12}$

17 An organic compound, X, reacts with chlorine gas to give two products. Which of these statements is true?
- **A** X is an alkene and the reaction is a substitution reaction
- **B** X is an alkane and the reaction is an addition reaction
- **C** One product is a halogenoalkane and the other product is hydrogen gas.

D One product is hydrogen chloride and sunlight is needed for the reaction.

18 Which equation does **NOT** describe a characteristic reaction of ethanoic acid?

A $2CH_3COOH + 2Na \rightarrow 2CH_3COONa + H_2$

B $2CH_3COOH + Cu \rightarrow (CH_3COO)_2Cu + H_2$

C $2CH_3COOH + CuCO_3$
$\rightarrow (CH_3COO)_2Cu + H_2O + CO_2$

D $CH_3COOH + C_2H_5OH$
$\rightleftharpoons CH_3COOC_2H_5 + H_2O$

19 Which of the following statements is/are true about the fractions obtained from the fractional distillation of petroleum?

I   They may be used as fuels.

II   Each fraction is a pure alkane.

III   They contain both alkanes and alkenes.

IV   They are mixtures of alkanes.

A I only

B I and II only

C II and III only

D I and IV only

20 Which one of the following processes is similar to the chemical changes which take place in digestion?

A Conversion of amino acids to proteins

B Hydrolysis of starch

C Hydration of ethene

D Fermentation

24 Non-metals and their compounds

In this chapter, you will study the following:

- the solids carbon, sulphur and, to a lesser extent, silicon and phosphorus;
- the gases chlorine, nitrogen, hydrogen and oxygen.
- the natural occurrence of the elements;
- the relationship between their physical properties and their uses;
- their typical chemical reactions;
- the nature of the compounds formed when they react;
- the relative reactivity of the non-metals.

This chapter covers
Objective 5.1 of Section C of the CSEC Chemistry Syllabus.

In Chapter 4, we compiled a list of physical features by which we could identify an element as a non-metal or a metal. For example, a hard, shiny, heavy solid is more likely to be a metal than a non-metal and a substance that does not conduct electricity is more likely to be a non-metal than a metal.

But are the physical differences always that distinct? You can already answer this question from your study of the metals in Group II and the non-metals in Group VII of the Periodic Table (see Chapter 6). Calcium, for example, does not have the typical appearance of a metal, as it is soft, quite light and not very shiny. By contrast, the non-metal carbon could exist either as diamond, a hard shiny solid, or as graphite, which though dull in appearance conducts electricity. You may recall, too, that the non-metal iodine in Group VII is a shiny solid.

It seems, therefore, that we must consider whether an element possesses most of the physical properties listed for a metal or non-metal before we can classify it as either. Furthermore, we also need to examine the chemical properties of the element before making our conclusions.

24.1   Non-metals

Non-metals are elements that are not metals. They are not ionised by loose electrons. Therefore, non-metals do not have the same characteristics that metals have. Non-metals are located on the right-hand side of the diagonal in the Periodic Table. Non-metals are usually solids or gases.

H hydgen																	He helium
Li lithium	Be beryllium											B boron	C carbon	N nitrogen	O oxygen	F flourine	Ne neon
Na sodium	Mg magnesium											Al aluminium	Si silicon	P phosphorous	S sulfur	Cl chlorine	Ar argon
K potassium	Ca calcium	Sc scandium	Ti titanium	V vanadium	Cr chromium	Mn manganese	Fe iron	Co cobalt	Ni nickel	Cu copper	Zn zinc	Ga gallium	Ge germanium	As arsenic	Se selenium	Br bromine	Kr krypton
Rb rubidium	Sr strontium	Y yttrium	Zr zirconium	Nb niobium	Mo molybdenium	Tc technetium	Ru ruthenium	Rh rhodium	Pd palladium	Ag silver	Cd cadmium	In indium	Sn tin	Sb antimony	Te tellurium	I iodine	Xe xenon
Cs caesium	Ba barium	La lanthanium	Hf hafnium	Ta tantalum	W tungsten	Re rhenium	Os osmium	Ir iridium	Pt platinum	Au gold	Hg mercury	Tl thallium	Pb lead	Bi bismuth	Po polonium	At astatine	Rn radon
Fr francium	Ra radium	Ac actinium	Rf rutherfordium	Db dubnium	Sg seaborgium	Bh bohrium	Hs hassium	Mt meitnerium	Ds darmstadtium	Rg roentgenium							

▲ **Figure 24.1** The position of metals and non-metals in the Periodic Table

Some of the non-metal gases exist as diatomic molecules. The diatomic molecules are bromine, iodine, fluorine, chlorine, hydrogen, oxygen and nitrogen.

Properties of metals and non-metals

Table 24.1 highlights the properties of metals versus non-metals.

▼ **Table 24.1**   Properties of metals and non-metals

Metals	Non-metals
high melting and boiling points (except alkali metals)	low melting and boiling points (except carbon as diamond or graphite)
solids (except mercury)	gases and solids (except bromine)
lustrous (when freshly cut)	non-lustrous
sonorous (clang when hit)	not sonorous
malleable (can be made into thin sheets) and ductile (can be made into thin wires)	brittle
good conductors of heat and electricity	poor conductors, except graphite, which conducts electricity well
form positive ions	usually form negative ions
reducing agents	oxidising agents
form basic oxides (ZnO, Al_2O_3 and PbO are amphoteric)	form acidic oxides (CO, H_2O, NO are neutral)
low electronegativity	high electronegativity

24.2 The uniqueness of hydrogen and silicon

Hydrogen is a difficult element to classify.

- Hydrogen atoms form positive ions that migrate towards the cathode during electrolysis (see Section 13.2). This property would suggest that hydrogen is like a metal.

- Under standard conditions, hydrogen is a gas containing discrete diatomic molecules. Like the halogens, hydrogen forms anions, H^- ions, in combination with the more reactive metals such as sodium and magnesium. These properties would suggest that hydrogen is like a non-metal.

Because hydrogen can behave both like a metal and like a non-metal, it is usually placed on its own at the top of the Periodic Table.

Silicon can also be considered to have both metallic and non-metallic properties:

- It has some metallic properties, for example, it conducts electricity under some conditions.

- It has some non-metallic properties, for example, it has an acidic oxide.

- It can be classified as a metalloid, as its position in the Periodic Table would suggest (see Chapter 3).

Figure 24.2
Silicon ▶

▼ **Table 24.2**   The natural occurrence of some solid non-metals

Element	How it occurs free	How it occurs combined
Carbon	Occurs in different forms (allotropes), e.g. diamond and graphite.	Combined with oxygen as carbon dioxide in the air. Combined with hydrogen as hydrocarbons
Sulphur	Deposits of the element often found near petroleum deposits, e.g. in Texas. Sulphur is found in sulphur springs in St. Lucia.	Combined with metals as sulphides containing the S^{2-} anion. Examples are: zinc blende – ZnS Gaseous compounds of sulphur, such as SO_2 and H_2S, are usually present in polluted air.
Silicon	Does not occur uncombined.	Silicon is mainly found combined with oxygen as silica, SiO_2 (for example in sand), and in a variety of silicates (for example in clays)

▼ **Table 24.3** The natural occurrence of some gaseous non-metals

Element	How it occurs free	How it occurs combined
Chlorine	Does not occur uncombined.	Occurs combined in a number of chlorides, of which the most important is sodium chloride, rock salt. Salt ponds occur in Anguilla.
Hydrogen	Does not occur uncombined.	Found combined in many organic compounds and as the inorganic compound water.
Nitrogen	The major constituentof the air (approx. 78%). Apart from the noble gases (Group 0), nitrogen is the least reactive of the gaseous non-metals.	Oxides of nitrogen are found in polluted air. These arise from the oxidation of nitrogen, under high temperature conditions, in internal combustion engines of motor vehicles.
Oxygen	Oxygen occupies approximately 21% of air.	Occurs combined with hydrogen in water. Occurs with metals in a number of oxide ores. Occurs as oxides of the non-metal silicon (sand is silicon dioxide) and in carbonates, for example limestone, in soil.

24.3 Occurrence of the elements

Some elements are sometimes found uncombined in nature. More often, though, elements are found in combination with other elements. Highly reactive elements are more likely to occur in compounds than as the free elements. Tables 24.2 and 24.3 summarise the natural occurrence of selected non-metals.

Find out more

Some impure forms of carbon are charcoal, coal and coke.

- Charcoal, commonly called 'coals' in the Caribbean, is formed by heating wood or animal bones in a limited supply of air/oxygen. Charcoal is used as a fuel.
- Coal is mainly carbon and is formed by action of heat and pressure on plants over millions of years. Coal is used as a fuel.
- Coke, formed by heating coal in the absence of air, is used to reduce iron oxide to iron at high temperatures in the blast furnace.

24.4 How can non-metals be identified?

You have already encountered a number of physical properties which help us to identify non-metals. You will remember that these properties are closely associated with the structure of the elements and their type of bonding (see Chapter 4).

Here we will recap some of this information:

- Most non-metallic elements are covalently bonded molecules, the majority of which are simple and discrete. Note that the element carbon is different in this respect.

- Non-metal elements usually show some combination of the following physical properties:
  - They have low melting points and boiling points.
  - They are usually non-conductors of electricity.
  - They are not shiny.
  - They are usually softer than metals.
  - They generally have low densities.

Practice

1 For each of the physical properties listed above, give at least two examples of a non-metallic element that shows the property, and at least one example of a non-metallic element that is an exception. You may use the same element more than once in your answer.

Carbon and sulphur: allotropy, structure and bonding

You already know that diamond and graphite are allotropes of carbon. Sulphur also exhibits allotropy. The crystalline allotropes of sulphur are:
- rhombic sulphur (also known as alpha sulphur);
- monoclinic sulphur (also known as beta sulphur).

Figures 24.3 and 24.4 show the shapes of the two types of crystals.

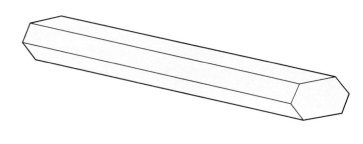

▲ **Figure 24.3** Rhombic sulphur   **Figure 24.4** Monoclinic sulphur ▲

(a)  (b)

Figure 24.5
The crystalline allotropes of sulphur: (a) rhombic sulphur and (b) monoclinic sulphur ▶

You can produce these two crystal forms in the laboratory.

Experiment 24.1   Making the two crystal forms of sulphur

Procedure

1 Place some crystals of rhombic sulphur (alpha sulphur) in an evaporating dish.

2 Heat the dish over a Bunsen burner until it melts to a pale gold, runny liquid.

3 Pour approximately half of the molten sulphur into an empty evaporating dish. Place the dish in an ice bath.

4 Place the rest of the liquid in an evaporating dish on the laboratory bench to cool naturally.

5 Examine the underside of the crust that forms in contact with the molten sulphur for the formation of crystals. Look at the crystals closely with a magnifying glass. Draw the crystals.

Questions

Q1 Describe how the crystals formed in steps 3 and 4 differ.

Q2 Which crystals are rhombic sulphur (alpha sulphur) and which are monoclinic sulphur (beta sulphur)?

Sulphur also exists in a non-crystalline (amorphous) form as a pale yellow powder. This is easily obtained in the laboratory by adding dilute acid to sodium thiosulphate:

$$Na_2S_2O_3(aq) + 2HCl(aq) \rightarrow 2NaCl(aq) + H_2O(l) + S(s) + SO_2(g)$$

Plastic sulphur is another allotrope that can be prepared in the laboratory.

Find out more

Plastic sulphur

A soft brown solid is formed when liquid sulphur, close to its boiling point, is poured into water. Plastic sulphur is elastic because it contains long chains of sulphur atoms. The chains are formed from S_8 rings, which were split at high temperatures and then joined together. The allotrope is unstable and can revert to crystals of rhombic sulphur.

The shape of the S_8 rings found in sulphur are shown in Figure 24.6.

You should note that there is a difference in the type of allotropy shown by carbon and sulphur.

- Diamond and graphite can co-exist under the same conditions of temperature and pressure.

- By contrast, rhombic sulphur is stable at temperatures below 96 °C and monoclinic sulphur must be formed above 96 °C. This temperature is known as the transition temperature of the allotropes.

Differences in the physical properties of carbon and sulphur are related to the bonding and structure in these elements. Table 24.4 compares some of the physical properties of carbon and sulphur.

▼ **Table 24.4** Some important physical properties of the allotropes of carbon and sulphur.

Physical property	Allotropes of carbon		Allotropes of sulphur	
	Diamond	Graphite	Rhombic sulphur (alpha)	Monoclinic sulphur (beta)
physical state at room temperature	solid	solid	solid	unstable solid
melting point (°C)	3 730	3 570	119	112
boiling point (°C)	4 830	4 200	444	444
electronegativity	2.5	2.5	2.5	2.5
electrical conductivity	non-conductor	conductor	non-conductor	non-conductor
density (g cm⁻³)	3.51	2.22	2.06	1.96
hardness	hardest natural material	soft and flaky	soft and crystalline	soft and crystalline
lustre	+	−	+	+

Diamond and graphite are giant molecular structures. Sulphur on the other hand, has a simple molecular structure in which eight sulphur atoms join in a ring and then adopt a chair-like shape (Figure 24.6).

Figure 24.6
The chair-like structure of sulphur (S_8). These sulphur molecules are stacked differently to give the two types of crystals. ▶

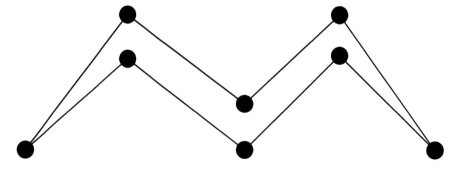

It is how these S_8 molecules pack in the crystals that give either an octahedral shape (rhombic sulphur) or a needle-like shape (monoclinic sulphur). Although there is strong covalent bonding within the S_8 molecule, the intermolecular van der Waals forces are weak. As a result, sulphur is softer and has much lower melting and boiling points than diamond and graphite.

Note that phosphorus also has allotropes: white, red and black phosphorus. White phosphorus is the most reactive and black phosphorus is the least reactive. Molecules of white phosphorus contain four phosphorus atoms arranged tetrahedrally.

Combustion is the combination of a substance with oxygen, releasing energy as heat and often as light.

Practice

2 Equal masses of pure rhombic sulphur and pure monoclinic sulphur are burnt in an excess of oxygen to form sulphur trioxide gas.
  a Will the volume of gas produced at RTP be the same for both allotropes?
  b Give a reason for your answer to part (a).

24.5 Physical properties of the gaseous non-metals

Chlorine, nitrogen, hydrogen and oxygen are all gases. They are all colourless and odourless, with the exception of chlorine – which is a green-yellow gas with a choking smell. Because they are gases, they all have low densities, hydrogen having the lowest and chlorine the highest. Their melting and boiling points are lower than those of the solid non-metals since intermolecular forces are much weaker between the smaller diatomic molecules of these non-metals. Table 24.5 summarises the physical properties of these gaseous non-metals.

▼ **Table 24.5** Physical properties of the gaseous non-metals

Element	Colour	Odour	Density	Melting point (°C)	Boiling point (°C)
chlorine	green-yellow	sharp	3.21 g dm⁻³	−101	−34
nitrogen	colourless	odourless	1.25 g dm⁻³	−210	−196
hydrogen	colourless	odourless	0.09 g dm⁻³	−259	−253
oxygen	colourless	odourless	1.43 g dm⁻³	−219	−183

Structure and bonding

All of these gases (chlorine, nitrogen, hydrogen and oxygen) exist as discrete diatomic molecules. When writing equations involving these gases, they should be written as diatomic molecules rather than as single atoms (see the equations in Table 24.6).

24.6 Chemical reactions of non-metals
1 Reaction with oxygen

Non-metals react with oxygen to form covalently bonded oxides that show mainly acidic properties. A few non-metal oxides are neutral.

Carbon and sulphur

Carbon reacts on heating in a limited supply of oxygen to form carbon monoxide. If oxygen gas is plentiful, carbon dioxide is formed. The reaction is described as a combustion reaction.

When sulphur burns in oxygen, the gaseous oxide sulphur dioxide (sulphur(IV) oxide) is formed. Sulphur dioxide has a characteristic 'choking smell'. If you live near a sulphur spring or a volcano you will know what sulphur dioxide smells like.

When a mixture of dry sulphur dioxide and oxygen is passed over a heated catalyst, such as vanadium(V) oxide, a reversible reaction occurs and sulphur trioxide (sulphur(VI) oxide) is formed. Both gases must be dry, since sulphur trioxide readily dissolves in water to form sulphuric acid. Only small quantities of sulphur(VI) oxide are formed because the reaction is reversible and the equilibrium position is to the left (see Section 7.2).

The reactions of oxygen with sulphur and with carbon are summarised in Table 24.6.

▼ **Table 24.6** The reactions of carbon and sulphur with oxygen

Element	Reaction with oxygen	Equation	Description of the oxide
Carbon	Heated in a limited supply of oxygen	$2C(s) + O_2(g) \rightarrow 2CO$ carbon monoxide	A neutral oxide; toxic; combines irreversibly with haemoglobin, reducing the ability of the latter to pick up oxygen.
	Heated in excess oxygen	$C(s) + O_2(g) \rightarrow CO_2(g)$ carbon dioxide	An acidic oxide; dissolves in water forming carbonic acid. A by-product of fermentation.
Sulphur	Heated in air	$S(s) + O_2(g) \rightarrow SO_2(g)$ sulphur dioxide	An acidic oxide with a choking smell. Functions as both an oxidising agent and a reducing agent. Dissolves in water to form sulphurous acid.
		$2SO_2(g) + O_2(g \rightleftharpoons 2SO_3(g)$ sulphur trioxide	The acid anhydride of sulphuric acid.

The gaseous non-metals

The reactions of the gases chlorine, nitrogen and hydrogen with oxygen, and the nature of the oxides formed are summarised in Table 24.7.

▼ **Table 24.7** The reactions of the gaseous non-metals with oxygen

Element	Reaction with oxygen	Equation	Description of the oxide
Chlorine	When chlorine reacts with oxygen, the oxides formed are highly unstable (explosive)	$2Cl_2(g) + O_2(g) \rightarrow 2Cl_2O(g)$ $Cl_2(g) + 2O_2(g) \rightarrow 2ClO_2(g)$	• Cl_2O is an orange gas. • ClO_2 is a yellow gas. • Cl_2O_6 is a red liquid. • Cl_2O_7 is a colourless liquid. All of the oxides are acidic and react with alkalis to form salts.
Nitrogen	In lightning storms	$N_2(g) + O_2(g) \rightarrow 2NO(g)$ nitrogen(II) oxide	A neutral oxide; colourless gas that combines instantly with the oxygen of the air to form the brown gas nitrogen dioxide.
		$2NO(g) + O_2(g) \rightarrow 2NO_2(g)$ nitrogen(IV) oxide	A mixed acid anhydride; reacts with water to form nitrous acid and nitric acid.
Hydrogen	When ignited	$2H_2(g) + O_2(g) \rightarrow 2H_2O(l)$ water	A neutral oxide; colourless liquid.

2 Reaction with metals

Carbon and sulphur

Carbon does not normally react with metals.

Sulphur, however, reacts on heating to produce metallic sulphides, which are ionic compounds:

$$Mg(s) + S(s) \rightarrow MgS(s)$$
magnesium sulphide

The gaseous non-metals

The reactions of chlorine, nitrogen, hydrogen and oxygen with metals, and the nature of the products formed are summarised in Table 24.8.

▼ **Table 24.8** The reactions between metals and the gaseous non-metals

Element	Example of reaction with metal	Equation	Description of product
hydrogen	alkali metals	$2Na(s) + H_2(g) \rightarrow 2NaH(s)$ sodium hydride	ionic compound containing Na^+ and H^- ions
oxygen	reacts with all metals	$2Ca(s) + O_2(g) \rightarrow 2CaO(s)$ calcium oxide	ionic compound containing Ca^{2+} and O^{2-} ions
nitrogen	reacts with lit magnesium	$3Mg(s) + N_2(g) \rightarrow Mg_3N_2(s)$ magnesium nitride	ionic compound containing Mg^{2+} and N^{3-} ions
chlorine	direct combination with most metals	$2Fe(s) + 3Cl_2(g) \rightarrow 2FeCl_3(s)$ iron(III) chloride	ionic compound with the metal ion in its highest oxidation state (for transition metals)

3   Oxidising and reducing properties

Carbon and sulphur

See Section 12.7 for more information about oxidation states.

Carbon and sulphur react with oxygen to form oxides in which the element exists in a higher oxidation state. These reactions show that carbon and sulphur have reducing properties. These equations show how the oxidation numbers change:

$$2C^0 + O_2^0 \rightarrow 2C^{+2}O^{-2}$$
$$C^0 + O_2^0 \rightarrow C^{+4}O_2^{-2}$$

The reducing properties of carbon and sulphur are also demonstrated in their reactions with the oxidising acids, concentrated sulphuric acid and concentrated nitric acid:

$$S(s) + 2H_2SO_4(aq) \rightarrow 3SO_2(g) + 2H_2O(l)$$
$$3S(s) + 4HNO_3(aq) \rightarrow 3SO_2(g) + 4NO(g) + 2H_2O(l)$$
$$C(s) + 2H_2SO_4(aq) \rightarrow CO_2(g) + 2SO_2(g) + 2H_2O(l)$$
$$C(s) + 4HNO_3(aq) \rightarrow CO_2(g) + 4NO_2(g) + 2H_2O(l)$$

Carbon is also used to reduce some metal oxides (ores) to the corresponding metal:

$$ZnO(s) + C(s) \rightarrow Zn(s) + CO(g)$$

Non-metals that show reducing properties usually have lower electronegativities than those which show oxidising properties.

Understand it better

Why do sulphur and carbon show reducing properties?
Typically, non-metals are electronegative elements and they tend to accept electrons from metals to form anions. When this happens, the non-metals behave as oxidising agents (oxidising agents accept electrons).

When non-metals react with each other by sharing electrons, the more electronegative non-metal behaves as the oxidising agent, and the less electronegative non-metal is forced to behave as the reducing agent. When non-metals react with more powerful oxidising agents, such as concentrated nitric and sulphuric acids, they show reducing properties.

Q1   How are the redox properties of non-metals different from those of metals?

Practice

3 a Indicate which non-metal is the oxidising agent and which is the reducing agent in each of the following equations:

   (i) $C(s) + 2S(s) \rightarrow CS_2(l)$
   (ii) $H_2(g) + S(s) \rightarrow H_2S(g)$
   (iii) $H_2(g) + Cl_2(g) \rightarrow 2HCl(g)$
   (iv) $N_2(g) + 3H_2(g) \rightleftharpoons 2NH_3$

   b To what extent do your answers agree with the relative electronegativity values shown in Table 24.9?

24.7 Relative reactivity of the non-metals

The reactivity of non-metals is linked to their electronegativity (see Section 3.3). The greater the **electronegativity** of the non-metal the more reactive it is. Electronegativity depends on the size of the atom (its atomic radius) and the nuclear charge. Small atoms with a high nuclear charge are highly electronegative.

▼ **Table 24.9** Atomic radii and electronegativity values for some non-metals.

Non-metal	Atomic radius (nm)	Electronegativity value
carbon	0.077	2.5
sulphur	0.102	2.5
hydrogen	0.037	2.1
oxygen	0.073	3.5
nitrogen	0.075	3.0
chlorine	0.099	3.0

Practice

4 Using their chemical symbols, arrange the elements in Table 24.9 in order:
   a of decreasing atomic radius;
   b of increasing electronegativity.

5 Comment on the order of the elements in the two lists in question 4.

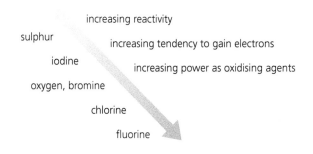

The reactivity series of non-metals arranges selected non-metals on the basis of the ease with which they accept electrons, that is, on their oxidising power. The most powerful oxidant (fluorine) is placed at the bottom of the list.

▲ **Figure 24.7** The reactivity series of the non-metals.

Practice

6 Use the electronegativity values of 4 for fluorine and 2.5 for iodine, together with the relevant values from Table 24.9, to place hydrogen, oxygen and carbon into their relative positions in the reactivity series for non-metals.

The order of the non-metals in the reactivity series can be determined experimentally. For the halogens, for example, the order of their reactivity can be determined by the displacement experiments discussed in Section 6.2. In these experiments, the more reactive halogen displaced the less reactive halogen from a solution of its halide ions.

The results of these and other displacement reactions of non-metals are shown in Table 24.10.

▼ **Table 24.10** Reactions of some non-metals with solutions of non-metal compounds.

Non-metals	Solutions of non-metal compounds			
	Sodium chloride	Sodium bromide	Sodium iodide	Sodium sulphide
chlorine (Cl_2/H_2O)		+	+	+
bromine (Br_2/H_2O)	–		+	+
iodine in KI(aq)	–	–	–	+
oxygen from H_2O_2	–	–	+	+
sulphur	–	–	–	

+ indicates that a displacement reaction occurred. – indicates that no reaction occurred.

From the data in Table 24.10, it can be seen that chlorine is the most reactive of the non-metals studied here, whereas sulphur is the least reactive.

24.8 Uses of the non-metals

It is best to use elements in everyday life in such a way that there is a good correlation between their special properties and their intended functions. Tables 24.11 and 24.12 list the uses of the non-metals and link these uses to their properties.

▼ **Table 24.11** Uses of the solid non-metals, linked to their properties

Solid non-metal	Uses	Properties to which uses are linked
carbon, as graphite	lubricant, also used as the 'lead' in lead pencils.	weak forces between the layers allow plates of graphite to slide past each other
carbon, as diamond	used for drill bits and as gemstones	strong three-dimensional covalent bonding makes diamond very hard; it reflects light
sulphur	production of chemicals, e.g. sulphuric acid, which is widely used in the manufacture of fertilisers and detergents	can be oxidised to sulphur dioxide, then to sulphur trioxide, which is the acid anhydride of sulphuric acid
	component of matches and gunpowder	burns easily and quickly
	sulphur drugs	antibacterial property
	treatment of pulp and paper	bleaching action
	vulcanising rubber	forms links between polymer chains

continued

▼ **Table 24.11** *continued*

Solid non-metal	Uses	Properties to which uses are linked
silicon	used in electronic devices, such as calculators, transistors and microcomputers	semi-conducting property
	silica, as sand, is used in the manufacture of glass; silicates in clay are used in the brick and cement industry	silica, SiO_2, is an acidic oxide
phosphorus	used to make flares and fireworks red phosphorus is used on the striking surface of safety matches	spontaneous inflammable nature
	phosphates are used in fertilisers and pesticides, cleaning agents and water softeners	an essential element for plant growth; particle binding properties

▼ **Table 24.12** Uses of the gaseous non-metals, linked to their properties

Gaseous non-metal	Uses	Properties linked to uses
chlorine	disinfection of water, bleaches cotton and paper manufacture of polychloroethene	chlorine is a powerful oxidant adds readily to alkenes
nitrogen	in handling of explosive mixtures, to anneal metals at high temperatures	the inertness of nitrogen prevents premature explosions
	flushing out of boilers and pipes during non-use periods	reduces the chances of corrosion
	protecting foods from spoilage	bacteria cannot survive in an atmosphere of nitrogen
hydrogen	in liquid form as a coolant and to freeze-dry some food stuffs	low freezing point
	manufacture of ammonia extraction of metals	combines chemically with nitrogen
	manufacture of methanol	has reducing properties
	hardening of oils	adds to double bonds in oils
	to make fuel cells	undergoes reaction with oxygen, which releases energy
	used in rocket fuel	very high energy density (kJ per g), product of combustion is water

continued

▼ **Table 24.12** *continued*

Gaseous non-metal	Uses	Properties linked to uses
oxygen	in the manufacture of steel	oxidises impurities in wrought iron to gaseous products
	to produce methanol	
	to weld and cut metals	produces a very hot flame when mixed with ethyne
	as a component of rocket fuel	releases energy on reaction
	in medicine, on aircraft and in deep-sea diving	oxygen is needed for proper breathing and respiration

▼ **Table 24.13** Properties and uses of non-metal compounds

Non-metal compounds	Uses	Properties
rubber	producing tyres	a high coefficient of friction with the ground that helps in acceleration, braking and cornering
		flexible and absorbs the impact of driving
		does not heat up and burn as easily, plus it is air and water tight
		Charles Goodyear accidentally discovered that by mixing sulfur and rubber, the rubber became tougher, more resistant to heat and cold, and increased in elasticity
insecticides	used to kill insects	the most prominent classes of insecticides are organochlorines, organophosphates, carbamates, and pyrethroids
		may affect non-pest insects, people, wildlife and pets
		most chemical insecticides act by poisoning the nervous system of insects
		the chemical structure of organochlorines is diverse, but they all contain chlorine, which places them in a larger class of compounds called chlorinated hydrocarbons organochlorines, which include DDT (dichlorodiphenyltrichloroethane), demonstrate many of the potential risks and benefits of insecticide use
		Organophosphates were initially developed in the 1940s as highly toxic biological warfare agents (nerve gases).
matches	used to make fires	the coated end is the head, which is ignited by frictional heat created by striking the match against a suitable surface
		the head contains phosphorus or phosphorus sesquisulphide (P_4S_3) as the active ingredient and gelatin as its binder

continued

▼ **Table 24.13**   *continued*

		the head contains phosphorus or phosphorus sesquisulphide (P4S3) as the active ingredient and gelatin as its binder
fertilisers	used for growing crops	improper or excessive use can lead to nitrate pollution of ground or surface water and eutrophication
		chemical fertilisers are nutrient specific; specific chemical fertilisers are chosen to supply specific nutrients, for example, superphosphate is used when soil is deficient in phosphorus
		the basic non-metal minerals supplied in fertilisers are nitrogen are phosphorus, which helps plants grow

▲ **Figure 24.8**   A truck tyre at a works in Kingston, Jamaica.

Summary

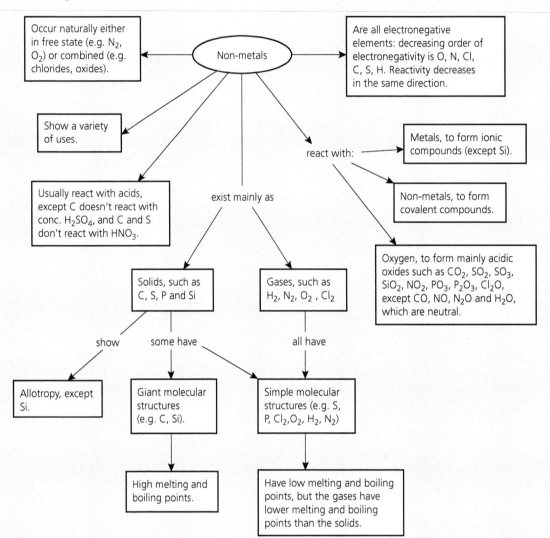

Occur naturally either in free state (e.g. N_2, O_2) or combined (e.g. chlorides, oxides).

Non-metals

Are all electronegative elements: decreasing order of electronegativity is O, N, Cl, C, S, H. Reactivity decreases in the same direction.

Show a variety of uses.

react with:

Metals, to form ionic compounds (except Si).

Usually react with acids, except C doesn't react with conc. H_2SO_4, and C and S don't react with HNO_3.

Non-metals, to form covalent compounds.

exist mainly as

Oxygen, to form mainly acidic oxides such as CO_2, SO_2, SO_3, SiO_2, NO_2, PO_3, P_2O_3, Cl_2O, except CO, NO, N_2O and H_2O, which are neutral.

Solids, such as C, S, P and Si

Gases, such as H_2, N_2, O_2, Cl_2

show

some have

all have

Allotropy, except Si.

Giant molecular structures (e.g. C, Si).

Simple molecular structures (e.g. S, P, Cl_2, O_2, H_2, N_2)

High melting and boiling points.

Have low melting and boiling points, but the gases have lower melting and boiling points than the solids.

End-of-chapter questions

1 Which one of the following does not occur as a diatomic molecule in its elemental form?
A Hydrogen
B Chlorine
C Oxygen
D Sulphur

2 When compared to metals, non-metals:
A Are better conductors of heat
B Are more sonorous
C Are more electronegative
D Have higher melting and boiling points

3 The following question are concerned with the elements A–D below. Each letter may be used once, more than once or not at all.
A Sodium
B Chlorine
C Carbon
D Nitrogen
a Which of the elements is a non-metal and a powerful oxidising agent?
b Which of the elements has a low melting point and conducts electricity?
c Which of the elements forms an acid oxide?
d Which of the elements forms a neutral oxide?

4 What happens to non-metals when they chemically combine with metal atoms?
A They gain electrons to become negative ions.
B They lose electrons to become positive ions.
C They remain electrically neutral.
D They share electrons to achieve a stable octet configuration.

5 How many non-metals are diatomic?
A 4
B 5
C 6
D 7

6 Which of the following statements is correct?
A Some non-metals are ductile.
B Non-metals are gases and solids (except bromine).
C All metals are ductile.
D Non-metals are poor conductors of heat and good conductors of electricity, respectively.

7 Which of the following non-metals is a green-yellow gas with a sharp odour?
A Chlorine
B Nitrogen
C Hydrogen
D Oxygen

8 Using carbon and sulphur as examples, discuss what is meant by the term allotropy.

9 Account for the differences in the melting points of:
a graphite and sulphur;
b sulphur and hydrogen.

10 Give evidence to justify the statement 'both carbon and sulphur are non-metals'.

11 Write equations to show how sulphuric acid reacts with:
a a metal of your choice;
b the non-metal carbon;
c a base of your choice.

12 Non-metallic oxides are usually acidic or neutral.
a Give two examples of non-metallic oxides that are acidic.
b Show by means of equations how each of the oxides from part (a) react with:
(i) water;
(ii) sodium hydroxide solution.

c Give one example of:

    (i) a neutral oxide that has solvent properties;

    (ii) a neutral oxide that can be formed by partial oxidation of a solid non-metal.

d Write an equation to show the reducing property of the oxide in part (c) (ii).

13 What would be the justification for placing hydrogen in (a) Group I or (b) Group VII?

14 Indicate the property of hydrogen on which the following commercial uses are based:

  a Filling balloons

  b The production of margarine from vegetable oils

  c The extraction of a metal from its oxide

15 Identify the following elements, giving a reason for each of your decisions:

  a The gaseous non-metal that may be present in a packet of potato chips

  b The gaseous non-metal that is essential to deep-sea diving

  c The solid non-metal that is found near to volcanoes

  d The solid non-metal that is included in the head of a match

  e The gaseous non-metal that is oxidised during lightning storms

  f The solid non-metal that forms an oxide which is the main component in sand

  g The solid non-metal used in many medicines

  h The solid non-metal with semiconducting properties

  i Two non-metal oxides that contribute to acid rain

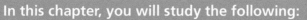

25 Preparation of gases and acids

In this chapter, you will study the following:

- the laboratory preparations, properties and uses of a few gases;
- the industrial preparation and uses of chlorine, ammonia, sulphuric acid and nitric acid;
- the choice of reaction conditions in industrial processes in terms of reversible reactions;
- the tests for cations, anions and gases.

This chapter covers
Objectives 5.2–5.3 and 6.1–6.3 of Section C of the CSEC Chemistry Syllabus.

Previous knowledge
You may have seen how some gases and some acids are prepared in your school laboratory. The quantities of gases and acids prepared in school are small, as these substances are prepared mainly to illustrate certain chemical principles to students. However, some of these substances are in demand on a larger scale, either to be used as they are or to be used further in manufacturing, in what are commonly called 'downstream' industries. Ammonia, for example, is used in the production of ammonium sulphate or urea. Larger quantities of these chemicals must be prepared industrially in chemical plants. Sometimes, the chemical industry is sufficiently important to make significant contributions to the economy of a country.

25.1 Laboratory preparation of gases

Here we will describe the preparation of the gases oxygen, hydrogen, chlorine, hydrogen chloride, carbon dioxide, sulphur dioxide and ammonia.

There are a number of requirements that are necessary in order to prepare and collect gases in the laboratory. Before you begin, you might want to consider several aspects of the procedure for preparing gases.

1 Reactants and conditions:
- What are the reactants needed to prepare the gas?
- Is a catalyst needed?
- Are the reactants to be heated or left at room temperature?

2 Properties of the gas to be prepared:
- Is the gas soluble in water or not?
- Is the gas more or less dense than air?
- Is the gas acidic or alkaline?
- Is the gas an irritant?
- Is the gas poisonous?

3 Method of collection:
- How will the gas be collected? Dependent on its properties, such as solubility, acid/alkaline nature and density, the gas may be collected (i) over water, (ii) by downward delivery (upward displacement of air) or (iii) by upward delivery (downward displacement of air).
- What type of drying agent will be used to obtain a dry gas?

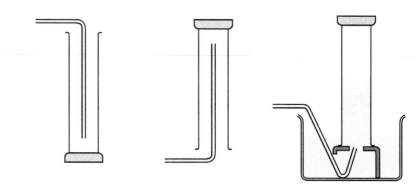

Figure 25.1
Methods of collecting
gases ▶

4 Safety considerations:
- Are special precautions necessary?
- Must the preparation be carried out in a fume chamber or not?
- Careful consideration of all of the above will guide you, not only with regards to the procedure for the preparation of the gas, but also to choice of the appropriate apparatus for the preparation.

5 Choice of apparatus:
- Comments on the function of the parts of the equipment used in this section will guide you in this respect.

6 Identity of the gas:
- Having prepared the gas you need to test its identity.
- For each gas there is a test or combination of tests which separates it from all other gases.

As you study or carry out the preparation of the gases, the areas mentioned above must be addressed.

The laboratory preparation of oxygen

Oxygen:
- is a colourless, odourless gas;
- does not burn but supports combustion;
- re-lights a glowing splint;
- is neutral – has no effect on litmus;
- is slightly soluble in water.

The most convenient method of preparing oxygen involves the catalytic decomposition of hydrogen peroxide:

$$2H_2O_2(l) \rightarrow 2H_2O(l) + O_2(g)$$

If left to itself, hydrogen peroxide is unstable and slowly decomposes to produce oxygen and water. However, if the hydrogen peroxide is allowed to drop onto manganese(IV) oxide, which acts as a catalyst, there is a rapid evolution of oxygen. Figure 25.2 shows the apparatus needed for the preparation.

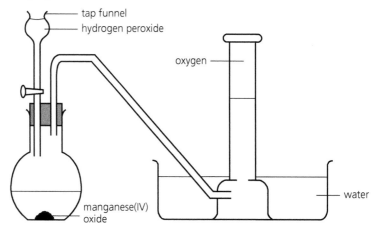

tap funnel
hydrogen peroxide
oxygen
water
manganese(IV) oxide

Figure 25.2
The apparatus used for
the preparation of
oxygen ▶

- A thistle (tap) funnel is used to introduce liquid into the reaction vessel.
- A flat-bottomed flask is used where heat is not required.
- Oxygen is collected over water as it is sparingly soluble.
- If dry gas is needed, insert a wash bottle containing concentrated sulphuric acid or a vessel containing solid silica gel or anhydrous calcium chloride or calcium oxide after the reaction vessel. Also, replace the trough and gas jar containing water with an empty gas jar. The gas will be collected by downward delivery.

Oxygen can also be prepared by the thermal decomposition of oxygen-rich substances such as sodium nitrate and potassium chlorate(V):

$$2NaNO_3(s) \xrightarrow{\text{heat}} 2NaNO_2(s) + O_2(g)$$

$$MnO_2 \longrightarrow \text{catalyst}$$
$$2KClO_3(s) \longrightarrow 2KCl(s) + 3O_2(g)$$

Uses of oxygen

Oxygen is a tasteless gas that has no smell or colour. Oxygen gas makes up 22% of the air around us. Some of the uses of oxygen gas include the following:

- It is used in industry for cutting, welding and melting metals. Oxygen gas can generate temperatures of 3 000 °C. These temperatures are required for oxy-hydrogen and oxy-acetylene blow torches.
- It is used to produce energy in industrial processes, generators and ships. Oxygen gas is also used in airplanes and cars. Spacecraft burn liquid oxygen for thrust.
- Oxygen supplies are kept in stock in health-care institutions. Astronauts, mountaineers and scuba divers use breathing apparatus that contain oxygen gas.
- Oxygen gas is used to treat victims of carbon monoxide poisoning.
- Oxygen gas is used to destroy bacteria.
- All living things use oxygen in a process called aerobic respiration.

The laboratory preparation of hydrogen

Hydrogen:
- is a colourless, odourless gas;
- is less dense than air;
- is insoluble in water;
- is highly flammable (handle with care!);
- gives a slight explosion when mixed with air and ignited; this is a test for hydrogen.

As you can see in Table 21.2, some metals react with water or steam to produce hydrogen. Other metals, for example zinc and aluminium, also react with aqueous sodium hydroxide to produce hydrogen. However, the most convenient method of producing hydrogen involves the reaction between a moderately active metal, such as magnesium or zinc, and dilute hydrochloric acid or dilute sulphuric acid. The arrangement of apparatus is shown in Figure 25.3.

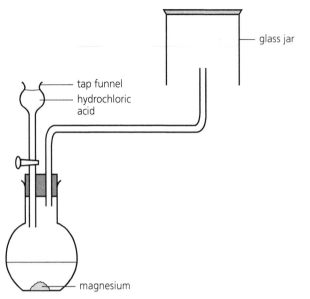

▲ **Figure 25.3** The apparatus used for the preparation of hydrogen

- Upward delivery is used here because the gas is less dense than air.
- A graduated syringe can replace the gas jar if the volume of gas released is to be measured.
- A drying agent can be introduced as needed.

Uses of hydrogen

Hydrogen gas is the lightest and most common element that exists. It is one of the components of water and is vital to life. Some of the uses of hydrogen include the following:

- Hydrogen is used as a hydrogenating agent to change unhealthy, unsaturated fats to saturated oils and fats.
- Because hydrogen is light, it is used in weather balloons that are fitted with equipment to record information necessary to study the climate.
- Hydrogen fuel cells are used to generate electricity from oxygen and hydrogen. Hydrogen fuel cells are considered environmentally friendly because they generate only water vapour.
- Hydrogen is used in fertilisers, food, and chemical and paint industries.

The laboratory preparation of chlorine

Chlorine can be prepared in the laboratory in several ways, all involving the oxidation of concentrated hydrochloric acid. Among the oxidising agents that can be used are lead(IV) oxide (PbO_2), potassium manganate(VII) and manganese(IV) oxide. The apparatus is shown in Figure 25.4.

Chlorine:
- is a green-yellow gas;
- is soluble in water;
- is more dense than air;
- is acidic, so turns blue litmus red;
- has bleaching properties, and so turns red litmus colourless;
- is an oxidising agent;
- turns aqueous potassium iodide from colourless to yellow.

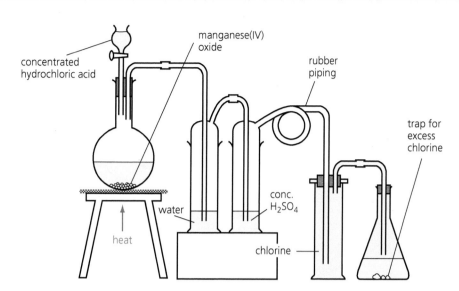

▲ **Figure 25.4**   The apparatus used for the preparation of chlorine from manganese(IV) oxide

This reaction should be carried out in a fume chamber.

- A round-bottom flask is used as the reaction vessel because the reactants must be heated.
- The water in the first wash bottle removes acid fumes from the gas.
- The drying agent in the second wash bottle removes moisture (water) from the gas.
- Note that the delivery tubes bring the gas in direct contact with the washing or drying agents, but the tubes through which gases leave are not in contact with≈liquids.
- Collection is by downward delivery. Chlorine is denser than air.
- The trap prevents escape of the poisonous gas.

The laboratory preparation of hydrogen chloride

Hydrogen chloride:
- is a colourless gas;
- has a pungent smell;
- fumes in moist air;
- fumes (forms a fine white solid) with ammonia:

$HCl(g) + NH_3(g)$
$\rightarrow NH_4Cl(s)$

- is strongly acidic, so turns moist blue litmus red.

Hydrogen chloride gas can be prepared by direct combination of hydrogen and chlorine. It can be prepared in the laboratory by the action of concentrated sulphuric acid on sodium chloride (rock salt).

The equation for the reaction is:

$$NaCl(s) + H_2SO_4(l) \rightarrow NaHSO_4(aq) + HCl(g)$$

The apparatus used to prepare hydrogen chloride is similar to that for the preparation of chlorine gas except:

- the gas is not bubbled into water before drying it;
- the trap used in preparing chlorine gas is not included.

However, some precautions are necessary since hydrogen chloride gas can have an irritating effect, even though it is not poisonous.

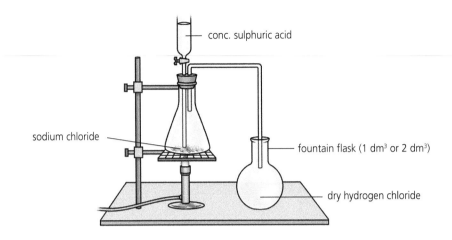

Figure 25.5
Preparation of hydrogen
chloride ▶

Carbon dioxide:

- is a colourless, odourless gas;

- does not burn;

- is more dense than air;

- is weakly acidic, so turns blue litmus pink;

- gives a white precipitate with lime water:

$Ca(OH)_2(aq) + CO_2(g)$
$\rightarrow CaCO_3(s) + H_2O(l)$

- the precipitate dissolves on reaction with more CO_2:

$CaCO_3(s) + CO_2(g) + H_2O(l)$
$\rightarrow Ca(HCO_3)_2(aq)$

- heating the solution reforms the precipitate:

$Ca(HCO_3)_2 \rightarrow CaCO_3(s)$
$+ H_2O(l) + CO_2(g)$

The laboratory preparation of carbon dioxide

Acids react with carbonates and hydrogencarbonates to give carbon dioxide. This is the basis for the preparation of carbon dioxide by the action of dilute hydrochloric acid on calcium carbonate:

$$CaCO_3(s) + 2HCl\,(aq) \rightarrow CaCl_2(aq) + H_2O(l) + CO_2(g)$$

The apparatus is shown in Figure 25.6.

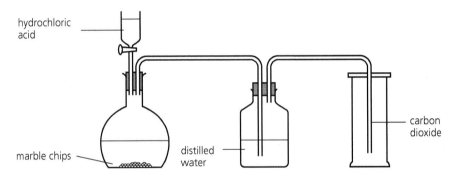

▲ **Figure 25.6**   The apparatus used for the preparation of carbon dioxide.

Uses of carbon dioxide

◀ **Figure 25.7**
Carbon dioxide gas is forced into liquid under pressure and so it dissolves. When the can is opened, the pressure is lowered and the solubility of the gas is decreased. The carbon dioxide comes off with a characteristic fizz.

- Used to give drinks fizz.
- In fire fighting, especially where the use of water is not advised. The high density of the gas allows it to cover potentially hazardous material. It prevents oxygen reaching the burning material and so prevents combustion.
- As a refrigerant. Solid carbon dioxide, which sublimes at –78 °C, can be used to keep materials cold.
- In pressurising oil wells.
- As an aerosol propellant.
- In treating alkaline waste water.

Practice

This question is about the laboratory preparation of carbon dioxide.

1 Can you explain the following:
  a Why a flat-bottomed flask is used?
  b The need for a tap funnel?
  c The function of the distilled water?
  d Why the gas is collected as shown?
  e How the apparatus can be modified to collect dry carbon dioxide gas?

The laboratory preparation of sulphur dioxide

Sulphur dioxide (SO_2) can be prepared by the reaction of hot concentrated sulphuric acid on metallic copper. In this reaction, the concentrated sulphuric acid functions as an oxidising agent.

$$Cu(s) + 2H_2SO_4(aq) \xrightarrow{\text{heat}} CuSO_4(aq) + SO_2(g) + 2H_2O(l)$$

Sulphur dioxide:

- is a colourless gas;
- has a pungent smell of burning sulphur;
- is denser than air;
- is soluble in water;
- does not burn;
- is weakly acidic, and so turns blue litmus pink;
- has a bleaching property, so slowly turns red litmus colourless;
- is a reducing agent which decolorises acidified potassium manganate(VII) and turns orange acidified potassium dichromate green.

Sulphur dioxide can also be prepared by the action of dilute acid on a sulphite or a hydrogen sulphite:

$$Na_2SO_3(s \text{ or } aq) + 2HCl(aq) \xrightarrow{\text{heat}} 2NaCl(aq) + SO_2(g) + H_2O(l)$$

Note that the reaction mixture must be heated, since sulphur dioxide is very soluble in water and would remain dissolved in the dilute acid unless its solubility is decreased by heating.

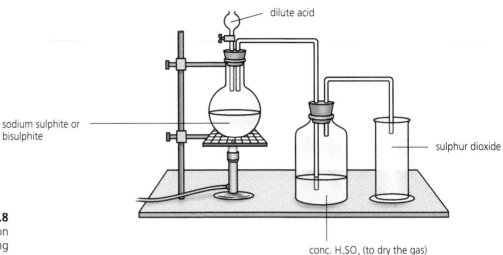

dilute acid

sodium sulphite or
bisulphite

sulphur dioxide

conc. H_2SO_4 (to dry the gas)

Figure 25.8
Laboratory preparation
of sulphur dioxide using
a sulphite. ▶

Practice

Carefully read the description of the preparation of sulphur dioxide (above).

2 Draw a simple line diagram of the apparatus you would use to prepare it in the laboratory.

3 a Explain the pieces of equipment you used in the apparatus.

  b Explain the method of collection you used.

Uses of sulphur dioxide

• A bleaching agent (e.g. For wool and wood pulp)

• A food preservative

• In detergents

• During manufacture of dyestuffs

• In the manufacture of sulphuric acid (this is the major use).

The laboratory preparation of ammonia

Ammonia:

• is a colourless gas;

• has a pungent smell;

• is very soluble in water;

• is weakly alkaline, so turns red litmus blue;

• gives dense white fumes with hydrogen chloride.

Ammonia is a weak base. A stronger base can therefore displace ammonia from an ammonium salt. This is the basis for the laboratory preparation of ammonia, which involves heating a mixture of solid calcium hydroxide and solid ammonium sulphate:

$$(NH_4)_2SO_4(s) + Ca(OH)_2(s) \longrightarrow CaSO_4(s) + 2NH_3(g) + 2H_2O(l)$$

Any other strong alkali and any other ammonium salt can be used. As ammonia is very soluble in water and is less dense than air, it is collected by upward delivery. To obtain a dried sample of the gas, the drying agent used is calcium oxide, as both sulphuric acid and anhydrous calcium chloride react with ammonia. The apparatus is illustrated in Figure 25.10.

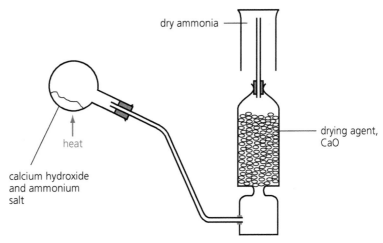

▲ **Figure 25.9** The apparatus used for the preparation of ammonia in the laboratory

Practice

These questions are about the laboratory preparation of ammonia.

4 Notice the downward slant of the reaction vessel. What explanation can you offer for this?

5 The drying agent is calcium oxide, which is an alkali. Why can't concentrated sulphuric acid be used as the drying agent?

6 Anhydrous calcium chloride forms a complex compound with ammonia. Will you use this drying agent in the preparation of the gas?

7 Justify the method of collection.

8 Ammonia gas is colourless. How can you tell when the gas jar is filled with the gas?

25.2 Industrial production of gases and acids

Find out more

Some important considerations in the industrial production of chemicals

The commercial production of a gas or any other substance is subject to quite different considerations from the laboratory preparation.

You need to consider some or all of the following factors:

- Availability of raw materials and suitable sources of energy.
- Cost of production – including cost of raw materials, energy, equipment and labour. The company that uses the best combination of routes and raw materials is normally able to produce a commodity most economically.
- Conservation of resources – efforts to conserve resources could include recycling of materials and the use of alternative sources of energy.
- Location – the best site for the plant should be selected taking the following into consideration, for example:
  - ease of getting to raw materials and to markets;
  - proximity to a pool of workers needed by the industry.

- Environmental impact – the industrial process and wastes produced may disrupt important ecosystems.
- Safety – the equipment and materials used in the industrial process may be hazardous to the health of workers and exhausts from the plant may affect the members of the community.

Q1 Select an industrial chemical plant in your country. Find out how each of the above factors has been taken into account in setting up the plant.

The commercial production of ammonia

Find out more

People have long recognised the need to replenish or improve soil fertility. For much of our history, we have used the droppings of animals to achieve this (and in many parts of the world manure is still the most used method of fertilising the soil).

In the early part of the 19th century saltpetre (sodium nitrate) was imported from South America to enrich the soils of Europe. In the latter part of the 19th century, ammonium sulphate became available as a by-product of certain chemical works, but the quantities were not sufficient to satisfy demand. A cheap method of producing ammonia, and hence of ammonium compounds, was urgently required. Such a method, the Haber process, was developed in Germany in 1913. Today, huge ammonia-producing plants have become a familiar part of the landscape of many countries. There are two ammonia production plants in Trinidad and Tobago.

In the Haber process, nitrogen is made to react with hydrogen to form ammonia gas:

$$N_2(g) + 3H_2(g) > 2NH_3(g)$$

Nitrogen and hydrogen are compressed to a pressure of more than 200 atm and passed over an iron catalyst at about 400 °C. Under these conditions, approximately 25% of the gaseous mixture is changed into ammonia. The unreacted nitrogen and hydrogen are passed over the catalyst again. (See Section 12.6).

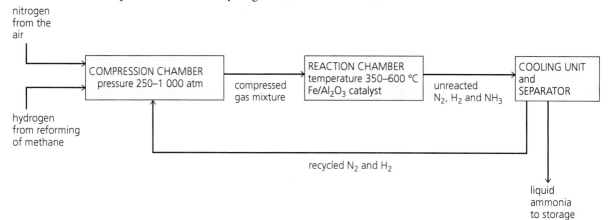

▲ **Figure 25.10** The large-scale manufacture of ammonia

Ammonium fertilisers

Ammonia produced in the process shown in Figure 25.11 is used mainly in the production of fertilisers. These can be prepared by:

- Neutralisation: The alkaline ammonia is treated with an acid such a nitric, sulphuric or phosphoric acid, for example.

  $$NH_3(g) + HNO_3(aq) \rightarrow NH_4NO_3(aq)$$
  $$2NH_3(g) + H_2SO_4(aq) \rightarrow (NH_4)_2SO_4(aq)$$
  $$3NH_3(g) + H_3PO_4(aq) \rightarrow (NH_4)_3PO_4(aq)$$

- Double decomposition (ionic precipitation): the ammonia is converted to an aqueous solution of ammonium carbonate which reacts in a double decomposition reaction with a saturated solution of calcium sulphate.

  $$(NH_4)_2CO_3(aq) + CaSO_4 \rightarrow (NH_4)_2SO_4(aq) + CaCO_3(s)$$

- Alternatively, liquid anhydrous ammonia may be applied directly to the soil. This is a highly potent fertiliser (contains 82% nitrogen) when compared with urea (46% N) and ammonium nitrate (35% N).

Understand it better

How are conditions chosen to ensure the best yield of ammonia in the Haber process?

The equation for the reaction is:

$$N_2(g) + 3H_2(g) \rightleftharpoons 2NH_3(g)$$

There are two points to note about the reaction:
- The reaction is reversible (see Section 12.6).
- The forward reaction is exothermic.

After the reaction has proceeded for a while, the reaction mixture will contain both reactants (nitrogen and hydrogen) and the product (ammonia), and the rate of the forward and backward reactions will be equal.

Any change in reaction conditions that favours a move from left to right will increase the amount of ammonia in the reaction mixture.

The volume of gas decreases from left to right. Since the volume of a gas decreases with increasing pressure (see Section 1.4), high pressures will favour the move from left to right, that is, increased production of ammonia. Therefore, in the Haber process, high pressures are used.

The forward reaction is exothermic and consequently the backward reaction is endothermic. Lowering the temperature always favours an exothermic reaction. However, lowering the temperature will also decrease the reaction rate (see Section 14.3). In order to make the production of ammonia economically viable, moderately high temperatures are used in the Haber process and a catalyst is introduced to speed up the reaction.

The commercial production of chlorine

The flowing mercury cell is the main method for the manufacture of chlorine (Figure 25.11). Sodium hydroxide and hydrogen are co-products. This is essentially the electrolysis of concentrated brine using a continuously flowing mercury cathode and graphite or titanium anodes.

- The raw material for this process is a concentrated sodium chloride solution.
- The power requirement is approximately 4.5 V, 300 000 A.
- The process is carried out at a pH of about 4.5.

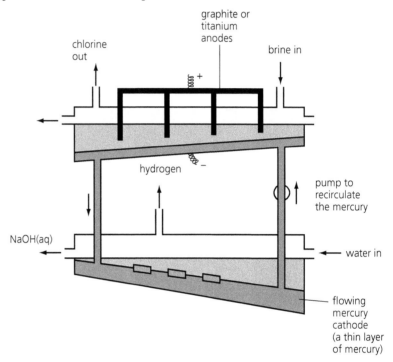

Figure 25.11
The flowing mercury cathode cell used to manufacture chlorine ▶

The primary products of this electrolysis are sodium and chlorine. The chlorine is removed (bubbled off) at the anodes, where the reaction is:

$$2Cl^-(aq) - 2e^- \rightarrow Cl_2{}^0(g)$$

At the cathode, sodium 'dissolves' in the mercury to form an amalgam which is treated with water as it flows out of the cell to form sodium hydroxide and hydrogen gas:

$$Na^+(l) + 1e^- \rightarrow Na(l)$$

$$Na(l) + Hg(l) \rightarrow Na/Hg \text{ (amalgam)}$$

$$2Na/Hg(l) + 2H_2O(l) \rightarrow 2NaOH(aq) + H_2(g) + 2Hg(l)$$

The flowing mercury cathode cell is one of the main uses of mercury, which is both expensive and toxic. The cells are so constructed that the two gaseous products – chlorine and hydrogen – are produced in separate tanks.

Chlorine, sodium hydroxide and hydrogen are obtained when sodium chloride is electrolysed in the mercury cathode cell. Figure 25.13 shows that many of the products we use every day come originally from sodium chloride (common salt).

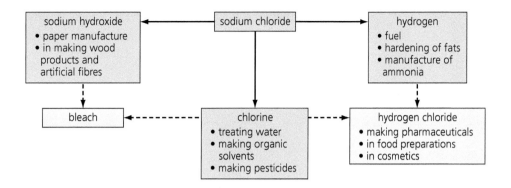

Figure 25.12
A flow diagram
illustrating the many
uses of the products
made from sodium
chloride ▶

The commercial production of acids

Sulphuric acid

Most of the world's supply of sulphuric acid is produced by the Contact Process. The stages in the process are summarised in Figure 25.13.

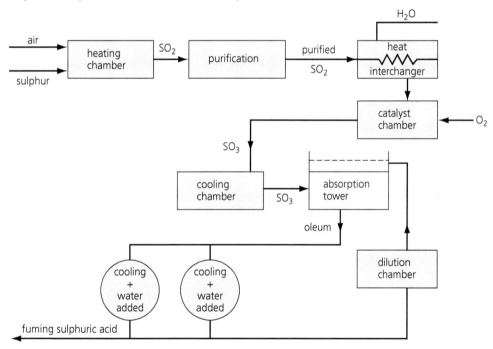

Figure 25.13
The manufacture of
sulphuric acid by the
Contact Process ▶

The raw materials needed for the Contact Process are sulphur dioxide, air and a catalyst, usually vanadium oxide.

The sulphur dioxide can be obtained by various methods:

* Burning sulphur in air at about 1 000 °C:
$$S(s) + O_2(g) \rightarrow SO_2(g)$$

* Roasting suitable sulphide ores:
$$4FeS_2(s) + 11O_2(g) \rightarrow 8SO_2(g) + 2Fe_2O_3(s)$$

Sulphur dioxide is cooled to 450 °C and passed through hot gas filters, where dust particles, which can 'poison' the catalyst, are removed by electrostatic precipitation.

$$2SO_2(g) + O_2(g) \rightleftharpoons 2SO_3(g) \qquad \Delta H = -98 \text{ kJmol}^{-1}$$

Since the reaction of sulphur trioxide and water is highly exothermic, the sulphur trioxide is not directly absorbed into water. Instead it is absorbed into concentrated sulphuric acid:

$$SO_3(g) + H_2SO_4(l) \rightarrow H_2S_2O_7(l)$$
oleum

The oleum is then diluted, slowly and during cooling, with water:

$$H_2S_2O_7(l) + H_2O(l) \rightarrow 2H_2SO_4(l)$$

Uses of sulphuric acid

Car batteries contain quite strong sulphuric acid. Like all acids, it is highly corrosive and will burn holes through human skin as well as other materials!

Sulphuric acid is one of the major industrial chemicals. Figure 25.14 summarises the major uses of sulphuric acid.

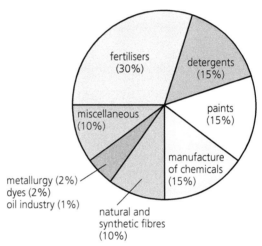

Figure 25.14
The major uses of sulphuric acid ▶

Key reactions of sulphuric acid

Concentrated sulphuric acid acts as an oxidising agent:

$$Cu^0(s) + H_2SO_4(l) \rightarrow CuSO_4(aq) + SO_2(g) + H_2O(l)$$

and as a **dehydrating agent**:

$$CuSO_4.5H_2O(s) \longrightarrow 5H_2O(l) + CuSO_4(s)$$

blue crystals   water   white
removed   powder
by sulphuric
acid

Dilute sulphuric acid also acts as a **typical acid**:

$$Mg(s) + H_2SO_4(aq) \rightarrow MgSO_4(aq) + H_2(g)$$
$$ZnO(s) + H_2SO_4(aq) \rightarrow ZnSO_4(aq) + H_2O(l)$$
$$2NaOH(aq) + H_2SO_4(aq) \longrightarrow Na_2SO_4(aq) + 2H_2O(l)$$
sodium sulphate

$$NaOH(aq) + H_2SO_4(aq) \longrightarrow NaHSO_4(aq) + H_2O(l)$$
sodium hydrogen sulphate

25.3 Common tests for gases, anions and cations

Tests for gases

▼ **Table 25.1** The common tests for gases

Gas	Colour/smell	Test	Results
oxygen	none	hold glowing splint in gas	splint glows brighter, or relights
hydrogen	none	hold lighted splint in gas	flame is extinguished with a slight explosion
carbon dioxide	none	bubble the gas through lime water, $Ca(OH)_2(aq)$	lime water turns milky initially; may turn back clear
hydrogen chloride	irritating smell; colourless	hold moist blue litmus paper in the gas;	litmus turns red (gas is acidic)
		bring ammonia in contact with gas	dense white fumes of ammonium chloride (NH_4Cl) form
nitrogen dioxide	irritating smell; reddish brown	hold moist blue litmus paper in the gas	litmus turns red, but is not bleached
sulphur dioxide	colourless; choking smell of burning sulphur	hold filter paper soaked in acidified potassium dichromate(VI) in gas	colour changes from orange to green
		hold moist blue litmus paper in the gas	litmus turns red, may also be bleached
water vapour	none	hold blue cobalt chloride paper in vapour	cobalt chloride paper turns from blue to pink
		bring anhydrous copper(II) sulphate in contact with gas	there is colour change from colourless to blue
ammonia	colourless pungent smell	hold moist red litmus paper in gas	litmus turns blue (ammonia is the only common alkaline gas)
		bring HCl(g) in contact with the gas	dense white fumes of ammonium chloride form
chlorine	pale green; irritating smell	hold moist blue litmus paper in the gas	litmus turns red, and is then bleached
		hold moist starch–iodide paper in the gas	starch–iodide paper turns blue-black

Tests for anions

▼ **Table 25.2** Tests for anions, using heat and acids.

Anion	Action of heat	Action of dilute HNO_3 or dilute HCl	Action of concentrated H_2SO_4
hydrogencarbonate, HCO_3^-	CO_2(g) and steam evolved	CO_2(g) evolved	CO_2(g) evolved
carbonate, CO_3^{2-}	CO_2(g) evolved (except Na, K)	CO_2(g) evolved	CO_2(g) evolved
chloride, Cl^-	–	–	HCl(g) evolved (also some Cl_2(g) if an oxidising agent is present)
bromide, Br^-	–	–	Br_2(g) and HBr(g) evolved
iodide, I^-	–	–	I_2(aq) or (s) formed; H_2S(g) may be formed
nitrate, NO_3^-	NO_2(g) and O_2(g) evolved (only O_2(g) if $NaNO_3$ or KNO_3)	–	NO_2(g) and HNO_3(g) on warming (colour deepens on addition of Cu(s))
nitrite, NO_2^-	–	NO_2(g) evolved	NO_2(g) evolved
sulphate, SO_4^{2-}	SO_2(g) and SO_3(g) evolved from some	–	–
sulphite, SO_3^{2-}	SO_2(g) evolved	SO_2(g) evolved	SO_2(g) evolved
sulphide, S^{2-}	H_2S(g) evolved	H_2S(g) evolved	H_2S(g) evolved

▼ **Table 25.3** Tests for anions, using $BaCl_2$ or $Ba(NO_3)_2$, $AgNO_3$ and $Pb(NO_3)_2$

Anion	$BaCl_2$(aq) or $Ba(NO_3)_2$(aq) followed by dilute HCl(aq) or dilute HNO_3(aq)	$AgNO_3$(aq) followed by NH_3(aq)	$Pb(NO_3)_2$(aq) followed by dilute HNO_3(aq)
HCO_3^-(aq)	white precipitate; soluble in acid	white precipitate; turns brown on heating; soluble in dilute HNO_3	white precipitate; soluble in dilute HNO_3
CO_3^{2-}(aq)	white precipitate; soluble in acid (CO_2(g) evolved)	white precipitate, turns brown on standing; soluble in dilute HNO_3	white precipitate; soluble in dilute HNO_3

continued

▼ **Table 25.3** *continued*

Anion	BaCl$_2$(aq) or Ba(NO$_3$)$_2$(aq) followed by dilute HCl(aq) or dilute HNO$_3$(aq)	AgNO$_3$(aq) followed by NH$_3$(aq)	Pb(NO$_3$)$_2$(aq) followed by dilute HNO$_3$(aq)
Cl$^-$(aq)	–	white precipitate; soluble in NH$_3$(aq)	white precipitate; soluble on heating; re-precipitates on cooling
Br$^-$(aq)	–	white precipitate; slightly soluble in NH$_3$(aq)	white precipitate; soluble on heating; re-precipitates on cooling
I$^-$(aq)	–	pale yellow precipitate; insoluble in NH$_3$(aq)	insoluble in dilute HNO$_3$; bright yellow precipitate
NO$_3^-$(aq)		–	–
NO$_2^-$(aq)	–	turns black	–
SO$_4^{2-}$(aq)	white precipitate; insoluble in acid	–	white precipitate; insoluble in dilute HNO$_3$
SO$_3^{2-}$(aq)	white precipitate; soluble in acid (SO$_2$ evolved)	turns black	white precipitate; soluble in dilute HNO$_3$
S^{2-}(aq)	–	black precipitate	black precipitate; insoluble in dilute HNO$_3$

Tests for cations

▼ **Table 25.4** Tests for cations

Cation	Addition of NaOH(aq)	Addition of NH$_3$(aq)
NH$_4^-$	ammonia evolved on warming (see test for NH$_3$(g) in Table 25.1)	solutions mix; no precipitate forms
Na$^-$, K$^-$	no precipitate	no precipitate
Ca^{2-}	white precipitate; insoluble in excess	no precipitate
Al^{3-} excess	white precipitate; soluble in excess	white precipitate; insoluble in excess
Fe^{2-}	dirty green precipitate; turns brown on exposure to air; insoluble in excess	dirty green precipitate; turns brown in air; insoluble in excess
Cu^{2-}	red-brown precipitate; insoluble in heating	blue precipitate; dissolves in excess to deep blue solution
Zn^{2-}	white precipitate; soluble in excess	white precipitate; soluble in excess
Pb^{2-}	white precipitate; soluble in excess	white precipitate; insoluble in excess

Figure 25.15 shows a flow diagram. Follow this diagram when you want to detect cations using aqueous ammonia.

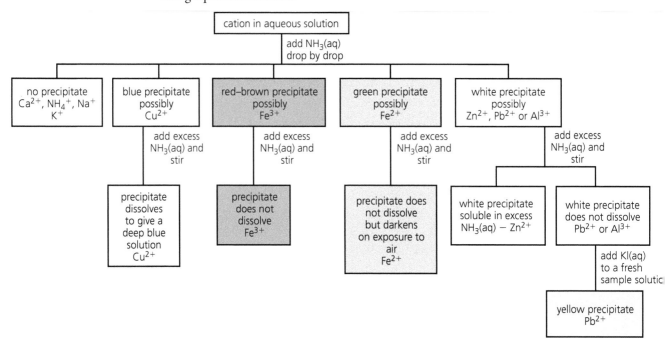

▲ **Figure 25.15**   Testing for cations – aqueous ammonia

Figure 25.16 shows another flow diagram. Follow this diagram when you want to detect cations using aqueous sodium hydroxide.

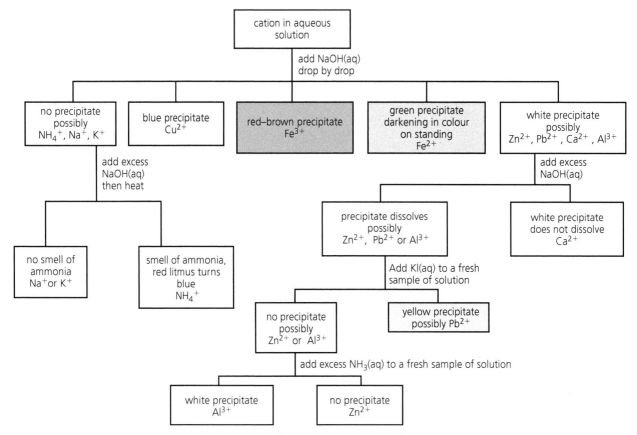

▲ **Figure 25.16**   Testing for cations – aqueous sodium hydroxide

Summary

- When preparing gases in the laboratory, consider: the reactants and conditions; the properties of the gas formed; collection methods; safety; the apparatus; how to test the identity of the gas.

- Some general observations such as the colour of a gas, its effect on moist red or blue litmus and glowing or lighted splints help to identify gases in the laboratory.

- More specific reagents such as acidified oxidising agents, potassium manganate(VII) and potassium dichromate(VI) help to identify reducing gases, e.g. SO_2 and H_2S.

- Gases that are insoluble in or unreactive with water may be collected over water. Examples are O_2 and H_2.

- If an acid is one of the reactants used to prepare the gas, the acid must first be removed by bubbling the gas through water before drying the gas.

- Gases with low densities, such as NH_3 and H_2, are collected by upward delivery.

- More dense gases, such as CO_2 and Cl_2, are collected by downward delivery.

- Poisonous gases, such as Cl_2, NO_2 and SO_2, should be prepared in a fume cupboard.

- When producing chemicals industrially, the following factors are considered: the availability of the materials needed, the cost, conservation of resources, location, environmental impact and safety.

- Sulphuric acid (a major industrial chemical used in fertilisers, detergents and other uses) is obtained commercially by the Contact Process.

- Conditions of temperature and pressure are carefully chosen and a catalyst is used to obtain a good yield of acid and to make the process economical.

- The Haber process is used to obtain ammonia. Some of the ammonia produced is used to make fertilisers; some is converted to nitric acid industrially.

- Chlorine is produced industrially by the electrolysis of brine. This process produces important by-products such as hydrogen gas and sodium hydroxide solution.

End-of-chapter questions

1 A variety of methods are used in the laboratory to prepare gases:

  a dilute acid + solid → H_2, CO_2, SO_2

  b concentrated acid + solid → HCl, Cl_2

  c non-acid liquid + solid → O_2, NH_3

  d mixture of solids → NH_3, (heated) O_2

  e single solid heated → O_2, CO_2, NO_2, NH_3, HCl

  For each of the methods, state the reagents needed to prepare each gas and write a balanced chemical equation for each reaction.

2 Oxygen is a very reactive non-metal. Write balanced chemical equations to show how oxygen reacts with:
  a hydrogen;
  b nitrogen;
  c magnesium;
  d iron;
  e sodium;
  f sulphur.
  You need to write more than one equation for some elements.

3 In the laboratory, how would you test for the presence of the following compounds?
  a Ammonia
  b Carbon dioxide
  c Chlorine
  d Oxygen
  e Sulphur dioxide

4 Various solids are heated in the laboratory. What conclusions can you draw about the solids when the following gases are liberated?
  a Oxygen gas
  b A mixture of oxygen gas and nitrogen dioxide gas
  c Carbon dioxide gas
  d Ammonia gas

5 a Describe what you will expect to see if a sample of ammonium chloride is heated in the laboratory.
  b What differences would be expected if sodium hydroxide is added to the ammonium chloride before heating?

6 a Identify the gases described here.
    (i) Gas X is very soluble in water, is alkaline and has a low density.
    (ii) Gas Y is neutral, reacts with oxygen in air explosively.
    (iii) Gas Z turns acidified potassium dichromate green and bleaches red litmus.
  b (i) Predict the method of collection and drying agent that can be used for gas X.
    (ii) Predict the reagents that may be used to generate gas Y in the laboratory.
    (iii) Predict the anions that are formed when gas Z reacts with water.

7 a Describe the laboratory preparation of a sample of carbon dioxide.
  b How does carbon dioxide react with the following compounds?
    (i) $Ca(OH)_2(aq)$
    (ii) $Na_2CO_3(aq)$
    (iii) Heated magnesium
  c Write a balanced equation for each of the reactions in part (b).
  d How can carbon dioxide be converted to carbon monoxide?
  e Account for the use of carbon dioxide in fire fighting and keeping perishables cold.

8 The chief constituents of the air, nitrogen and oxygen, have important roles in a number of industrial processes.
  a Discuss the role of nitrogen in the Haber process.
  b Discuss the role of oxygen in the Contact Process.

9 What is the importance of commercially produced ammonia to crop production?

10 Explain why the following procedures are necessary in the industrial production of sulphuric acid:
  a Moisture is excluded in the conversion of sulphur to sulphur dioxide.

b Dust is removed from sulphur dioxide gas.

c Sulphur trioxide is dissolved in concentrated sulphuric acid rather than in water.

11 a What do you understand by the following terms?

(i) Reversible reaction

(ii) Equilibrium

(iii) Percentage yield of product

b Illustrate the meaning of each term in part (a) by referring to the industrial production of ammonia.

12 A flow diagram for the manufacture of sulphuric acid is shown at the top of the next page.

a Give the name of a possible starting material for the production of sulphur dioxide.

b Why must the sulphur dioxide be purified? How is this purification achieved?

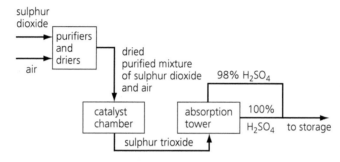

c Name a suitable drying agent for both sulphur dioxide and air.

d What experimental conditions are normally employed in the catalyst chamber?

e Write balanced equations for all reactions that occur in this process.

f Why is sulphur trioxide not dissolved directly in water?

g Sulphur dioxide is an atmospheric pollutant. State two ways in which sulphur dioxide gets into the atmosphere, giving balanced equations for reactions that occur.

13 a Outline the manufacture of sulphuric acid from sulphur.

b Describe one reaction in which sulphuric acid behaves as:

(i) an acid;

(ii) a dehydrating agent.

c Sulphuric acid is sometimes described as a dibasic (or diprotic) acid.

(i) What is meant by the term 'dibasic'?

(ii) In which of the following reactions is sulphuric acid behaving as an oxidising agent?

$$NaCl(s) + H_2SO_4(l) \rightarrow NaHSO_4(aq) + HCl(g)$$

$$Mg(s) + H_2SO_4(aq) \rightarrow MgSO_4(aq) + H_2(g)$$

$$2HBr(aq) + H_2SO_4(aq) \rightarrow Br_2(g) + SO_2(g) + 2H_2O(l)$$

14 Review the industrial production of chlorine by the electrolysis of brine and then answer the following questions.

a Write the ionic half-equation for the reaction at:

(i) the cathode;

(ii) the anode.

b Write an equation for the reaction between the product at the cathode and the mercury.

c What are the by-products of this industrial process? Write an equation for their formation.

d What are the main uses of chlorine produced in this industrial process?

26 Non-metals in the environment

- the harmful effects of non-metals on living systems and the environment;
- the adverse effects of a number of compounds on the environment;
- the problem associated with disposal of solid waste;
- how some elements and compounds are essential to the proper functioning of living systems;
- whether these materials are needed in relatively small or large quantities;
- the composition of the atmosphere;
- the effects of the intake of excessive amounts of non-metals on living systems;
- how elements in the environment are made available to living systems.

This chapter covers
Objectives 5.4–5.5 of Section C of the CSEC Chemistry Syllabus.

The environment is everything in our surroundings. Our surroundings include: the atmosphere, oceans, soil and all living things. The environment means all of the conditions that are needed for growth and for development.

26.1 The environment

An **ecosystem** is a community of living organisms (plants, animals and microbes) together with the non-living components (air, water and mineral soil) of their environment, interacting as a system.

The environment is everything in our surroundings and creates the conditions that we need to survive. **Pollutants** are materials that are toxic and can interfere with the food chain or change the rate of growth of animals and plants. Pollutants damage the environmental ecosystem. Pollutants can be in any physical state: solids, liquids or gases. We can classify pollutants in the following ways:

- By their effects: Does the pollutant cause local damage or does it damage the entire **ecosystem**?

Figure 26.1
Oil slick on Red Mangroves, Port Royal, Jamaica ▶

• By their properties: Are they toxic, and how long is their **half-life**.

Half-life is the time required for half the mass of a substance to break down (decay).

Figure 26.2
Solid waste, mainly plastic, can be seen on the Kingston Harbour shoreline. Plastic has a very long half-life ▶

• Natural or artificial: Carbon dioxide can be produced both artificially by combustion and also naturally from volcanoes.

Figure 26.3
Carbon dioxide can be produced naturally from volcanoes. ▶

• By how easy it is to control: Dust is easy to filter out of the air in factories, but gases are harder to remove.

▲ **Figure 26.4** in 2013, the staff of the RJR Communications Group on Lyndhurst Road in Kingston, Jamaica were forced into a mandatory evacuation because of certain fumes. The fumes also affected other businesses in the area.

Pollution is any damage caused to the environment by a pollutant. There are many sources of pollution, for example, the burning of fossil fuels; the burning of forests to clear land; the use of pesticides and fertilisers; waste materials from industrial processes such as mining, quarrying and smelting ores to make metals; the disposal of plastic and other non-biodegradable waste minerals, and sewage polluted water.

26.2 The composition of non-metals and their compounds in the atmosphere

The composition of the air is thought to have remained largely unchanged for the past 200 million years. The approximate percentages (by volume) of the main gases present in unpolluted, dry air are given in Table 26.1. There are also very small amounts of the other noble gases in air.

It is important to understand that the figures in Table 26.1 apply to dry air. Air can contain anywhere between 0% and 4% of water vapour.

▼ **Table 26.1** Percentage of gases in the unpolluted atmosphere

Non-metal	Composition (%)
nitrogen	78.1
oxygen	21.0
argon	0.9
carbon dioxide	0.03

26.3 Non-metals and the environment

Phosphates, nitrates and sulphates

Phosphates, nitrates and sulphates are found in syndets (to increase cleaning power), fertilisers, untreated sewage and adenosine triphosphate (ATP) molecules. The energy produced from respiration is stored in ATP to be released when the body requires it.

When there is a high concentration of phosphates and nitrates in water bodies (for example, rivers, stream and seas) they cause excessive growth of algae. This excessive growth is called eutrophication. The concentration of oxygen available to aquatic animals is reduced when dead plants decay. The surface growth also reduces the amount of sunlight that reaches underwater plants, which helps to oxygenate the water.

▲ **Figure 26.5** Kingston Harbour, Jamaica, is the third largest port in the Caribbean and Latin America. The rich biodiversity of the harbour is under threat from pollution and alien species brought there by shipping. The very feature that makes Kingston Harbour a safe and sheltered haven, contribute to its vulnerability to pollution: it is largely landlocked, with a relatively small openings on the southwestern end. The waters of Kingston Harbour are considered one of the most highly polluted in the Caribbean.

Hydrogen sulphide

Hydrogen sulphide (H_2S) gas smells of rotten eggs. Hydrogen sulphide gases are released when sulphur-containing compounds in volcanic gases and some mineral springs decay.

Hydrogen sulphide reacts with a number of cations to form insoluble sulphides. Some of these cations are needed in small amounts for metabolic activities, and for the formation of bone, pigment and haemoglobin. As a result, inhaling H_2S is toxic to animals.

Sulphur dioxide

Sulphur dioxide, which is produced during the extraction of metals from sulphide ores (for example, copper from copper pyrite), is an acidic anhydride. Sulphur dioxide is also released into the air when coal is burned and crude oil is refined.

SO$_2$ mixes with smoke and water vapour to form smog, which damages the lining of the respiratory tract. SO$_2$ accumulates in the stomata of leaves and kills the cells. Acid rain is created when SO$_2$ dissolves in rain. Acid rain damages leaves and sculptures, as well as acidifies the soil.

Chlorofluorocarbons (CFCs)

Aerosol sprays, foam plastics and some refrigerants contain CFCs. CFCs are odourless, non-toxic, heat resistant, inert, non-flammable and can be easily liquefied. Consequently, these properties allow CFCs to escape into the air without being detected and move up into the atmosphere.

Scientists believe that CFCs might be responsible for destroying the protective ozone layer. The ozone layer prevents harmful ultraviolet (UV) radiation from reaching the earth. UV light splits CFC molecules, producing chlorine atoms, which cause a chain reaction that destroys ozone.

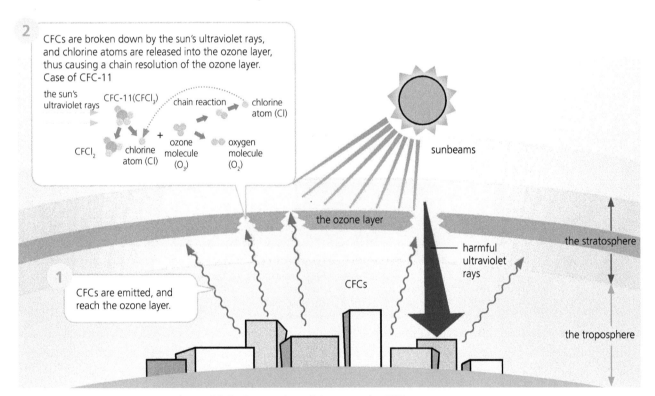

▲ **Figure 26.6** Destruction of the ozone by CFCs

Carbon monoxide

Carbon monoxide (CO) is formed from the incomplete combustion of carbon or carbon compounds. This includes decay of organic matter, burning of fossil fuels and reduction of metallic oxides by carbon.

Carbon monoxide combines more readily with haemoglobin than oxygen. Thus, CO reduces the oxygen concentration of the blood. As a result, the cells become starved of oxygen. Therefore, prolonged exposure to CO can lead to death.

Carbon dioxide

Carbon dioxide is a product of all combustion reactions and it is used during photosynthesis. This means that the increase in the use of fossil fuels, deforestation and population explosions lead to the increase in the carbon dioxide concentration in the atmosphere.

When radiation from the sun penetrates the Earth's atmosphere, only some of the radiation escapes back into space; the rest is absorbed by gases in the atmosphere. Gases such as CO_2, methane, CFCs and nitrous oxide (N_2O) increase the amount of radiation trapped. This trapping of solar radiation is called the **greenhouse effect**. The radiation heats up the air and causes **global warming** and changes in rainfall patterns. This leads to the polar icecaps melting, which causes flooding of lowlands and reduced harvests.

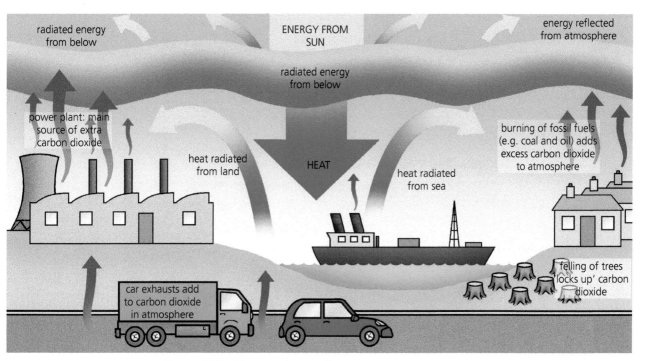

▲ **Figure 26.7**　The greenhouse effect

26.4　Non-metals in living systems

Non-metals are also essential to the development and maintenance of healthy organisms. The following tables explain the functions of the non-metals nitrogen, phosphorus and sulphur in plants.

▼ **Table 26.2**   The primary plant nutrients

Nutrient	Source	Comments
Nitrogen	Becomes available through nitrogen fixation and from dead organisms and animal wastes. Supply replenished by organic manure and artificial fertilisers.	Necessary part of proteins. Part of chlorophyll. Produces lush green growth and increases plant vigour. Increases fruit and seed production. A deficiency causes yellowing of leaves and reduced branching.
Phosphorus	Phosphorus often comes from fertiliser bone meal and superphosphate.	Needed especially by plants with bulbs and flowering plants and non-leafy vegetables. Important in plant processes such as photosynthesis. A deficiency causes leaves and stems to develop a purple coloration.

▼ **Table 26.3**   The secondary plant nutrients

Nutrient	Availability/source	Comments
Potassium	Potassium is supplied to plants by soil minerals, organic materials and fertiliser.	An enzyme activator: its level determines the rate of some reactions. A deficiency causes stunted growth and premature loss of leaves.
Calcium	Available as free Ca^{2+} ions or in complex ions. Provided by calcium hydroxide, calcium carbonate and calcium sulphate.	A deficiency leads to growth abnormalities in the tips of shoots and roots.
Magnesium	Available as free Mg^{2+} ions or in complex ions. Important in chlorophyll formation. Provided by magnesium sulphate and magnesium carbonate.	A deficiency leads to growth abnormalities in the tips of shoots and roots.
Sulphur	Provided by the pure element and ionic sulphates (SO_4^{2-}).	A constituent of some proteins, enzymes and glycosides. Deficiency leads to stunted. yellowish plants.

▼ **Table 26.4** Plant micronutrients

Nutrient	Availability/source	Comments
Iron	Provided by iron(II) sulphate.	An important component of the catalyst involved in chlorophyll formation. Deficiency leads to a yellowing of leaves. (If too much lime or phosphate is added to soils, the iron becomes less available.)
Boron	Provided by disodium tetraborate (borax).	Only trace amounts needed. Associated with the biochemical processes that lead to cell division. Toxic, if applied in excess of a critical maximum.
Zinc	Provided by zinc sulphate.	Involved in the synthesis of chlorophyll. Soya bean and sesame plants, for example, have high demands for zinc.
Manganese	Provided by manganese(II) sulphate.	Deficiency leads to poor flavours in some crops, e.g. the potato.
Copper cereals.	Provided by copper(II) sulphate.	Necessary for the healthy growth of cereals. Involved in the production of vitamin C and carotene.

Macronutrients are elements found in the soil that are required in relatively large amounts by plants.

Micronutrients are elements found in the soil that are only needed in minute (micro) quantities.

Non-metal elements are combined in compounds such as proteins and carbohydrates that are synthesised in plants and pass into animals through the food chain. Proteins are important for growth and carbohydrates are important in storage and supply of energy in living organisms.

Deficiencies in **macronutrients** and **micronutrients** can lead to diseases in plants and animals. Health problems can also arise if ions are taken in larger quantities than are necessary. These problems occur especially in detoxifying organs such as the liver and kidneys.

26.5 Three important natural cycles

Many of the elements needed by plants and animals are taken in as compounds and most of these are recycled naturally through the environment.

Oxygen, carbon dioxide and nitrogen play important roles in the biochemical cycles that sustain living matter. These cycles keep important materials in circulation. Without this recycling, important materials would eventually be used up.

The oxygen/carbon dioxide cycle

Two main processes are featured in the oxygen/carbon dioxide cycle – **photosynthesis** and **respiration**.

The oxygen/carbon dioxide cycle is summarised in Figure 26.8.

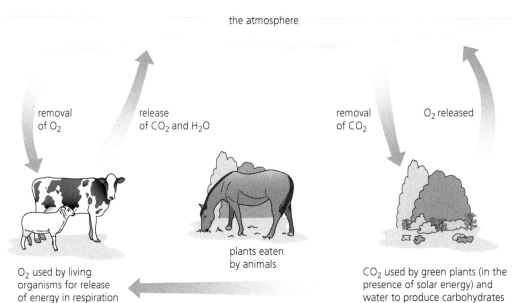

the atmosphere

removal
of O_2

release
of CO_2 and H_2O

removal
of CO_2

O_2 released

plants eaten
by animals

O_2 used by living
organisms for release
of energy in respiration

CO_2 used by green plants (in the
presence of solar energy) and
water to produce carbohydrates
in photosynthesis

Figure 26.8
The oxygen/carbon
dioxide cycle ▶

Carbon dioxide and water combine in photosynthesis to form the simple carbohydrate, glucose. Photosynthesis involves the green plant pigment chlorophyll and sunlight.

$$6CO_2(g) + 6H_2O(l) \xrightarrow[\text{chlorophyll}]{\text{light energy}} C_6H_{12}O_6(aq) + 6O_2(g)$$

Plants usually convert simple glucose molecules into complex polymers, one of which is starch. When animals eat plants, the starch is digested in stages to reform the simple sugar glucose.

Glucose can readily release energy when it is burnt (oxidised) in the presence of oxygen. Oxygen is taken in through the lungs when animals breathe in air. The process by which glucose is oxidised to form carbon dioxide, water and energy is known as respiration. The equation for the reaction is:

$$C_6H_{12}O_6(aq) + 6O_2(g) \rightarrow 6CO_2(g) + 6H_2O(l) + \text{energy}$$

Note that this equation is the reverse of the equation for photosynthesis.

The importance of the oxygen/carbon dioxide cycle

Respiration and photosynthesis help to maintain the concentration of CO_2 in the atmosphere at an approximately constant level. Presently, the carbon dioxide concentration in the air is increasing, mainly because of the burning of carbon-based fuels to supply energy for transportation and for industrial processes.

The importance of the oxygen/carbon dioxide cycle to living things is clear. Both animals and plants need the oxygen released from plants during photosynthesis for respiration. Respiration supplies energy that plants and animals need for the many chemical reactions that occur in living organisms. Respiration also supplies the carbon dioxide needed by plants for photosynthesis.

The **nitrogen cycle** is the flow of nitrogen from the atmosphere back to the atmosphere, via atmospheric fixation, industrial fixation and biological fixation.

The nitrogen cycle

Animals and most plants cannot utilise the element nitrogen directly. The nitrogen must first be fixed, i.e. combined with oxygen and hydrogen, before being taken up by plants. In turn, the nitrogen contained in plants is consumed by animals.

Nitrogen is converted (fixed) into compounds which can be used by plants and animals by:

- atmospheric fixation;
- industrial fixation;
- biological fixation.

Atmospheric fixation

Atmospheric fixation is achieved by lightning. The electrical discharge in lightning provides sufficient energy for nitrogen and oxygen to combine to form nitrogen monoxide:

$$N_2(g) + O_2(g) \rightarrow 2NO(g)$$

The nitrogen monoxide is further oxidised by oxygen in the air to form nitrogen dioxide, which then reacts with rainwater to produce nitric acid:

$$2NO(g) + O_2(g) \rightarrow 2NO_2(g)$$
$$2NO_2(g) + H_2O(l) \rightarrow HNO_2(aq) + HNO_3(aq)$$

This nitric acid forms nitrates on reacting with soil materials. These nitrates can be incorporated into plant proteins.

Industrial fixation

Industrial fixation involves the direct synthesis of ammonia from nitrogen and hydrogen, using the Haber process (see Section 25.2).

The ammonia is subsequently converted to ammonium salts and urea, which are then used as nitrogen sources (fertilisers) for plants.

Biological fixation

In biological fixation, specialised micro-organisms (nitrogen-fixing bacteria) convert nitrogen in the air into compounds such as ammonia which plants can utilise. Nitrogen-fixing bacteria can be found on the root nodules of leguminous plants (clover, peas, beans and the like).

Continuing the cycle

When plants and animals die they return nitrogen to the soil. The returned nitrogen may be recycled through a new generation of plants and animals, or it may be returned to the atmosphere.

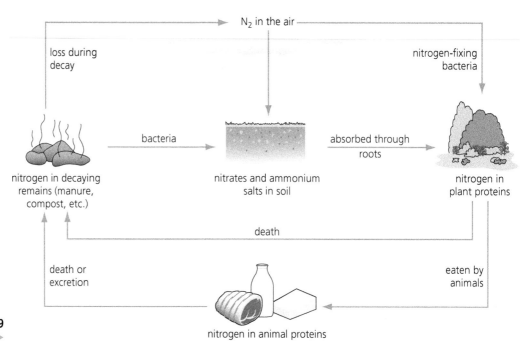

Figure 26.9
The nitrogen cycle ▶

The importance of the nitrogen cycle

The percentage of nitrogen that is recycled is quite small. The nitrogen cycle functions mainly to control the quantity of nitrogen in the soil that is available for use by the plant.

Without the nitrogen cycle, plants will not be able to produce proteins. Animals too will be deprived of proteins because they use both plants and animals as a source of proteins.

The water cycle

In the water cycle, water is continually exchanged between the surface of the Earth and the atmosphere. The exchange is made possible by energy from the sun and the pull of gravity. Unlike the nitrogen and carbon cycles, in which chemical changes occur, the processes in the water cycle are mainly physical (see Section 1.5).

The essential processes in the water cycle are shown in Figure 26.11. They are:

*   evaporation;
*   precipitation;
*   run-off.

Practice

3 'Unlike the nitrogen and carbon cycles, in which chemical changes occur, the processes in the water cycle are mainly physical'. Justify this statement by using suitable examples.

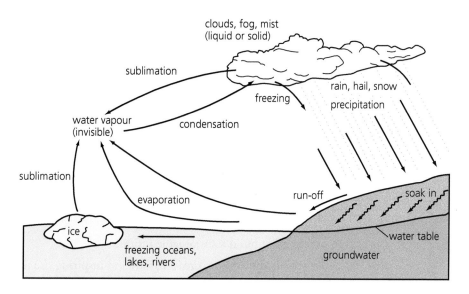

Figure 26.10
The water cycle ▶

The importance of the water cycle

Sources of water

The water that is recycled in the water cycle come from several sources. These sources can be broadly classified as surface sources and groundwater.

- **Surface sources**, for example, rivers, lakes and ponds, are on the Earth's surface. This does not include the water in the oceans or water locked within the Earth. Sometimes rivers are dammed to serve as collection centres. We should use surface water with great care because it is still the major source of water for our needs.

- **Groundwater**, or sub-surface water, includes soil moisture, groundwater within a 0.8 km depth and groundwater lying deep within the Earth. Underground water reserves are much larger than those on the surface, but as they are unseen we tend to underestimate them.

Water conservation

Modern society places high demands on our water reserves. We use water:

- in homes for washing, drinking and cooking;
- in industries such as paper-making, production of soft drinks, in beer production and other chemical industries.

Our population has grown and is still growing and our lifestyle is more sedentary than nomadic communities of long ago. Our manufacturing industries have expanded. The tourism industry, on which many Caribbean islands depend for foreign exchange, uses large quantities of water. In addition, long periods of drought can deplete our water resources.

Individuals and organisations can help improve the situation by:

- installing water-saving devices on showers in the home;
- using collected rainwater rather than tap water for gardens and lawns;
- increasing the tariff on water that is used excessively;
- recycling water where practical, e.g. recycling waste water.

26.6   Green Chemistry

Green chemistry, also called sustainable chemistry, is defined as the design of chemical products and processes through chemical research and engineering that reduce or eliminate the use and generation of hazardous waste.

Green chemistry helps to maintain the balance of nature by:

* Emulating nature through the use of renewable materials that are readily biodegradable in the environment.
* Reducing the adverse impact of chemistry on the environment through the prevention of pollution at its source and mitigating the use of fewer natural resources.
* Using materials with more efficiently less energy.
* Helping to build a sustainable future.
* Fostering innovation, creating jobs and inspiring the next generation of chemists.

Green chemistry allows for a clean and sustainable way in which new discoveries does not harm the planet. The practise of green chemistry is also economically beneficial with many positive social impacts.

Examples of Green Chemistry

Several companies throughout the world have developed materials that minimises the release of hazardous waste on the environment. These companies have developed:

* Biodegradable plastics which are made from renewable biodegradable sources. For instance the scientists, of NatureWorks in Minnesota, discovered a method where microorganisms convert corn starch into a resin that is just as strong as the rigid petroleum-based plastic currently used for containers such as water bottles and yogurt pots. Also Baden Aniline and Soda Factory (BASF) which is the largest chemical company in the world has developed a biodegradable bag that completely disintegrates into water, carbon dioxide and biomass in industrial composting system. The use of these bags instead of conventional plastic bags for kitchen and yard waste will easily degrade in municipal composting systems.

▲ **Figure 26.10**   Many supermarkets across the Caribbean region are now using biodegradable plastic bags and reusable green bags.

- Sherwin-Williams has developed a water based acrylic alkyd paint with low volatile organic chemicals (VOCs) that can be made from recycled soda bottle plastics (polyethylene terephthalate – PET), acrylics and soyabean oil. In 2010, Sherwin-Williams manufactured enough of these new paints to eliminate over 800,000 pounds, or 362,874 kgs, of VOCs.

The twelve principles of green chemistry

The twelve principles of green chemistry are:

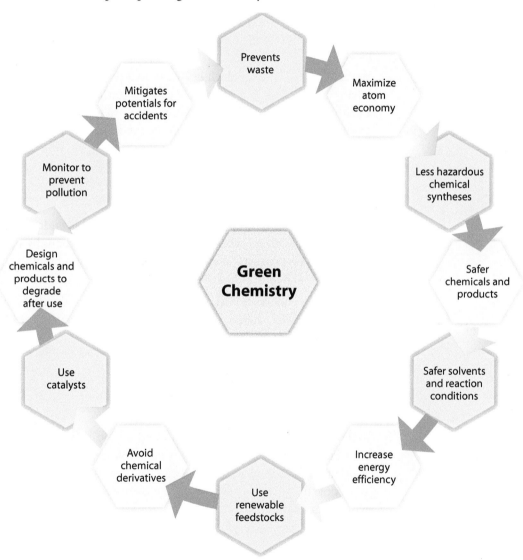

1 **Prevention** – the reduction and prevention of waste generation creates a more sustainable environment which minimizes hazards associated with waste treatment, transportation and storage.

2 **Atom economy** – this is a concept that evaluates the efficiency of a chemical transformation. Atom economy is a ratio of the total mass of atoms in the desired product to the total mass of atoms in the reactants, which is similar to percentage yield.

The reduction of haematite (Fe_2O_3), is reduced by carbon monoxide in the blast furnace, leads to an atom economy of

$$Fe_2O_3 \quad + \quad 3CO \quad \rightarrow \quad 2Fe \quad + \quad 3CO_2$$

haematite	carbon		iron
(55.85 x 2) +	monoxide		2(55.85)
(3 x 16)	3(12+16)		=111.7
=159.7	= 84		

Waste is minimized when the chemical transformations is designed to maximize the incorporation of all materials used in the process into the final product, resulting in few if any wasted atoms. The atom economy should be as close to 100% as much as possible.

For example, Ibuprofen which is used to reduce fever and treat pain or inflammation, became obtainable without prescription in the 1980s which utilized a method with an overall atom economy of 40% which translates into 60% waste products. This means that if 30 million pounds of ibuprofen is produced each year, then more than 35 million pounds of waste is generated which is not sustainable for the environment. Through the application of green synthesis there was atom economy of 99% which translates to less than 500,000 pounds of waste for the production of ibuprofen.

3 **Design less hazardous chemical synthesis** – use less hazardous reagents whenever possible and design processes that do not produce hazardous by-products to generate only benign by-products. These practices will generate substances that possess little or no toxicity to human health and the environment.

4 **Design safer chemicals and products** – the toxicity of the chemical products should be reduced when designed so that they only affect their desired function. New products can be designed through green chemistry that are safer for the environment.

5 **Use safer solvents/reaction conditions** – harmful solvents, catalysts and substances that are used in separation techniques should be minimized or not used to maintain a sustainable environment.

6 **Increase energy efficiency** – synthetic and purification methods should be designed to be performed at room temperature and pressure, if possible, so that energy costs associated with extremes in temperature and pressure are minimized. As much as possible the heat evolved in exothermic reactions should be used for another experiment and not released to the environment to cause heat pollution.

7 **Use renewable feedstocks** – raw materials and feedstocks that are renewable should be used whenever possible to prevent the depletion of natural resources. Feedstocks are natural occurring materials that have undergone some slight processing which is used as a starting material for a chemical process. Examples of depleting feedstocks are haematite – Fe2O3 – and magnetite – Fe3O4 (iron ores). Examples of renewable feedstocks include agricultural products.

8 **Avoid chemical derivatives** – the use of protecting or blocking groups tend to require additional reagents and can generate waste. Therefore synthetic transformations are more selective and will eliminate or minimize the need for protecting groups. Alternative synthetic sequences may eliminate the need to convert functional groups in the presence of other sensitive functionality.

9 **Use catalysts** – the use of catalysts enhance the selectivity of a reaction, reduce the temperature of a reaction, and enhance the extent of conversion to products and reduce-based waste. The reduction of temperature saves energy and removes the possibility of unwanted side reactions.

10 **Design for degradation** – the products of a chemical reaction should be designed so that they disintegrate at the end of their function and do not persist in the environment. The disintegrated products should not stay long in the environment.

11 **Monitor to prevent pollution** – the progress of the reaction should be monitored to detect the presence of pollutants such as unwanted by-products. The analytical methods should be developed to control and minimize the formation of hazardous substances.

12 **Minimize the potential for accidents** – the potential for chemical accidents may be reduced by selecting reagents and solvents that minimize the possibility of explosions or fires. The physical state (solid, liquid or gas) or composition of the reagents may be altered so as to lessen the risks associated with these types of accidents.

Practice

4 Calculate the percentage atom economy for the named product in the reactions shown below:

a) sodium in the reaction $\qquad$ $2NaCl \rightarrow 2Na + Cl_2$

b) hydrogen in the reaction $\qquad$ $Zn + 2HCl \rightarrow ZnCl_2 + H_2$

c) calcium oxide in the reaction $\qquad$ $CaCO_3 \rightarrow CaO + CO_2$

d) sulfur trioxide in the reaction $\qquad$ $2SO_2 + O_2 \rightarrow 2SO_3$

e) oxygen in the reaction $\qquad$ $2H_2O_2 \rightarrow 2H_2O + O_2$

5 Comment on the percentage waste product for the calculations in question four above.

Summary

- Non-metals are important to living things for respiration (O_2), photosynthesis (CO_2), nutrients and cell structures (C, H, and O).

- Some non-metals, usually in large amounts, pollute the environment and may be toxic to living things. Such impacts include global warming (CH_4 and CO_2), asphyxiation (CO), acid rain and photochemical smog (SO_2 and NO_2) and eutrophication (nitrate and phosphate fertilisers).

- Some common non-metals can be identified from their peculiar properties. For example, H_2 gives an small explosion when lit with a splint, O_2 relights a glowing splint, CO_2 forms a white precipitate with lime water, NH_3 turns moist pink litmus blue, Cl_2 bleaches moist blue litmus, SO_2 bleaches manganate (VII) paper, NO_2 is an acidic brown gas that seems to relight a glowing splint, and H_2O turns blue anhydrous cobalt chloride pink.

- Biochemical cycles such as the nitrogen and carbon cycles sustain living matter.

- In the nitrogen cycle, nitrogen from the air is fixed through reactions with both oxygen and hydrogen.

- Fixation of nitrogen takes place in the air, in industry and in living organisms.

- The changes that occur in the carbon cycle involve two main biochemical processes: photosynthesis and respiration.

- Photosynthesis and respiration help to maintain the concentration of carbon dioxide and oxygen in the air.

- The burning of fuels can upset the balance of photosynthesis and respiration.

- The physical changes in the water cycle are also essential to maintenance of life.

- In the water cycle water evaporates from seas, rivers, lakes and from leaves of plants. This water later condenses and falls as rain.

- The water cycle keeps us supplied with fresh water.

- A range of elements are needed for the normal growth of green plants.

- The essential plant nutrients are divided into:
  - non-mineral elements (from the air and water) – C, O and H;
  - mineral elements (from the soil).

- The mineral elements are divided into:
  - macronutrients (needed in relatively large amounts);
  - micronutrients (needed in minute amounts) – Fe, B, Zn, Mn, Cu.

- Macronutrients are further divided into:
  - primary nutrients – N, P, K;
  - secondary nutrients – Ca, Mg, S.

End-of-chapter questions

1 Which is true of solid sulphur?
A It has a high melting point.
B It has a giant molecular structure.
C The intramolecular covalent bonds are weak.
D The intermolecular van der Waals forces are weak.

2 Which of the properties listed below does graphite share with most metals?
I It conducts electricity.
II It has a high melting point.
III It forms a giant molecular structure.
IV It combines with oxygen to form an acidic oxide.
A I and II only
B I and III only
C II and IV only
D II and III

3 Chlorine can be prepared by adding manganese(IV) oxide to concentrated hydrochloric acid. The function of the manganese(IV) oxide is to:
A Dry the gas
B Neutralise the acid
C Oxidise the acid
D Act as a catalyst

4 Which set of conditions favours a high yield of ammonia in the Haber process?
A High pressure and moderately high temperature
B Low pressure and the use of a catalyst
C Very low temperature and low pressure
D Very high temperature and low pressure

5 Chlorine may be obtained by an electrolytic process using a flowing mercury cathode and graphite anodes. Which of the following is/are not true about this process?
I the electrolyte is molten sodium chloride.
II the anode wears away with time.
III an alloy is formed at the cathode.
IV sodium hydroxide is a by-product of the reaction.
A I only
B I and II only
C I and III only
D III and IV only

6 Which of the following are involved in some way in making nitrogen available to plants?
I The Haber process
II The Contact Process
III Denitrifying bacteria
IV Lightning storms
A I and II only
B II and III only
C I and IV only
D I, II and IV only

7 A gas which is acidic to litmus and also decolorises acidified potassium manganate(VII) is:
A Chlorine
B Sulphur dioxide
C Nitrogen dioxide
D Carbon dioxide

8 All of the following processes occur in the water cycle except:
A Hydrolysis
B Evaporation
C Condensation
D Crystallisation

9 All of the following equations represent reactions found in the carbon cycle except:
A $C(s) + O_2(g) \rightarrow CO_2(g)$
B $C_6H_{12}O_6(aq) + 6O_2(g)$
$\rightarrow 6CO_2(g) + 6H_2O(l)$
C $H_2O(g) + CH_4(g) \rightarrow 3H_2(g) + CO(g)$
D $CO_2(g) + H_2O(l) \rightarrow H_2CO_3(aq)$

10 Which of the following all pollute the environment?
A Calcium carbonate, carbon dioxide, silica
B Chlorofluorocarbons, carbon dioxide, rock salt
C Sulphur dioxide, nitrogen dioxide, carbon monoxide
D Soaps, hydrogen sulphide, iron sulphide

11 What is meant by each of the following terms?

a Polar molecule   **c** Hardness of water

b Ionic lattice   **d** Surface tension

12 List four unique properties of water.

13 Why is water essential for life?

14 List four reactions that lead to rainwater being acidic.

15 What is the link between the acidic nature of rain water and the hardness of water from natural sources?

16 Distinguish between soft and hard water.

17 Discuss methods for softening hard water.

18 List the advantages of soft and hard water.

19 Describe the water cycle, commenting on its importance to life.

20 Why is water from many natural sources not suitable for drinking?

21 How is water from natural sources made suitable for drinking?

22 Is there enough water to meet the needs of:

a your community?

b the world?

23 List reasons why some places have water shortages.

24 List the ways in which we pollute our water supplies.

25 Find out what is meant by the term 'algal bloom'. Why are algal blooms undesirable?

26 Thermal pollution is the warming of natural water supplies above normal temperatures. Identify possible sources of thermal pollution in your country. How does thermal pollution affect the quality of the water supply? What steps can be taken to reduce thermal pollution from the source(s) which you have identified?

27 Lead is a chemical which pollutes some of our water supplies.

a What are the possible sources of lead contamination?

b What steps can be taken to reduce contamination from lead?

28 a Nitrogen is described as a 'primary plant nutrient'. Distinguish between 'primary' and 'secondary' plant nutrients.

b Explain how nitrogen is important to plant growth.

c Name and describe two natural processes by which nitrogen is returned to the soil.

29 Soils to which ammonium fertilisers have been repeatedly applied become acidic.

a Give an explanation for this observation.

b Describe how you would determine the pH of a soil sample.

c Describe how acidic soils can be treated. (Include equations in your answers where applicable).

30 Many farmers are choosing organic farming methods to cultivate their crops. Two practices they may employ are the use of organic fertilisers and the use of biological pest control. Compare the advantages and disadvantages of using:

a organic fertilisers over inorganic fertilisers;

b biological pest control over chemical pest control.

27 Water

> **In this chapter, you will study the following:**
>
> - the properties of water, which include: density changes, solvent properties, specific heat capacity and volatility;
> - how to relate the unique properties of water to its functions in living systems;
> - the difference between hard and soft water;
> - the temporary and permanent hardness of water;
> - the consequences of the solvent properties of water;
> - the methods used in the treatment of water for domestic purposes;
> - leaching;
> - chemical equations for softening water.
>
> **This chapter covers**
> Objectives 5.6–5.8 of Section C of the CSEC Chemistry Syllabus.
>
> Water is essential to all life on Earth.

Mona Reservoir

▲ **Figure 27.1** Mona reservoir supplies water for the whole of Kingston, Jamaica

27.1 The quality of water

Potable water is water that is safe for drinking.

Water quality is affected by the inorganic and organic materials it contains. **Potable** water is water that comes through the tap and is safe for drinking. Potable water is free from undesirable colours, odours tastes, sediments and pathogens. Potable water is created when natural sources of water are treated. As rain falls onto soil, the water becomes contaminated with soil materials such as minerals, bacteria and any pollutants.

27.2   Water: A unique commodity

Water is essential to all forms of life. The unique properties of water and how it impacts on living things and the environment are described in Table 27.1.

▼ **Table 27.1**   The impact of water's properties on living things and the environment

Property	Explanation	Impact on living things and the environment
High specific heat (the heat requried to raise the temperature of 1 g of water by 1 °C)	This means that water absorbs plenty of heat and yet its temperature will rise only slowly. On the other hand when water cools, it releases a lot of heat.	Gradual temperature changes mean a more stable environment for many aquatic species. Many of the chemical reactions in living things need a liquid environment and water remains a liquid even after absorbing a lot of heat. The temperature of large bodies of water remains relatively constant.
High heat of vaporisation (the regions – when amount of heat needed to change a substance from liquid to gas)	A lot of heat is required for liquid water to change to a gas, due to the presence of the hydrogen bonds.	As water evaporates it takes heat up from warmer regions of Earth and transfers it to cooler regions – when it condenses and releases the stored heat. In this way the Earth's heat is spread out keeping the temperatur more moderate.
Relatively high melting and boiling points	Water has higher melting and boiling points than other similar substances. It remains as a liquid over the important temperature range of 0–100 °C.	Liquid water has solvent properties which is essential for chemical reactions of living organisms. Much of the Earth's surface is covered with water.
Water is a very good solvent	The polarity of water allows it to dissolve ionic substances as well as polar covalent substances.	Many reactions in living things take place in solution. Water provides a suitable medium for these reactions. Plants get their nutrients from solutions in the soil. Too much water results in: • leaching of nutrients from the soil; • washing out of fertilisers into water bodies causing eutrophication; • run off of pesticides into rivers and streams; • acid rain.
Cohesion and high surface tension: the surface tension of water is greater than that of all liquids except mercury*	Hydrogen bonds hold water molecules together. Water surfaces tend to have a 'skin'.	Water molecules stick together, thus water can move as a column up the stems of plants and move through small blood vessels in animals. Water surfaces can support some animals. Some insects, for example, can 'walk' across the water surface.

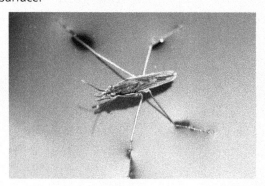

continued

▼ **Table 27.1**   *continued*

Density: below 4 °C the density of water decreases instead of increasing	Above 4 °C, water behaves like other substances, that is, as the temperature increases, it expands and its density decreases. Below 4 °C, however, the density of water decreases as temperature decreases. This means that ice is less dense than liquid water and floats on liquid water.	When the surface temperature of lakes, for example, falls below 0 °C, ice forms on the surface. The ice insulates the water below so that it does not lose heat and freeze. Living things can thus survive.

*Surface tension is that property of a liquid that makes it behave as if its surface is enclosed in an elastic skin. Surface tension arises from forces between the molecules of the liquid. Cohesion refers to the force of attraction between like molecules. It reflects the tendency of a substance to hold together.

Understand it better

Why can water dissolve both ionic and polar covalent substances?

The ability of water to dissolve both ionic and polar covalent substances is linked to the polarity of the water molecule. To dissolve a solid ionic compound, the water molecules must attract the ions from the crystal lattice and then surround them.

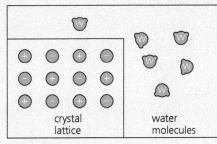

Stage 1
Separate ionic crystal and water molecules

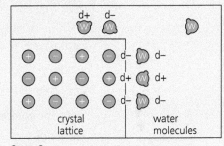

Stage 2
The following interactions are taking place:
1 the attraction of water molecules to water molecules
2 the attraction between oppositely charged ions
3 the attraction between ions and water molecules

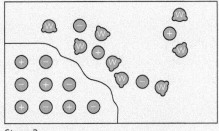

Stage 3
The crystal lattice is broken up because ion–water attractions are stronger than ion–ion attractions or water–water attractions put together

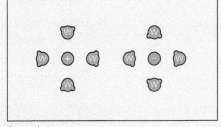

Stage 4
'Free' hydrated ions

▲ **Figure 27.2**   Water dissolving an ionic substance.

Water can dissolve a polar covalent substance such as ethanol (Figure 27.4).

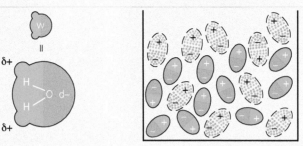

▲ **Figure 27.3** The structure of water

▲ **Figure 27.4** Water mixes completely with ethanol because of mutual attractions between dipoles. Ethanol is also dipolar.

Practice

Table 27.2 summarises the solvent properties of water.

▼ **Table 27.2** The solvent properties of water

Substance	Solubility in water (g per 100 cm³) and comments	Particles in aqueous solution and comments
ionic		
potassium nitrate	247	$K^+(aq)$ $NO_3^-(aq)$
sodium chloride	39	
magnesium sulphate	74	$Mg^{2+}(aq)$ $SO_4^{2-}(aq)$
polar covalent		
ethanol		
sucrose		no ions present
non-polar covalent		
tetrachloromethane		
cyclohexane	two distinct layers observed	

Use the data presented in the table to answer the following questions.

1 Define the term 'solubility'.
2 Which of the ionic compounds has the highest solubility in water?
3 Complete the table by filling in the spaces.

27.3 Hardness of water

We say that water is hard when it contains minerals such as calcium and magnesium salts. Hard water leaves white fur or scale, called limescale, in boilers and kettles. Hard water reduces the ability of soaps to lather by reacting with it to form scum, which is left in basins and baths. On the other hand, hard water contains minerals that promote strong bones and teeth.

Figure 27.5
Limescale ▶

Soap lathers easily with distilled water, but in some countries it is difficult to get a lather with tap water or water from natural sources. Water in which soap lathers easily is described as 'soft', whereas water in which soap does not lather easily is described as 'hard'.

Causes of hardness of water

The hardness of tap water or of water from some natural sources is a direct result of the excellent solvent properties of water. Hardness in water is caused principally by the presence of dissolved calcium and magnesium salts.

Water becomes hard in two ways.

- Rainwater running over gypsum and anhydrite rocks slowly dissolves some of the calcium and magnesium sulphate out of these rocks. Calcium and magnesium sulphate are the major sources of **permanent hardness** in water.

Permanent hardness is hardness which cannot be removed by boiling. This type of hardness is caused by dissolved calcium sulphate and magnesium sulphate.

- Rainwater contains dissolved carbon dioxide and other acidic gases and so rainwater has a pH of less than 7. Rain water falling on chalk, limestone or marble reacts with some of these materials producing a dilute solution of calcium hydrogen carbonate, according to the equation:

$$CaCO_3(s) + H_2O(l) + CO_2(g) \rightarrow Ca(HCO_3)_2(aq)$$

Temporary hardness is hardness which can be removed by boiling. This type of hardness is caused by dissolved calcium hydrogen carbonate and magnesium hydrogen carbonate.

Dissolved calcium hydrogen carbonate is the source of **temporary hardness** in water. This action is responsible for the formation of caves in limestone areas.

Find out more

How do stalactites and stalagmites form in caves?
It takes millions of years for stalactites to grow down from the ceiling of a cave. As the water drips from the roof of the cave, some of it evaporates, leaving behind a deposit of limestone, calcium carbonate. Limestone caves can only form in districts where the water entering the cave has run through limestone rocks.

Q1 Use suitable resource materials to find out how quickly stalactites and stalagmites are formed.

▲ **Figure 27.6**
Stalactites and stalagmites in Green Groto Cave, between Ocho Rios and Montego Bay, Jamiaca. The huge, labyrinthine limestone cave found here is 5 003 feet (1 525m) long and characterised by stalactites, stalagmites, overhead ceiling pockets, numerous chambers, light holes, and in its depths a subterranean lake. The caves have played an important role in Jamaican history: the island's original inhabitants, the Arawak Indians, used them for shelter; they were used as a hideout for Spaniards during the British take-over; they were a natural haven for runaway slaves; the Jamaican government used them to store barrels of rum during World War II and they were even used as a den for smugglers running arms to Cuba. Apparently they have also been used as a setting for scenes from a James Bond film.

Practice

4 Create a table comparing hard water and soft water.
5 a Explain why calcium carbonate does not cause hardness of water whereas calcium hydrogencarbonate does.
   b Explain why magnesium sulphate causes permanent hardness but calcium hydrogencarbonate causes temporary hardness.
6 If you were given a 1% soap solution, how would you compare the hardness of different water samples?

27.4 The treatment of water

The treatment of water is a process that reduces the amount of inorganic and organic materials it contains. We can treat water at home in the following ways:

- Boiling: Removes temporary hardness and some microbes.
- Filtering: Removes solid impurities and some microbes.
- Chlorinating: Used mainly for pools or water collected after a water-lock-off period to kill microbes.
- Softening: Removes hardness to conserve detergents and reduce scale formation in kettles and irons.

27.5 Removal of hardness

Hard water can be softened. The softening of water involves the removal of the salts of calcium and magnesium. If the salt is a carbonate, sulphate or chloride of calcium or magnesium, we say that the water has permanent hardness. Water that has permanent hardness cannot be softened by boiling.

However, if the salt is a bicarbonate, we say that the water has temporary hardness, that is, it can be softened by boiling. Water can be softened using the following methods:

- Boiling (removes only temporary hardness)
- Using sodium carbonate (washing soda)
- Using ion-exchange resins

▼ **Table 27.3** Methods used to remove hardness of water

Method of removal	Temporary hardness	Permanent hardness
Boiling	When water is boiled, the soluble calcium hydrogencarbonate is converted to the insoluble calcium carbonate: $Ca(HCO_3)_2(aq) \xrightarrow{heat} CaCO_3(s) + H_2O(l) + CO_2(g)$ Calcium carbonate is responsible for the scale or fur which builds up in kettles and boilers	Not affected
Use of sodium carbonate	This treatment removes both temporary hardness and permanent hardness. Dissolved carbonate magnesium ions or calcium ions are precipitated as the insoluble carbonate: $Na_2CO_3(aq) + Ca^{2+}(aq) \rightarrow CaCO_3(s) + 2Na^+(aq)$	
Use of ion-exchange resins	An ion-exchange resin (Permutit is the trade name of a widely-used one) contains adsorbed sodium ions. The calcium and magnesium ions in the water displace the sodium ions as they pass through the column. Calcium and magnesium ions now become adsorbed onto the resin: $Ca^{2+}(aq) + 2Na–Permutit \rightarrow Ca–Permutit + 2Na^+(aq)$	

Figure 27.7
An ion-exchange water softening device. When hard water is passed through an ion-exchange column, calcium ions and magnesium ions in the water are replaced by sodium ions, which do not cause hardness. ▶

Practice

7 Explain as fully as you can how water pipes get encrusted over time.
8 Remembering that rain leaves the clouds as chemically pure water, explain why much of our water supply is hard.

Leaching

Leaching is the removal of soluble substances as water percolates through the soil. Many of the nutrients in the soil can be leached out, making them less available to the plants. Soluble fertilisers can also be removed by leaching. This can lead to eutrophication.

Freshwater purification

Water from natural sources is usually contaminated. Water quality varies widely with environmental conditions and levels of human activity. Contaminated water must be treated to make it both potable and **palatable**.

Palatable water is water that is pleasant to taste.

The stages in water purification are shown in Figure 27.8.

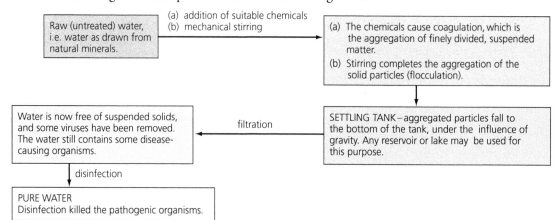

Raw (untreated) water, i.e. water as drawn from natural minerals.

(a) addition of suitable chemicals
(b) mechanical stirring

(a) The chemicals cause coagulation, which is the aggregation of finely divided, suspended matter.
(b) Stirring completes the aggregation of the solid particles (flocculation).

SETTLING TANK – aggregated particles fall to the bottom of the tank, under the influence of gravity. Any reservoir or lake may be used for this purpose.

filtration

Water is now free of suspended solids, and some viruses have been removed. The water still contains some disease-causing organisms.

disinfection

PURE WATER
Disinfection killed the pathogenic organisms.

▲ **Figure 27.8** Water purification

Find out more

What is waste water?
Water that has been used for domestic purposes and then discarded is described as sewage or waste water. Sewage or waste water, if not treated, poses a threat to health. In water treatment plants, bacteria break down the contaminants in the waste water to harmless products.

The liquid part of the sewage is then pumped out and allowed to percolate through filter beds. Organic matter is further broken down by aerobic bacteria. The water which emerges from the filter bed then flows into settling tanks where dead bacteria settle out. The treated water is then discharged into rivers, etc., or it can be recycled.

▲ **Figure 27.9** A sewage treatment plant

Find out if, and how, sewage is treated in your country.

Find out more

Why is purification of water necessary?

The quality of water deteriorates if:

- its dissolved oxygen is used up;
- toxic substances accumulate or are discharged into it.

Water pollutants include:

- sewage;
- pesticides;
- fertilisers;
- detergents, both soap and soapless ones;
- industrial waste.

Fertilisers get into the water supply either by run-off from farms or by percolation through the soil down to the water table. Fertilisers in the water supply provide nutrients, in the form of phosphates and nitrates, which are readily used by plants and algae. These aquatic plants and algae grow wildly and reach an undesirable condition of overgrowth called eutrophication. When the vegetation dies, its decomposition products are unsightly and have disagreeable odours. The rotting vegetation also uses up the dissolved oxygen supply in the water so it can no longer support fish and other aquatic animal life. Pollution by sewage and detergents can also result in eutrophication.

Both metallic and non-metallic elements and compounds have been implicated in polluting water:

- Mercury, which can accumulate in fish
- Lead compounds in paints, batteries and exhaust fumes of cars
- Carbon monoxide in exhaust fumes of motor vehicles
- Acidic anhydrides, such as carbon dioxide, sulphur dioxide, sulphur trioxide, and nitrogen dioxide
- Chlorofluorocarbons, which were used as aerosol propellants for many years

Q1 Find out the upper limits specified by the WHO for nitrate ions in drinking water. What are the negative effects of excess nitrate ions in potable water?

Disinfection of water

There are several ways to disinfect water, but chlorination is most frequently used in water treatment. As chlorine is a poisonous gas, very small amounts are added – only 5 ppm (parts per million). Disinfection is necessary because of water-borne diseases such as cholera, typhoid, paratyphoid, dysentery, infectious hepatitis and gastro-enteritis.

Raw water could have been taken through the stages as shown in the purification process, and still not be of an acceptable quality. World health standards specifying acceptable limits to the levels of contaminants in drinking water have been established and it is one of the functions of water authorities to ensure that these limits are not exceeded.

Desalination involves the removal of the dissolved salts, mainly sodium chloride and sodium iodide from sea water or brackish water.

Desalination

Where there is a depletion in fresh water supplies due to drought, for example in Antigua, or in cases where the demand for water is high, for example in Barbados (where the tourism industry places stress on water supplies), sea water or brackish water is used as the source for drinking water. The process of **desalination** is used.

Modern desalination plants use a process of **reverse osmosis**.

Find out more

What is reverse osmosis?

Reverse osmosis is the process where brackish water is forced under pressure through a thin filament or membrane. The membrane used must be semi-permeable (see Section 5.3).

If a salt solution is separated from fresh water by a semi-permeable membrane, movement of water from the pure water into the saline solution occurs by osmosis.

In reverse osmosis, the saline solution is forced through the semi-permeable membrane under pressure. Under these conditions, water leaves the brackish water and collects on the other side of the semi-permeable membrane. The pure water is collected in tanks and treated, and is then pumped into reservoirs which feed into the national supply.

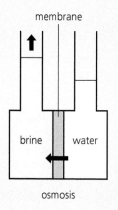
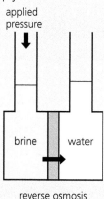

▲ **Figure 27.10** Osmosis and reverse osmosis

Summary

- Water has unique properties which make it essential to life.

- The unique properties of water include high melting and boiling points, high specific heat, high heat of vaporisation, high surface tension and a decrease in density with decreasing temperature below 4 °C. In addition, water is an excellent solvent.

- The solvent property of water is dependent on the fact that water is a polar molecule.

- During dissolution, polar water molecules are able to attract ions and polar molecules and surround them.

- Water is important biologically, because it is a good solvent and because it holds nutrients and other important biological materials in colloidal suspension.

- Natural water samples may be described as hard or soft.

- Soft water contains low levels of calcium and magnesium salts.

- Soft water lathers easily, it does not form much scum and it does not waste soap.

- Hard water contains significant levels of calcium and magnesium salts.

- Hard water can be softened by:
  - boiling (temporary hardness only);
  - addition of sodium carbonate (washing soda);
  - ion exchange resins.

- When rain falls, the water either soaks into the ground (groundwater) or runs into rivers and streams (surface water).

- Groundwater and surface water are sources of water for domestic and industrial uses.

- Water from natural sources must be treated before use.

- Treatment of water for domestic use must remove solid materials and kill disease-causing organisms.

- Water must be conserved because of increasing demands and reduced availability.

- Conservation measures focus on better water management through reduced wastage and recycling of waste water.

End-of-chapter questions

1 Which of the following is not a domestic home water purification method?
   A Filtration
   B Boiling
   C Chlorinating
   D Flocculation

2 Which of the following is not involved in water purification?
   A Crystallisation
   B Sedimentation
   C Filtration
   D Flocculation

3 Which compound is responsible for permanent water hardness?
   A $Ca(HCO_3)_2$
   B $MgCl_2$
   C $NaCl$
   D Na_2SO_4

4 Why is chlorine added to water?
   A To coagulate impurities
   B To increase the number of germs
   C To kill bacteria
   D To improve the taste

5 Which of the following allows two hydrogen atoms to bond to a single oxygen atom in a single molecule of water?
   A Hydrogen bonds
   B Non-polar covalent bonds
   C Polar covalent bonds
   D Van der Waals interactions

6 What is the name given to the attraction that allows the slightly negative charge of one end of a water molecule to be attracted to the slightly positive charge of another water molecule?

   A A hydrogen bond
   B A covalent bond
   C An ionic bond
   D A hydrophobic bond

7 The cohesiveness of water molecules is due to:
   A Hydrophobic interactions
   B Non-polar covalent bonds
   C Ionic bonds
   D Hydrogen bonds

8 Which property of water allows a water strider to walk across the surface of a small pond?
   A Cohesion and high surface tension
   B Density
   C High specific heat capacity
   D High heat of vaporisation

9 Why is water considered to be a good solvent?
   A Water has higher melting and boiling points than other similar substances.
   B The polarity of water allows it to dissolve ionic substances as well as polar covalent substances.
   C Hydrogen bonds hold water molecules together.
   D Above 4 °C, water behaves like other substances, that is, as the temperature increases, it expands and its density decreases.

10 Which pair of salts does **NOT** cause hardness of water?
   A Calcium carbonate and sodium sulphate
   B Calcium hydrogencarbonate and magnesium sulphate
   C Calcium carbonate and magnesium sulphate
   D Calcium sulphate and sodium carbonate

11 What is meant by each of the following terms?
  a Polar molecule
  b Ionic lattice
  c Hardness of water
  d Surface tension

12 List four unique properties of water.

13 Why is water essential for life?

14 List four reactions that lead to rainwater being acidic.

15 What is the link between the acidic nature of rainwater and the hardness of water from natural sources?

16 Distinguish between soft and hard water.

17 Discuss methods for softening hard water.

18 List the advantages of soft and hard water.

19 Describe the water cycle, commenting on its importance to life.

20 Why is water from many natural sources not suitable for drinking?

21 How is water from natural sources made suitable for drinking?

22 Is there enough water to meet the needs of:
  a your community?
  b the world?

23 List reasons why some places have water shortages.

24 Write an essay on waterborne diseases.

25 What steps should be taken to better conserve water:
  a in your household;
  b nationally;
  c internationally?

26 Identify some problems involved in the treatment of domestic and municipal waste water.

Index

abrasion 69
acid(s) 185, 186, 188, 194, 207
 anhydrides 188, 193, 211, 212
 anhydrous state 192
 basisity 210
 classifying 194
 concentrated 196
 definition 192
 dibasic 194, 203
 dilute 196
 hydrolysis 366
 indicators see indicators
 indigestion 189
 in living systems 189
 mineral 194
 monobasic 194
 non-oxidising 197
 organic 194
 oxidising 197
 preparation 194
 rain 188
 solids 197
 strong 194, 196-197, 211
 testing 197-199
 tribasic 194, 203
 in water 192, 196
 weak 194, 196-197, 212
acidity 189
 salts 189
activation energy 260
active
 ingredients 373
 metals 198
adenosine triphosphate (ATP) molecules 491
adhesives 349
aeroplane bodies 427
air 78
alcohol(s) 342
 breathalyser test 346, 357
 consumption 344, 355
  impact 344
aldehydes 298, 345
alkali metals 38
alkaline 185
 hydrolysis 364-368
alkalis 190, 201, 211
 conduct electricity 211
 strong 196, 212
 weak 196, 212
alkanes 296, 297, 300, 308, 324, 328, 337
 boiling point 325
 bonding 326
 branched 313
 butane 293
 chemical properties 328
 combustion reaction 328
 ethane 297
 homologous series 324, 325
 methane 297
 molecular formulae 324
 pentane 297
 physical states 324, 325
 propane 297
 reactions 313
 saturated compounds 333
 substitution reaction 327, 333
 unbranched 311
 uses 328
alkanoic acids 296, 298, 300-301, 338, 359-360
 boiling point 361, 362
 butanoic acid (butter) 359
 chemical reactions 359
 combustion 362
 conversion to esters 359
 functional groups 359

lactic acid (milk) 359
 neutralisation 363
 physical properties 339, 359, 360, 361
 reactions 363
  alkanols 363
  carbonates 362
  hydrogen carbonates 362
  hydroxides 362
  metals 362
  metal oxides 362
 soluble in water 361
 sources 336
 tartaric acid (grapes) 355
 uses 367-368
  preservative 367alkanols
alkanols 297, 299, 301, 335, 338, 344
 boiling point 348, 357
 chemical reaction 343
 combustion 347
 conversion to esters 347
 dehydration to alkenes 345
 fermentation process 336, 357
 flammable 338
 fuels 348
 functional groups 343
 organic acids 343
 oxidation 343
 physical properties 338-39
 reactions 354
 reactive metals 343
 sources 338
 uses 348
alkenes 298, 299, 232-233, 310, 316, 331-335, 336, 340
 addition reactions 333, 336
  hydrogen 333
 chemical properties 332
 combustion 333-334
 hydration 335
  direct 335
  indirect 335
 molecular formulae 331
 physical states 331, 332
 reactions 316, 332
 sources 332
 unsaturated compounds 336
 uses 337
alkyl groups 299, 233
alkynes 299, 331
allergic medicines 357
allotropes 63, 70
allotropy 45, 63, 70
 carbon 70
 phosphorus 70
 sulphur 70
alloys 78, 79, 428
 uses 428
alpha particles 27
alumina
 electrolysis 422-423
aluminium 20, 22, 43, 119-120, 122, 145, 182, 406, 407
 alloys 427, 428
 chloride 121
 conducting ability 121
 corrosion 250
 extraction 421-422, 429
 foil 153
 ore 418
 oxide 63, 121, 200, 347
 production 421-422
 reactions 122
  with chlorine 122
  with dilute acid 407
  with oxygen 122
  with water 122
 sulphate 132

Amazon River 82
amine 297
amino acids 95, 307, 390
 amine 301
 carboxylic acid functional group 306-307
 essential
 tests 95, 390
ammonia 49, 51, 52, 114, 156, 201, 219
 decomposition 160
 gas 4, 185, 201
 production 215
ammonium 133, 477
 chloride 53, 88, 139, 209
 fertilisers 477
 ion 133
 nitrate 283-284
 salts 199, 210
 sulphate 130, 147, 209
amperes 243
anions 44, 130, 235
 charges 130
anode 236
anodising 243
 aluminium 247
antacid 187
antihistamines 354
argon 119-120
arsenic
 agricultural insecticides 441
 groundwater 441
 old wood 441
 poisonous inorganic compounds 441
 smelting copper and zinc 441
ascorbic acid 173, 191
asphalt 316
aspirin 354, 357
astatine 113, 116, 117
 radioactive 115
atmosphere 490
 gases 490
 non-metals 490-491
 unpolluted 490
atmospheric fixation 497
atomic
 mass 16, 21
 numbers 16, 21, 31
 radius 39-40, 42
atom(s) 16, 31
 see also electrons; neutrons; protons
 charge 21
 composition 21
  nucleon number (mass) 21
  proton number (atomic) 21
 electronic configuration 19-20
 nucleus 18, 31
 polyatomic ions 222
 shells 31
 size 16-17
 structure 16-33, 43
 theories 17
Avogadro, Amedeo (1776-1856) 145, 155
Avogadro's
 Law 145, 155, 159, 158, 173, 176
  applications 157, 159
 number 145

baking
 powder 209, 213
 soda 191
balanced equations 137-138, 160, 178-179, 181
barium 106, 107
 chloride 175, 181, 182
 hydroxide 107
 reactions 107

sulphate 180, 181, 207
bases 189, 200-201, 211
 characteristics 201
 chemistry 191
 definition 200
 substances 201
basic oxides 189
bauxite 422, 423
Becquerel, Henri (1852-1908) 27
benzene 152
beryllium 106
beta particles 27
biological fixation 497
bitumen 317
blast furnaces 424-425, 430
blood 78
boiling 11-12
 point 54
bonding see chemical bonding
borax 191
brass 80, 236
bromine 32, 114-115, 116, 117, 292
 liquid 3, 115
 vapour 3
Brown, Robert (1773-1858) 5
Brownian motion 5-6
Buckminsrerfullerene 70

cadmium toxicity 441
calcination
calcium 105, 107, 344, 406, 416
calorimeter 281, 283, 284, 288
cancer 28
carbohydrates
carbon 19, 26, 54, 146, 311, 458
 allotropes 454
 -12 isotope 145
 -14 dating 29-30, 31
 bonding properties 291
 branched chain compounds 292-293
 dioxide 11, 49, 60, 62, 134, 147, 149, 157, 159, 177, 181, 256, 493
 monoxide 200, 492
 reducing properties 458
 saturation 292
  saturated 292
  unsaturated 29
 straight-chain molecules 290
carbohydrates 391, 393
carbonates 196, 197, 408
carbonic acid 192
carboxylic acids 359, 386
 see also alkanoic acids
catalysts 112, 266, 267, 270, 320
 biological (enzymes) 266
 in industry 270
cathode 236
cations 46, 109, 133
cement 204, 209
 Portland 209
chemical
 bonding 45-46, 57, 278, 286
 compounds 300
 equations 129, 133, 141, 174, 181, 184
 reactions 129, 160, 175, 213, 255, 260-261, 265, 272, 273, 275, 281-282, 287
chlorides 123
chlorine 22, 27, 34, 54, 113-115, 117, 119-120, 144, 144, 184, 241, 246
 atoms 220, 226
 disinfection of water 461
 gas 112, 115
chlorofluorocarbons (CFCs) 492
 molecules 492
 ozone layer destruction 492
chlorophyll 154

chromatography 93
cobalt 111
collisions 260
colloids 75, 83, 84, 85, 98
combination reactions 229
combustion 455
compounds 25, 45, 57, 97, 130, 132, 147, 221, 222, 289, 298-299
    ionic 131
    relative molecular mass 25
condensation 12
    polymerisation 377
    reactions 362
conduction 232
conversions 148
coordinate covalent bonding 52-53
copper 38, 57, 98, 110-111, 131, 144, 149, 178-179, 231, 247, 404, 407, 416
corrosion 247, 249, 250, 405, 426, 431
coulomb 245
covalent
    bonds 46, 49, 50, 52, 59
    compounds 129
    solids 63
cracking 319
crude oil 78, 312, 319, 334, 337
    fractional distillation 314, 315-317
crystallation 88
crystals 57
Curie, Marie 28

Dalton, John (1766-1844) 18
dative covalent bond see coordinate covalent bonding
decomposition reactions 215-216, 229
dehydration 347
detergents 370, 372-375, 480
diamond 45, 63-64, 236, 460
dichloromethane 92
diesel 316
diffusion 3-6
digestive enzymes 394
dimer 361
dipoles 61
direct combination reactions 214
displacement reactions 216, 229, 230, 282
distillation 78, 98
    fractional 89-90, 98
    simple 89, 98
'dot and cross' diagrams 46, 48
dry
    cells 274
    cleaning 79
    ice 11, 64
        see also carbon dioxide

ecosystems 488
elements 222
electric current 231, 254
electrical
    conductivity 232, 233
    conductors 67, 230-234
    non-conductors 232, 233
    semi-conductors 232
electricity 67, 245, 254, 276, 423
electrochemical series 240
electrodes 238, 239, 241, 242, 247
electrolysis 215, 232, 236, 237, 239-243, 246
electrolytes 229, 236, 237, 242, 243, 245, 246, 256
electrolytic cells 238
electronegativity 39, 42, 44
electron(s) 20, 223
electroplating 245, 249, 250-251
electrorefining 251, 255
elements 21-23, 36, 41, 57, 97, 103, 137, 219
    atomic structure 103
    periodicity 103

properties 57
    relative atomic mass 23-24
    structure 57
empirical formulae 149-150, 151, 152-153
end point see equivalence point
enthalpy see heat changes
environment 430, 431, 488
equilibrium reactions 137
esters 300, 346, 360, 366, 368, 374
esterification 364
ethane (alkane) 319, 326, 328, 337, 340, 348, 379
ethanoic acid 110, 140, 191, 192, 193, 195, 266, 354, 374, 376, 377
ethanol 54, 80, 231, 266, 285, 330, 342, 344, 345, 349-356
    commercial production 334, 347
    fermentation process 350, 354
    as a fuel 346, 353
    as a germicide 347, 348
    solubility in water 350
    as a solvent 349, 355
    toxic drug 352
ethene 311, 334, 336, 345
    direct hydration 334
ether 80
ethyl ethanoate 268, 365
eutrophication 491
evaporation 12
exothermic reactions 218

Faraday, Michael (1791-1867) 245
Faraday's constant 232, 245, 254
fats and oils 368
fatty acids 359
fertilisers 204, 463, 480
filtration 78, 86, 88
fluid catalytic cracking 320
fluorine 54, 113, 115, 117
formulae 129-132, 137, 294-296, 313
formula unit 47
fossil fuels 286, 316
fractional distillation process 90-91
fuels 274, 330
functional group 297, 311

galvanising 247
gamma radiation 31, 32
gasohol 347, 353
gasoline 79, 80, 316, 320, 336, 348-350
glucose 130, 140, 393, 394
graphite 45, 57, 63, 64, 68, 234, 236
green chemistry 500
    examples of 500
    principles of 501-3
greenhouse effect 493
groups
    Group 0 noble gases 39
    Group I alkali metals 39, 109, 222
    Group II alkali earth metals 39, 103, 126, 220
    Group III metals 37
    Group VII halogens 103, 112-113, 124

Haber process 476, 497
half-life 489
halide ions 116, 117
halogenoalkanes 329, 330
halogens 38, 113-116, 117, 247
heat 274
    changes 276-277, 280
    of fusion 58
heavy metal nitrates 408
helium 20, 23
hexane 68, 80
homologous series 291, 297, 306, 310
hydrocarbons 157, 291, 310, 311, 313
hydrochloric acid 108, 110, 166-167, 171-172, 174, 182, 183, 194, 197, 261, 281

hydroelectric power stations 423
hydrogen 26, 55, 144, 145, 155, 158, 179, 180, 191, 220, 222, 450
hydronium 190
hydroxide 140, 408

ice 62
indicators 187-188, 195, 196-197, 219
industrial fixation 497
inert gases see noble gases
inorganic
    carbon oxides 291
    compounds 291
insecticides 462
insulators 233
intermolecular forces 60, 61, 65, 66
intramolecular forces 60, 66
inter-particle forces 14
iodine 10, 43, 60-61, 62, 66, 79, 80, 86, 92, 113-114, 117, 223
ionic
    compounds 57, 58-59, 69, 129, 130
    crystals 57, 58
ions 24, 46, 66-67, 235
iron 108, 110, 130, 131, 143, 398, 421
    corrosion 251, 253
        see also rusting
    extraction 424, 429
    galvanising 253
isomers 315
isotopes 16, 27, 33

joules 282

ketones 2300
kerosene oil 316, 317
kilojoules 282
Kinetic Theory of Gases 2, 10

lactic acid 194, 195, 196
landfills 441-444
Law of Conservation of Matter 154, 159
lead 255, 407, 416
    bromide 235, 245, 246
    chloride 212
    iodide 141
    nitrate 217, 230
    oxide 202

macronutrients 495
magnesium 105, 107, 119-120, 144, 146, 180, 183, 233, 246, 261 406, 407
manganese 111
    (IV) oxide 223
marble 204
mass 152
melting point 11, 54
Mendeleev, Dmitri (1834-1907) 37
mercury 399
    dental amalgam 439
    environmental effects 437
    poisoning 438-439, 441
metal oxides 343, 374, 398, 400, 411
metallic
    bonding 54-55
    bromides 409
    carbonates 409, 410
    compounds 408
    hydroxides 409-410
    nitrates 408-409, 410
metalloids 38
metal oxides 196, 197, 411-412, 420
metals 37, 39, 55, 103, 140, 215, 220, 234, 416, 418
    alkaline earth 102, 105
    chemical properties 398, 400
    conductors of heat 399
    corrosion 250
    densities 399
    displacement reactions 3908, 403, 414

extraction 418, 421, 429
    melting points 399
    order of reactivity 414-415
    physical properties 39, 55-56, 103, 398-399, 415, 418
    reactivity 401-404, 405, 416, 419
    uses 410, 418-419
methane 50, 149, 283, 284, 316
methanoic acid (formic acid) 196, 359
methanol 344
methyl orange indicator 187-188, 205
methylbenzene 53, 191
micronutrients 495
mixtures 75, 77, 85, 87, 97
molar
    concentration 159, 160-161, 172, 181
    mass 142, 144, 156
    volumes 154-155
molarity see molar concentration
moles 143-147, 152, 159, 164, 174, 177, 182
molecular formulae 129, 149, 150, 151, 152
molecules 45, 48
monatomic ions 220
monomers 376, 377
Moseley, HGH (1887-1915) 37

naptha 319
neon 21
neutralisation 194, 229
neutrons 20-21
nickel 111, 249
nitrates 408
nitric acid 147, 162, 182, 191, 192, 193, 195, 226
nitrogen 20, 144, 158, 220, 223, 461
    cycle 497-498
    dioxide 177, 185, 200
    gas 157
    molecules 50, 145
    oxide 155, 177-178, 200
nitrous acid 192
noble gases 37, 44
non-bonded pairs 48
non-metals 38, 39, 215, 220, 448, 449, 464, 465
    chemical reactions 456
    compounds 448, 460
    electronegativity 459
    identification 451-452
    oxidising power 459
    physical properties 38, 449
    reaction with metals 218, 457
    reaction with oxygen 456, 457, 458
    reactivity series 459
    relative reactivity 459
non-polar covalent molecules 51-52
nuclear
    fission 31
    power 31
    reactor 30
nylon 383, 389, 395

oil refineries 316, 318
ores 418-420
    metal extraction 419-420
organic 311
    compounds 291, 294, 2959, 306, 307, 311, 323
osmosis 6-7
oxidation 149, 220, 225, 226
    numbers 23, 132, 219-220, 221, 224-2215, 2226
    -reduction see redox
    states 219, 221, 224, 225, 229
oxides 122, 187
    amphoteric 1857, 200
    basic 187, 200
    neutral 187, 200
    properties 123

oxidising
  acids see acids
  agents 223, 224, 225
oxygen 77, 142, 143, 175-176, 177, 211, 218, 237, 397, 454

paper chromatography 93-94, 95
paraffin 284
pentene 149
Period 3 103, 118, 123, 124-125
Periodic Table 36-44, 46, 104
perspex 378, 394
petrol 286, 316, 317
petroleum (crude oil) 313, 314
pH 188-189, 209, 211, 257
phosphoric acid 191, 192, 193, 195, 214
phosphorus 323, 62, 119-120, 145, 465
photochemical reactions 269
photography 269
photosynthesis 268, 271, 278, 286, 394
physical process 158
plants 392
plaster of Paris 204, 210
plastics 376, 391, 393, 394
  see also polymers
  disposal 376
  pollution 376
plutonium-238 32
polar covalent molecules 53, 54
polarity 53-54
pollen grains 5-6
pollutants 286
pollution 490
polyamides 388
polyatomic ions 130
polyesters 382, 386-387
polyethene 383, 384, 385-386
polymers 63, 378-379, 393, 395
  see also synthetic polymers
  addition 379-380, 385-386
  condensation 378, 379, 380
  and monomers 376, 377
  naturally occurring 378, 379
  properties 378, 385
  structure 378
  synthetic 380, 381

polysaccharides 391-394
polystyrene 385
potassium 83, 416
precipitation reaction 180
pressure 154
products and reactants 134, 139, 143, 160, 177, 185
propane 162, 322
propanoic acid 383
propanol 358, 365, 366, 369, 372, 372
propanone 54
proteins 390
proton(s) 20
pure substances 75-76, 78

radioactive
  isotopes 17, 30
  waste 32
radioactivity 29
  alpha particles 29
  uses 28, 29
radiocarbon dating see carbon-14 dating
radioisotopes 29-30, 32
radiotherapy 29
recrystallisation 95-96
redox reactions 214, 220-222, 230
reducing agents 225, 227, 228, 230
reduction 149, 224, 229, 419
reforming 320, 336
refractory materials 65
reverse osmosis 516
reversible reactions 219-220
Rontgen, Wilhelm (1845-1923) 28
room temperature and pressure (r.t.p.) 155, 171
rubidium 110
rusting 250-252, 253, 254

saponification 367
Scanning Tunnelling Microscope 20
semi-conductors 121
semi-permeable membranes 7
separating funnel 91, 99
sewage treatment plant 514

silica 63
  see also silicon dioxide
  gel 204
  structure 65
silicon 36, 119-120, 121, 450
silver 416, 418
soapless detergents 359
solder 428
solubility 67, 80, 92
solutes 79, 80, 97, 174
spectator ions 138
specific heat capacity 280
stalagmites 511
stalactites 511
starch 393, 394, 395
stearic acid 14
steel 425, 428, 429
stoichiometry 174
strontium 105, 108
structural isomerism 321
subatomic particles see protons; neutrons; electrons
sublimation 14, 86, 98
substitution reactions 216
sucrose 75, 96
sugar 79, 92, 98, 394
sulphates 108
sulphite 227
sulphur 40, 62, 119-120, 233, 224, 263, 452-453, 455, 458
sulphuric acid 26, 108, 146, 161, 172, 192, 193, 194, 200, 205, 226, 230, 239-240, 480
surface tension 371
surfactants 371, 362
suspensions 75, 83, 85, 98
synthetic polymers 378

tartaric acid 192, 193
technetium-99 29
temperature 75, 155
  change 276
thermal decomposition 215
thermometer 280
thermoplastics 393, 396
thermosets 393
thyroid gland 29

tin 131, 225, 253, 255
titration(s) 156, 166,167, 168, 172, 174, 182, 207
toluene 67
toothpastes 84, 189
tracers 29
transition metals 110-111, 113, 124, 224-225
trichloroethane 54

ultraviolet (UV) radiation 335, 339, 492
uranium-235 31

valence electrons 45, 55
vanadium 111, 112
Van de Waals forces 61, 337, 455
varnishes 347
vegetable oils 366
vibration 55
vinegar 191, 194, 238, 368-370
vitamin C (ascorbic acid) 193

washing soda 204
water 45, 50, 54-55, 66-67, 68, 98, 107, 149, 165, 179, 215, 280, 281, 340, 392, 405
  chlorinated 73, 510
  conservation 499, 513
  of crystallisation 204
  cycle 498-499, 450, 514
  desalination 520
  filtration 88
  hardness 510, 511, 517, 518
  leaching 513
  potable 507, 514, 515
  purification 515, 518
  softeners 507
  treatment 512
wrought iron 429

x-rays 28

zeolites 320
zinc 110, 135, 176, 229, 271, 278-279, 406, 416
zymase 336